Water Science and Technology Library

Volume 87

The aim of the Water Science and Technology Library is to provide a forum for dissemination of the state-of-the-art of topics of current interest in the area of water science and technology. This is accomplished through publication of reference books and monographs, authored or edited. Occasionally also proceedings volumes are accepted for publication in the series.

Water Science and Technology Library encompasses a wide range of topics dealing with science as well as socio-economic aspects of water, environment, and ecology. Both the water quantity and quality issues are relevant and are embraced by Water Science and Technology Library. The emphasis may be on either the scientific content, or techniques of solution, or both. There is increasing emphasis these days on processes and Water Science and Technology Library is committed to promoting this emphasis by publishing books emphasizing scientific discussions of physical, chemical, and/or biological aspects of water resources. Likewise, current or emerging solution techniques receive high priority. Interdisciplinary coverage is encouraged. Case studies contributing to our knowledge of water science and technology are also embraced by the series. Innovative ideas and novel techniques are of particular interest.

Comments or suggestions for future volumes are welcomed.

Vijay P. Singh, Department of Biological and Agricultural Engineering & Zachry Department of Civil Engineering, Texas A&M University, USA Email: vsingh@tamu.edu

More information about this series at http://www.springer.com/series/6689

Radin Maya Saphira Radin Mohamed
Adel Ali Saeed Al-Gheethi
Amir Hashim Mohd Kassim
Editors

Management of Greywater in Developing Countries

Alternative Practices, Treatment and Potential for Reuse and Recycling

 Springer

Editors
Radin Maya Saphira Radin Mohamed
Faculty of Civil and Environmental
  Engineering
Universiti Tun Hussein Onn Malaysia
  (UTHM)
Johor
Malaysia

Amir Hashim Mohd Kassim
Faculty of Civil and Environmental
  Engineering
Universiti Tun Hussein Onn Malaysia
  (UTHM)
Johor
Malaysia

Adel Ali Saeed Al-Gheethi
Faculty of Civil and Environmental
  Engineering
Universiti Tun Hussein Onn Malaysia
  (UTHM)
Johor
Malaysia

ISSN 0921-092X   ISSN 1872-4663   (electronic)
Water Science and Technology Library
ISBN 978-3-030-07980-2   ISBN 978-3-319-90269-2   (eBook)
https://doi.org/10.1007/978-3-319-90269-2

This Springer imprint is published by the registered company Springer International Publishing AG
part of Springer Nature
The registered company address is: Gewerbestrasse 11, 6330 Cham, Switzerland

Preface

The ever-increasing use of greywater has been associated with growing concerns over its ultimate destination and effects on the environment after discharge or when reused for irrigation. Greywater disposal is acknowledged as a serious worldwide problem due to the increased environmental awareness and stringent environmental standards that govern its use as set by various environmental protection agencies. In contrast, the utilization of greywater in agricultural production has been gaining increasing interest and attention in recent years. It offers economic and nutrient recycling advantages over the traditional disposal options. Nevertheless, potential risks derived from the accumulation of heavy metals and organic compounds, as well as pathogen contamination, must be taken into consideration. The developments in treatment technologies to produce high-quality greywater emphasize safe reuse. However, progress in the developing counties is a bit slower than in the developed countries. Nevertheless, there are a number of alternative technologies that have emerged, including phycoremediation processing and solar disinfection, which are more appropriate to the developing countries.

The current book is an effort to summarize the work of researchers in the field of greywater since 2011 who believe that this is the right time to write a reference book to discuss the current practice of greywater management. The main targets of this book are the academicians and researchers who seek to find more meaningful information on the management and recent treatment technologies of greywater in developed and developing countries.

This book aims to discuss the proper management of greywater in the developing countries. It provides a comprehensive review and referenced information on greywater in order to highlight its risk alongside the benefits. The book consists of 12 chapters: Chap. 1 describes the characteristics of greywater and quantities generated in the developing countries. This chapter highlights the main pollutants in greywater, such as heavy metals and infectious pathogens. Chapter 2 addresses the consequences of improper disposal of greywater into the environment, including the health risks associated with greywater, the potential of pathogenic organism to transmit into human and animals, and the adverse effects on the aquaculture organism in the water bodies which receive the discharged greywater. Chapter 3 reviews the isolation and assessment method of pathogens in greywater. Chapter 4 focuses on the greywater benefits for plants if it is reused in irrigation, the most appropriate technologies for the irrigation of plants with the greywater, and international regulations on the reuse of greywater for irrigation purpose. Chapter 5 reviews the occurrence and environmental impact of Xenobiotic Organic Compounds (XOCs) in greywater. Chapter 6 compares the management of greywater in developed and developing countries in order to give a better idea of the differences between countries in terms of treatment technologies level. Chapter 7 overviews the advantages and disadvantages of treatment technologies used for greywater and their ability to produce a higher quality of treated greywater without toxic by-products. Chapter 8 discusses the potential of phycoremediation as a green technology for removal of nutrients from greywater, as well as microbiological aspects of phycoremediation process. Chapter 9 highlights the biodegradation of XOCs by fungi and oxidative enzymes. Chapter 10 describes the most common technologies used for reduction of infectious agents in greywater before its disposal or reuse for irrigation purpose. Chapter 11 treats the recycling of greywater as a production medium for biomass and bioproducts, as well as economic aspects of greywater recycling. Chapter 12 discusses the transport systems for greywater in rural regions and their effects on the characteristics of greywater, as well as the application of nanotechnology in the treatment of greywater, challenges, and future directions. Finally, Chap. 13 focuses on the potential of microalgae to improve the quality of meat-processing wastewater for safe disposal.

Johor, Malaysia   Adel Ali Saeed Al-Gheethi

Contents

Contributors

Abd Halid Abdullah Department of Architecture and Engineering Design, Faculty of Civil and Environmental Engineering, Universiti Tun Hussein Onn Malaysia (UTHM), Batu Pahat, Johor, Malaysia

Adel Ali Saeed Al-Gheethi Micro-Pollutant Research Centre (MPRC), Department of Water and Environmental Engineering, Faculty of Civil and Environmental Engineering, Universiti Tun Hussein Onn Malaysia (UTHM), Parit Raja, Batu Pahat, Johor, Malaysia

Najeeha Mohd Apandi Micro-Pollutant Research Centre (MPRC), Department of Water and Environmental Engineering, Faculty of Civil and Environmental Engineering, Universiti Tun Hussein Onn Malaysia (UTHM), Parit Raja, Batu Pahat, Johor, Malaysia

A. Athirah Micro-Pollutant Research Centre (MPRC), Department of Water and Environmental Engineering, Faculty of Civil and Environmental Engineering, Universiti Tun Hussein Onn Malaysia (UTHM), Parit Raja, Batu Pahat, Johor, Malaysia

Siti Asmah Bakar Micro-Pollutant Research Centre (MPRC), Department of Water and Environmental Engineering, Faculty of Civil and Environmental Engineering, Universiti Tun Hussein Onn Malaysia (UTHM), Parit Raja, Batu Pahat, Johor, Malaysia

J. D. Bala Department of Microbiology, School of Life Sciences, Federal University of Technology, Minna, Niger State, Nigeria

Stewart Dallas Environmental Engineering, School of Engineering, Murdoch University, Perth, Australia

Norli Ismail Environmental Technology Division, School of Industrial Technology, Universiti Sains Malaysia (USM), George Town, Penang, Malaysia

Mohd Hairul Bin Khamidun Micro-Pollutant Research Centre (MPRC), Department of Water and Environmental Engineering, Faculty of Civil and Environmental Engineering, Universiti Tun Hussein Onn Malaysia (UTHM), Parit Raja, Batu Pahat, Johor, Malaysia

Anda Martin Environmental Engineering, School of Engineering, Murdoch University, Perth, Australia

Radin Maya Saphira Radin Mohamed Micro-Pollutant Research Centre (MPRC), Department of Water and Environmental Engineering, Faculty of Civil and Environmental Engineering, Universiti Tun Hussein Onn Malaysia (UTHM), Parit Raja, Batu Pahat, Johor, Malaysia

Amir Hashim Mohd Kassim Micro-Pollutant Research Centre (MPRC), Department of Water and Environmental Engineering, Faculty of Civil and Environmental Engineering, Universiti Tun Hussein Onn Malaysia (UTHM), Parit Raja, Batu Pahat, Johor, Malaysia

H. Nagao School of Biological Sciences, Universiti Sains Malaysia (USM), George Town, Penang, Malaysia

Efaq Ali Noman Faculty of Science, Technology and Human Developments, Universiti Tun Hussein Onn Malaysia (UTHM), Parit Raja, Batu Pahat, Johor, Malaysia

Norzila Othman Micro-Pollutant Research Centre (MPRC), Department of Water and Environmental Engineering, Faculty of Civil and Environmental Engineering, Universiti Tun Hussein Onn Malaysia (UTHM), Parit Raja, Batu Pahat, Johor, Malaysia

Fadzilah Pahazri Micro-Pollutant Research Centre (MPRC), Department of Water and Environmental Engineering, Faculty of Civil and Environmental Engineering, Universiti Tun Hussein Onn Malaysia (UTHM), Parit Raja, Batu Pahat, Johor, Malaysia

Junita Abdul Rahman Micro-Pollutant Research Centre (MPRC), Department of Water and Environmental Engineering, Faculty of Civil and Environmental Engineering, Universiti Tun Hussein Onn Malaysia (UTHM), Parit Raja, Batu Pahat, Johor, Malaysia

Balkis A. Talip Faculty of Applied Sciences and Technology (FAST), Universiti Tun Hussein Onn Malaysia (UTHM), 84000, KM11, Jalan Panchor, Pagoh Muar, Johor, Malaysia

A. S. Vikneswara Micro-Pollutant Research Centre (MPRC), Department of Water and Environmental Engineering, Faculty of Civil and Environmental Engineering, Universiti Tun Hussein Onn Malaysia (UTHM), Parit Raja, Batu Pahat, Johor, Malaysia

A. A. Wurochekke Micro-Pollutant Research Centre (MPRC), Department of Water and Environmental Engineering, Faculty of Civil and Environmental Engineering, Universiti Tun Hussein Onn Malaysia (UTHM), Parit Raja, Batu Pahat, Johor, Malaysia

Maizatul Azrina Yaakob Micro-Pollutant Research Centre (MPRC), Department of Water and Environmental Engineering, Faculty of Civil and Environmental Engineering, Universiti Tun Hussein Onn Malaysia (UTHM), Parit Raja, Batu Pahat, Johor, Malaysia

Abbreviations

2-DE	Two-Dimensional Electrophoresis
AOC	Assimilable Organic Carbon
AOPs	Advanced Oxidation Processes
API	Analytical Profile Index
APIs	Active Pharmaceutical Ingredients
ATP	Adenine Triphosphate
BCM	Billion Cubic Meters
BHI	Brain Heart Infusion Medium
BOD	Biochemical Oxygen Demand
CHIAT	Chemical Hazard Identification And Assessment Tool
CLSI	*Clinical And Laboratory Standards Institute*
CNCs	Cellulose Nanocrystals
COD	Chemical Oxygen Demand
CW	Constructed Wetlands
DAF	Dispersed Air flotation
DO	Dissolved Oxygen
DOE	Department of Environment
EDCs	Endocrine Disrupting Chemicals
ENPs	Engineered Nanomaterials
FC	Faecal Coliforms
FS	Faecal Streptococci
GC	Gas Chromatography
GC/MS/MS	Gas Chromatography–Tandem Mass Spectrometry
HABs	Harmful Algae Blooms
HAV	Hepatitis A Virus
HBC	Heterotrophic Bacterial Counts
HRT	Hydraulic Residence Time
ID	Infective Doses
L/p/d	Litre Per Day
LAC	Laccase

LC/MS/MS	Liquid Chromatography–Tandem Mass Spectrometry
LC-MS	Liquid Chromatography
LDPE	Low-Density Polyethylene
LPS	Lipopolysaccharide
MF	Membrane Filtration
MICs	Minimum Inhibition Concentrations
MPN	Most Probable Numbers
MRSA	Methicillin-Resistant *Staphylococcus Aureus*
MS	Mass Spectrometry
MSA	Mannitol Salt Agar
NA	Nutrient Agar
NH_3	Ammonia
NO_3^-	Nitrate
NVOC	Non-volatile Organic Carbon
OMPs	Organic Micro Pollutants
PAHs	Polyaromatic Hydrocarbons
PBT	Persistence–Bioaccumulation–Toxicity
PCR	Polymerase Chain Reaction
PFASs	Polyfluoroalkyl Substances
PFRs	Phosphorus-Containing Flame Retardants
PGP	Pathogen Growth Potential
PO_4^{3-}	Orthophosphate
POPs	Persistent Organic Pollutants
PPCPs	Pharmaceuticals and Personal Care Products
QMRA	Quantitative Microbial Risk Assessment
qPCR	Quantitative Polymerase Chain Reaction
RBC	Rotating Biological Contactor
RCRA	Resource Conservation And Recovery Act
RICH	Ranking and Identification of Chemical Hazard's Tool
SAR	Sodium Adsorption Ratio
SDW	Sterilized Distilled Water
SEM	Scanning Electronic Microscopy
–SH	Sulfhydryl Group
SMBR	Submerged Membrane Bioreactor
SODIS	Solar Disinfection
SPE	Solid Phase Extraction
STAATT	State And Territorial Association On Alternative Treatment Technologies
TC	Total Coliforms
TN	Total Nitrogen
TOC	Total Organic Carbon
TP	Total Phosphorus
TSS	Total Suspended Solids
UPLC-MS/MS	Ultra Performance Liquid Chromatography Coupled With Tandem Mass Spectrometry

USEPA	U.S. Environmental Protection Agency
VBNC	Available But Non-culturable
WA	Western Australia
WHO	World Health Organization
WRF	White Rot Fungi
WWTP	Wastewater Treatment Plant
XOCs	Xenobiotic Organic Compounds
ZTL	Suitable Zero Tension Lysimeter

Chapter 1
Qualitative Characterization of Household Greywater in Developing Countries: A Comprehensive Review

Adel Ali Saeed Al-Gheethi, Efaq Ali Noman,
Radin Maya Saphira Radin Mohamed, J. D. Bala
and Amir Hashim Mohd Kassim

Abstract Greywater is a type of wastewater generated from household activities which include bathing, laundry and kitchen activities. Greywater has a lower quality than potable water, but it is of higher quality than sewage. This article is a qualitative review presenting the quantities and qualitative characteristics of greywater in developed and developing countries. The chapter aims at highlighting the presence of nitrogen, phosphorus and pathogenic microorganisms. This paper provides a comprehensive review of greywater in order to understand the physiochemical and microbiological composition of greywater which represents the first step in choosing the most appropriate technology for the treatment process and to best evaluate the health risks associated with greywater discharge into the environment.

Keywords Chemical · Physical · Microbiological composition · Indicator bacteria

A. A. S. Al-Gheethi (✉) · R. M. S. Radin Mohamed (✉) · A. H. Mohd Kassim
Micro-Pollutant Research Centre (MPRC), Department of Water and Environmental Engineering, Faculty of Civil and Environmental Engineering, Universiti Tun Hussein Onn Malaysia (UTHM), 86400 Parit Raja, Batu Pahat, Johor, Malaysia
e-mail: adel@uthm.edu.my

R. M. S. Radin Mohamed
e-mail: maya@uthm.edu.my

E. A. Noman
Faculty of Science, Technology and Human Developments, Universiti Tun Hussein Onn Malaysia (UTHM), 86400 Parit Raja, Batu Pahat, Johor, Malaysia

J. D. Bala
Department of Microbiology, School of Life Sciences, Federal University of Technology, P.M.B 65, Minna, Niger State, Nigeria

© Springer International Publishing AG, part of Springer Nature 2019   1
R. M. S. Radin Mohamed et al. (eds.), *Management of Greywater in Developing Countries*,
Water Science and Technology Library 87,
https://doi.org/10.1007/978-3-319-90269-2_1

1.1 Introduction

The rapid increase in the total population in developing countries and their activities alongside the deficiency in clean water resources as well as the absence of advanced technology required to produce high quality treated wastewater leads to an increase in the level of natural water contamination. This is due to the direct discharge of wastewater into water bodies. In most developed countries, black water and greywater are treated separately. In contrast, these practices are common in the rural areas in developing countries in order to reduce the quantity of sewage discharged into the individual septic tank (IST) due to the absence of a central wastewater treatment plant. Besides, these practices are common in arid and semi-arid areas because they use greywater for irrigation purposes.

Research studies in greywater treatment and the reuse or recycling of greywater have started early since the 1990s and have increased significantly after 2005 (Fig. 1.1).

A review of these publications revealed that most of the studies were performed in developed countries. In the period between 1990 and 2000, the publications addressed the characteristics of greywater from different sources. However, in the last few years, they have shifted to the treatment and reuse of greywater for the production of biomass as well as for irrigation purposes. Most papers published in the Middle East region were produced in Jordan which has the best practice in the field of wastewater treatment among the developing countries in the Middle East region, and they are also related to arid and semi-arid weather in Jordan. Therefore, the greywater in

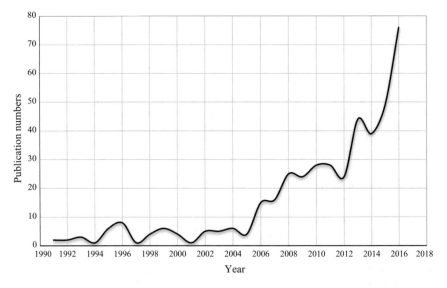

Fig. 1.1 Shows the distribution of 426 publications on greywater in the period between 1991 and 2016. The data were collected from Elsevier, Springer and Wiley publishers

Jordan represents an important alternative source of potable water. In Yemen, which has the least studies in the field of wastewater treatment, only three publications by Al-Mughalles et al. (2012) were found. In Malaysia, most papers on greywater were conducted by researchers working at the universities located in the suburban areas. Moreover, their work focused on the characteristics, treatment and reuse of greywater. The differences between developing countries in the Middle East in comparison to Malaysia are related to the presence of research facilities. There is some reliable information pertaining to the composition of greywater, treatment technologies as well as the potential to reuse and recycle greywater (Eriksson et al. 2002; Jefferson et al. 2004; Palmquist and Hanæus 2005). This gap offers researchers a greater opportunity to explore the qualitative characteristics of greywater which play an important role in the proper management of greywater to be used as an alternative source of fresh water. This chapter aims at viewing the physical, chemical as well as microbiological characteristics of greywater to provide a comprehensive idea of greywater composition.

1.2 Definition of Greywater

Greywater is a general term referring to different types of wastewater generated from household activities. Greywater is quite different from toilet wastewater which is known as black water. Greywater has a lower quality than potable water, but it is of higher quality than blackwater (Prathapar et al. 2005; Jamrah et al. 2006). The term of greywater refers to the colour change in water to grey during storage, but laundry greywater is grey even without the storage period. In some references, greywater is defined as light wastewater, diluted wastewater (Ledin et al. 2001) and reclaimed water (Gregory and Hansen 1996). Wilderer (2004) has classified the wastewater generated from houses into six categories including brown water (wastewater with faeces), yellow water (urine), blackwater (containing both urine and faeces), greywater (containing mainly detergents), green water (contains food particles) and storm water (rainwater). The main difference between black water and greywater is the high level of organic material, nutrients and infectious agents which are available in high concentrations in black water (Klammer 2013; Atiku et al. 2016), while they are lesser by 90% in greywater. In most definitions, kitchen and dishwasher wastewater are excluded from the category of greywater because it contains high levels of microbial loads which might be associated with high levels of organic matter resulting from grease, oil and detergents (Wurochekke et al. 2016). However, Al-Gheethi et al. (2016a, b) mentioned that those dishwasher, kitchen and restaurant wastewaters are closer to greywater than black water in terms of their characteristics. The source of microbial organisms in dishwasher, kitchen and restaurant wastewater might be caused by the washing of vegetables, fruits and meat. The microbial loads of these items are low but they may multiply in the kitchen greywater due to the presence of nutrients necessary for their growth (Friedler 2004). In contrast, the microbial loads in black water are very high even without the reproduction process. These differences might explain the exclusion of kitchen and restaurant wastewater from the black water category. Nonetheless, more clarifications are needed to justify the classification of these wastes within the greywater definition. This is because the quality of wastewater generated from cooking and dishwashers (both in kitchen and restaurant shops) is also similar to wet market and meat processing wastewater in terms of microbial loads and organic content as well as its nutrient content. Indeed, the classification of wastewater from household activities makes the creation of a separate network transport system for each type more complicated especially in developing countries in which all household wastewater is discharged without separation. Therefore, in order to overcome the confusion between greywater from baths and wastewater from the kitchen, Bodnar et al. (2014) used the term 'light greywater' to represent greywater from baths and dark greywater generated by laundry and cooking. In a report published by Morel and Diener (2006), 12 out of 15 references included kitchen wastewater within the greywater definition. Hence, in this chapter, the concept of greywater will include all the wastewater from household activities except for black water.

1.3   Sources and Quantity of Greywater in Developing Countries

The main source of greywater comes from household activities, including baths, showers, laundry as well as dishwashing. The percentage of greywater generated from household activities represents 50–80% of the total water usage. So far, quantities of greywater depend on domestic water consumption. The data depicted in Fig. 1.2 show that the maximum water consumption is utilized for laundry and shower activities (33 and 23% respectively). The toilets consumed around 20% while dishwashing and cooking consumed 11% of the total water usage.

The percentage of greywater compared to black water in developing countries is more than that in developed countries. For instance, in the UK, the quantities of greywater and sewage produced are equal, while greywater represents 70–80% of total domestic wastewater in Jordan and Oman (Prathapar et al. 2005; Jamrah et al. 2008). In South Africa, greywater is ranging between 65 and 85% of the total household water consumption (Carden et al. 2007). These differences would be also explained based on lifestyle. This indicates that the greywater quantity does not vary greatly.

Al-Mughalles et al. (2012) mentioned that the utilization of Arabic toilets might be associated with high generation of greywater, while the utilization of Western toilets with a flushing system might lead to increased black water production. Indeed, the estimation of greywater quantities in developing countries especially those located in the Middle East region is difficult because in most of those countries, the greywater is discharged with the sewage. Therefore, accurate information about the percentage of greywater in black water is unavailable. However, a more acceptable reason to

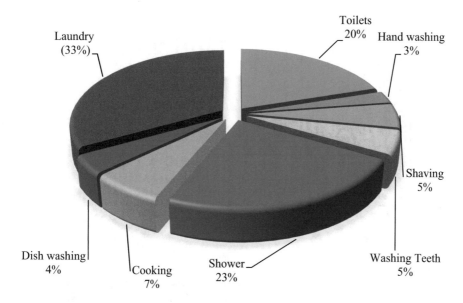

Fig. 1.2 Domestic water consumption per capita per day. (Adopted from Howard et al. 2003)

increase the percentage of greywater in Middle East countries is the ablution water which is generated from the ablution process and discharged along with greywater (Efaq et al. 2016; Mohamed et al. 2016a).

Greywater quantity depends on the number of household members, their age, nature of living, demographics and level of occupancy, geographical location, social habits and water usage pattern and time (Prathapar et al. 2005; Jamrah et al. 2008). Hence, the quantity of greywater produced differs in developed countries compared to developing countries (Fig. 1.3). Bodnar et al. (2014) also indicated that the amount of generated greywater in large cities (120–130 L/p/d) is higher than that in small villages (50–70 L/p/d).

Based on the quantities of greywater in different countries presented in Fig. 1.3, it can be noted that the quantities of greywater generated from household activities depend on the level of development in that particular country. The maximum quantity of greywater is recorded in the USA (281 L/p/d), while the lowest quantities were noted in South Africa (20 L/p/d).

The quantities of greywater in the countries which face a scarcity in freshwater resources such as Yemen (35 L/p/d), Syria (33.8 L/p/d) and Mali (30 L/p/d) are less compared to countries which have several water resources such as Malaysia (125 L/p/d). However, this situation depends on the level of the country's degree of economic development and the availability of facilities in the countries. For instance, in Jordan, the total amount of greywater generated from households is 50 L/p/d. In contrast, in Oman which is located in the same geographical area, the quantity of greywater generated is 184 L/p/d. This might be related to desalinated seawater.

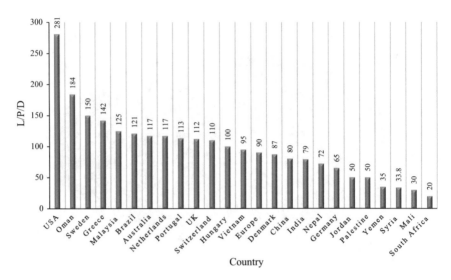

Fig. 1.3 Quantities of greywater (L/p/d) in selected countries. *Source* Faruqui and Al-Jayyousi (2002), Adendorff and Stimie (2005), Alderlieste and Langeveld (2005), Helvetas (2005), Busser et al. (2006), Jamrah et al. (2008); Mourad et al. (2011), Al-Mughalles et al. (2012), Harikumar and Mol (2012), Antonopoulou et al. (2013), Mohamed et al. (2013)

In Saudi Arabia, 15.1 billion cubic metres of desalinated seawater is provided per day (BCM) compared to 1.2 BCM per day of non-renewable groundwater resources (Al-Saud 2010). In Malaysia, greywater generated is estimated to be 100–150 L/p/d (Mohamed et al. 2016b). The common practices among the community might play an important role in the quantities of discharged greywater. Adendorff and Stimie (2005) stated that the low production of greywater in South Africa (20 L/p/d) is due to the lifestyle of the people who use rivers or lakes for washing clothes, utensils and to clean themselves. In comparison to developed countries, the average greywater volume is 113 L/p/d in Australia, 110 L/p/d in Switzerland (Helvetas 2005), 100 L/p/d in Hungary (Bodnar et al. 2014) and between 35 and 150 in Europe (Boyjoo et al. 2013).

The main resources for greywater include bathroom, laundry and kitchen. Greywater from baths constitutes the main percentage of the greywater. It has been reported that the greywater from bathrooms makes up 55% of the total greywater, followed by laundry wastewater (34%) and kitchen wastewater (11%) (Katukiza et al. 2015; Laghari et al. 2015). According to Mara and Cairncross (1989), bathroom greywater is the main greywater source with 54% of the total greywater generated by houses, while laundry greywater makes up 38% of the total greywater generated. In a study conducted by Ghaitidak and Yadav (2013), the percentages of bathroom, laundry and kitchen greywater were 47, 26 and 27% respectively. Mohamed et al. (2016b) revealed that the maximum generation of greywater in Malaysia was recorded from bathing activity, with an average of 50% of total greywater. It was due to frequent bathing among the occupants of more than three times daily. This is also related to the hot and humid weather in Malaysia. The study also found that laundry greywater contributed 22% of the total greywater. In contrast, the kitchen greywater quantity represented 28% of total greywater. The quantity of greywater during weekends is higher than greywater produced during weekdays (Mohamed et al. 2016b). These differences are due to more household activities on weekends, especially cooking activities. On weekdays, most people have their meals at their workplace thus less kitchen grey water is produced in the absence of cooking.

1.4   Qualitative Characteristics of Household Greywater

1.4.1   General Composition

Greywater is mixed with soap, shampoo, toothpaste, food waste, cooking oil and detergent depending on the utilization of resources by each household (Mohamed et al. 2016b). The main composition for different types of greywater is presented in Table 1.1. The quality of greywater depends mainly on the health status and personal hygiene of the users (Jamrah et al. 2008; Laghari et al. 2015).

The main parameters of greywater include biochemical oxygen demand (BOD), chemical oxygen demand (COD), suspended solids, turbidity, total nitrogen (TN), total phosphorus (TP), pH, alkalinity, electrical conductivity, heavy metals, disinfectants, bleach, surfactants and detergents (Adendorff and Stimie 2005). The quality of organic matter in greywater is similar to that in domestic wastewater but differs in terms of concentration. Moreover, the utilization of low quantities of water for household activities is associated with heavy pollution while high quantities of greywater show that pollutants are subjected to dilution. Previous studies have revealed that the pollutant loads of greywater from developing countries are higher than that reported in developed countries (Carden et al. 2007). This might be explained based on the quantities of greywater in developed countries which are higher in comparison to that in developing countries. Thus, more dilution for pollutants takes place with

Table 1.1 Composition of household greywater from three main resources

Source of greywater	General composition
Bathing	Represents real greywater and contains soaps, toothpaste, shampoos, body care products, shaving waste, body fats, skin, hair, lint, urine and faeces (which are common if there are children in the house), hair dyes, toothpaste, bacteria
Laundry	Sodium, phosphorous, surfactants, nitrogen, foam, bleaches, suspended solids, oils and grease, paints, solvents, fibres, bacteria, viruses, lint
Cooking and dishwashing	Food residues, oil and fat, detergents, suspended solids, nutrients, salt, bacteria, foam

high quantities of greywater. In some developed countries, the term 'diluted waste' is used to represent greywater.

1.4.2 Physical Characteristics

The physical characteristics of greywater include temperature, colour, turbidity, suspended solids and total dissolved solids. These characteristics depend upon the greywater source. Some of the physical characteristics such as temperature and colour vary depending on the environmental climate and conditions. For instance, wastewater temperature in cold regions ranged from 7 to 18 °C, while in warmer regions it is between 13 and 24 °C. The variations in their values affect the wastewater composition due to their role in chemical and biological reactions. The variations in temperature are associated with the changes in pH, conductivity and saturation level of gases. Besides, the higher temperature of greywater might provide a suitable environmental condition for infectious agents which have a temperature between 20 and 40 °C (Eriksson et al. 2002; Al-Gheethi et al. 2013). Therefore, these parameters should also be considered for the assessment of the quality of greywater. The colour parameter is an indicator of the level of total suspended solids and dyes in wastewater Dubey et al. (2013). In fact, greywater parameters interact and the evaluation for each parameter separately does not reflect greywater quality and the expected effects on the environment as well as the selection of treatment technology. Besides, the composition of wastewater varies seasonally and geographically based on the local activities and total population in the specific area, as well as the presence or absence of the wastewater transport system.

The main physical parameters of greywater are presented in Table 1.2. It can be noted that greywater has a wide range of physical characteristics. Turbidity is a physical parameter which reflects the presence of suspended solids resulting from clothes washing and cleaning dirty floors and leads to cloudiness in greywater.

In contrast, the 'total dissolved salts' is a term used for the expression of total concentration of dissolved substances (organic and inorganic) in greywater. The turbidity and TSS are high in greywater generated from laundry and the kitchen due to the presence of solids, fabric softeners and laundry detergent residues. Jong et al. (2010) revealed that the turbidity in mixed greywater from washing, cooking, bathroom and shower ranged from 152 to 4400 NTU, this value is associated with high TSS ranging from 72.5 to 4250 mg L^{-1}. The authors also indicated that kitchen greywater has a high content of TDS ranging between 412.57 and 1232.14 mg L^{-1} (Bodnar et al. 2014).

In a comparison between developed and developing countries, the concentrations of physical characteristics in greywater from developing countries are higher than those in developed countries with some exceptions. For example, TSS in greywater from the USA and the EU as well as Australia ranged from less than 10 to a maximum of 400 mg L^{-1}. These concentrations are also shown in greywater from India, Malaysia and Yemen, but it was more than 1500 mg L^{-1} in Uganda, Jordan

Table 1.2 General characteristics of bathroom greywater and related sources

Sources	Country	TSS	TDS	Turbidity (NTU)	References
Bathroom and laundry	Greece	7–250	ND	ND	Fountoulakis et al. (2016)
Bathroom and laundry	Malaysia	23.87 ± 11.01	ND	50.83 ± 6.39	Mohamed et al. (2016b)
Bathroom	Malaysia	81	ND	ND	Teh et al. (2015)
Cleaning and sink activities	Jordan	16–59	280–350	73–173	Bani-Melhem et al. (2015)
Bathroom	Malaysia	78–163	ND	ND	Mohamed et al. (2014)
Shower/bathtub, laundry and kitchen	Hungary	13.67–181.2	412.57–1232.14	25.45–357.39	Bodnar et al. (2014)
Mixed greywater	Uganda	2093–732, 828–735	ND	ND	Katukiza et al. (2014)
Bathroom	Nigeria	633.2	ND	ND	Nnaji et al. (2013)
Men hostel	India	564	ND	ND	Gokulan et al. (2013)
Showers and bathroom sinks	Spain		ND	50–158	Santasmasas et al. (2013)
Bathroom and laundry	Malaysia	88–110	ND	ND	Mohamed et al. (2013)
Bathroom	France	125	ND	ND	Chaillou et al. (2011)
Washing, kitchen, bathroom and shower	Korea	72.5–4250	ND	152–4400	Jong et al. (2010)

(continued)

A. A. S. Al-Gheethi et al.

Table 1.2 (continued)

Sources	Country	TSS	TDS	Turbidity (NTU)	References
Village houses	Jordan	23–358	ND	ND	Al-Hamaiedeh and Bino (2010)
Mixed greywater	Jordan	573–1679	ND	ND	Halalsheh et al. (2008)
Showers and hand basins	Denmark	7–207	ND	ND	Eriksson et al. (2002)
Bathroom	Australia	48–120	ND	ND	Christova-Boal et al. (1996)
Domestic greywater	Australia	45–330	ND	22–200	Jeppersen and Solley (1994)
Shower/bathtub	Greece	63 ± 38	601 ± 152	ND	Antonopoulou et al. (2013)
Hand basin		61 ± 37	578 ± 193		
Kitchen sink		299 ± 324	595 ± 202		
Total greywater	Jordan	213–803	1113–2930	ND	Ammari et al. (2014)
Bathroom greywater	Malaysia	101.3 206	ND	70.7 160.3	Oh et al. (2016)
Total greywater	Canada	313–543	ND	ND	Finley et al. (2009)
Municipal greywater	Egypt	ND	313–597	ND	Abdel-Shafy et al. (2014)
Bath/shower	India	148	277	ND	Vakil et al. (2014)

(continued)

Table 1.2 (continued)

Sources	Country	TSS	TDS	Turbidity (NTU)	References
Washbasin		48	237	ND	
Kitchen		308	245	ND	
Laundry		1852	1060	ND	
Total greywater	India	13.3	7.5	161.3	Saumya et al. (2015)
Hostel greywater	India	ND	355 ± 33	143 ± 54	Patil and Munavalli (2016)
Primary settling tank	Norway	155–160	ND	ND	Karabelnik et al. (2012)
Bathroom sinks	Brazil	40–105	ND	10.7–16.4	do Couto et al. (2013)
Kitchen sinks		67–208	ND	24.7–123.7	
Showers		64–143	ND	26.52	
Bathroom sinks	Bangladesh	74–84	432–532	74–102	Abedin and Rakib (2013)
Kitchen sinks		1253–2414	615–684	303–345	
Laundry		854–1160	832–1573	386–410	
Total greywater	Yemen	337.1–510.8	ND	266.24–618.60	Al-Mughalles et al. (2012)

ND (Non detected)

and Bangladesh (Al-Mughalles et al. 2012; Abedin and Rakib 2013; Katukiza et al. 2014). Similar trends were observed for turbidity and TDS but with little differences between the greywater in developed and developing countries. However, the data in Table 1.2 show that in some countries such as Korea, the turbidity ranges from 152 to 4400 NTU whereas in Yemen it was between 266.24 and 618.60 NTU (Jong et al. 2010; Al-Mughalles et al. 2012). Finally, it has to be mentioned that the physical properties of greywater might give a general idea of greywater while the real composition needs to be evaluated based on the chemical and microbiological characteristics.

1.4.3 Chemical Characteristics

The chemical characteristics of greywater are illustrated in Table 1.3. These parameters including pH, chemical oxygen demand (COD), biochemical oxygen demand (BOD), total nitrogen (TN) and total phosphate (TP) provide more details on the nature of these wastes in terms of organic and inorganic constitutes. Greywater has a more alkaline pH value ranging from 5 to 11 compared to black water which has a pH between 6 and 7.7 (Schäfer et al. 2006). The data in Table 1.3 indicate that there is no clear correlation between pH values and the development level for greywater from different countries.

However, it was noted that the greywater in developed countries has a neutral pH whereas those from developing countries ranged from less than 6 as that reported in Bangladesh, Jordan, Egypt and Yemen (Abedin and Rakib 2013; Ammari et al. 2014; Abdel-Shafy et al. 2014) to more than pH 9 as reported in India (Vakil et al. 2014). Meanwhile, Fountoulakis et al. (2016) found that the pH value in greywater from Greece ranged from pH 6.4 to 10. Nonetheless, pH values in greywater depend on the source of greywater. For bathing and kitchen greywater, the pH value is on average 7, while those generated from laundry has a pH more than 9 due to the presence of high concentrations of surfactants/detergents (Bodnar et al. 2014).

COD parameter is one of the main chemical characteristics which reveals real organic pollutants in greywater (Jais et al. 2016). The concentration of COD depends on the chemical reaction level between organic substances in greywater. Therefore, the increasing levels of COD in greywater indicate active chemical reactions with high consumption of free oxygen available in water. In contrast, BOD is an indicator for the biological oxidation of organic compounds in the presence of molecular oxygen as an oxidizing agent to produce carbon dioxide and water. This process takes place in the microbial cells and uses dissolved oxygen available in water. High concentrations of BOD in greywater are an indirect indicator for high density of microorganisms.

From the concentrations of COD and BOD_5 in previous studies presented in Table 1.3, greywater appears to have higher COD than BOD. Therefore, the $COD:BOD_5$ ratio (4:1) of greywater is higher than that of sewage. This would be due to the high levels of xenobiotic organic compounds (XOC) in detergent

Table 1.3 Chemical characteristics of bathroom greywater and related sources

Sources	Country	pH	COD	BOD$_5$	TN	TP	References
Bathroom and laundry	Greece	6.4–10	26–645	ND	3.6–21	0.1–1	Fountoulakis et al. (2016)
Bathroom	Malaysia	6.1	445	349	NR	NR	Teh et al. (2015)
Bathroom	Malaysia	6.1–6.4	445–621	40–105	NR	NR	Mohamed et al. (2014)
Bathroom	Nigeria	7.7	67.6	60.23	NR	NR	Nnaji et al. (2013)
Men hostel	India	7.48	986	279.6	NR	NR	Gokulan et al. (2013)
Bathroom and laundry	Malaysia	6.8–7.8	180–291	90–130	5.3–30	0.2–6.7	Mohamed et al. (2013)
Bathroom	France	7.6	399	240	3.8–17.0	0.1–2.0	Chaillou et al. (2011)
Mixed greywater	Jordan	6.9–7.8	92–2263	110–1240	38–61	NR	Al-Hamaiedeh and Bino (2010)
Showers and hand basins	Denmark	7.6–8.6	77–240	26–130	3.6–6.4	0.28–0.779	Eriksson et al. (2002)
Washing, kitchen, cooking, bathroom	Korea	7.02–7.86	119–3740	23.5–392.4	NR	NR	Jong et al. (2010)
Bathroom	Uganda	6–8	4699–8427	929–1861	NR	NR	Katukiza et al. (2014)
Bathroom sinks, baths and showers	UK	NR	33–138	8–34	NR	NR	Winward et al. (2008a, b, c)
Mixed greywater	Jordan	NR	92–668	NR	NR	NR	Bani-Melhem et al. (2015)

(continued)

A. A. S. Al-Gheethi et al.

Table 1.3 (continued)

Sources	Country	pH	COD	BOD$_5$	TN	TP	References
Bathing greywater	Sweden	NR	NR	NR	3.72 ± 2.21	3.73 ± 2.65	Günther (2000)
Total greywater	Jordan	NR	NR	1056	75	19.5	Halalsheh et al. (2008)
Shower, laundry and washbasins	Jordan	NR	366 (165)	150 (31)	11.4–12.6	10.2–11.8	Abu Ghunmi et al. (2010)
Total greywater	Sweden	NR	350–500	495–682	8.0–11	4.6–11	Palmquist and Hanæus (2005)
Shower, laundry and teeth-brushing	Australia	NR	225	76	0.3	40.8	Chin et al. (2009)
Shower/bathtub	Greece	NR	228–582	NR	8.4 ± 12.6	0.4 ± 0.6	Antonopoulou et al. (2013)
Hand basin		7.07 ± 0.25	335 ± 207	NR	2.6 ± 2.9	0.7 ± 0.9	
Kitchen sink		6.72 ± 0.40	775 ± 363	NR	4.0 ± 4.8	0.4 ± 0.4	
Total greywater	Jordan	5.44	816–2560	600–1710	54.30–155.03	5.20–9.14	Ammari et al. (2014)
Bathroom greywater	Malaysia	6.3–6.73	251–507.5	81–270.8	NR	NR	Oh et al. (2016)
Total greywater	Canada	6.7–7.6	278–435	NR	NR	NR	Finley et al. (2009)
Municipal greywater	Egypt	5.77–7.96	301–526	240–410	7.5–9.2	8.4–12.1	Abdel-Shafy et al. (2014)
Mixed greywater	South African	NR	280–310	NR	206 ± 5.8	40–69	Rodda et al. (2011)

(continued)

Table 1.3 (continued)

Sources	Country	pH	COD	BOD$_5$	TN	TP	References
Bath/shower	India	NR	461	NR	2.1–2.6	0	Vakil et al. (2014)
Washbasin		7.5	225	43	1.6–2.5	0	
Kitchen		6.2	602	293	4.7–11.4	5.3	
Laundry		9.4	824	269	10.7–79	18	
Hostel greywater	India	7.36 ± 0.32	146–168	NR	12 ± 1.5	1.33 ± 0.15	Patil and Munavalli (2016)
Primary settling tank	Norway	6.6–7	640–750	NR	13–16.5	6.8–7.5	Karabelnik et al. (2012)
Bathroom sinks	Brazil	7–7.5	68.4–123.5	45.7–79.7	4.5–7.2	0.7–1.4	do Couto et al. (2013)
Kitchen sinks		6.2–7.4	228.3–912.2	95.2–613.0	8.5–12.4	0.8–8.4	
Showers		7.2–7.8	131–159.0	82.590.7	12.7–32.5	1.2–2.4	
Total greywater	Croatia	NR	433 ± 4	75.1 ± 4.1	NR	NR	Grčić et al. (2015)
Bathroom sinks	Bangladesh	NR	560–804	320–420	NR	NR	Abedin and Rakib (2013)
Kitchen sinks		5.4–6.1	1104–2510	560–600	NR	NR	
Laundry		7.3–7.4	1056–1599	25–110	NR	NR	
Total greywater	Yemen	6 ± 0.35	1200–2000	186–518	7.84–11.28	16.10–29.5	Al-Mughalles et al. (2012)

NR (non-reported)

products used in the shower, washing process and dishwashing as well as the absence of organic matter in greywater generated from shower and laundry and the low concentrations of macro-nutrients such as nitrogen and phosphorus in greywater compared to sewage (Jefferson et al. 1999). Greywater has 90% less nitrogen than that available in sewage due to the absence of urine and faeces as well as toilet paper which leads to an increase in organic content (Al-Gheethi et al. 2016a, b). The high ratio in COD/BOD_5 indicates the presence of high concentrations of non-biodegradable compounds in greywater.

The difference in the type of detergents used for each process might have a significant effect on the concentrations of COD. The results in Table 1.3 indicate that the COD concentrations varied based on the source of greywater. Abedin and Rakib (2013) revealed that the greywater from kitchens has more COD than that from laundry (1104–2510 vs. 1056–1599 mg L^{-1}, respectively). However, Vakil et al. (2014) found that the COD in greywater from laundry was higher than that of kitchen greywater (824 vs. 602 mg L^{-1}). Moreover, COD in greywater from baths is less than that in laundry and kitchen greywater. So far, the reports in literature revealed that the concentrations of BOD_5 in kitchen greywater are more than that in bathing and laundry greywater due to high organic content.

The nutrients in greywater refer to total nitrogen and total phosphorous. Nitrogen in greywater is usually present in organic forms such as ammonia (NH_4^+, NH_3^-N) and inorganic forms such as nitrate (NO_3^-) and nitrite (NO_2^-). Some studies in the literature indicated that ammonium represents the main source of nitrogen more than nitrate and nitrite. Orthophosphate (PO_4^{3-}) is a common form of TP in greywater which results from the utilization of detergent builders and hygiene products (Eriksson et al. 2002).

Both TN and TP are usually determined by the authors, since they reflect the quantity of nutrients. The concentrations of TN and TP in greywater depend on the source. Many authors indicated that TN and TP are high in kitchen greywater while they are present in low concentrations in shower and laundry greywater due to the absence of organic matter. However, detectable concentrations of TN and TP in shower and laundry greywater have been revealed. Vakil et al. (2014) found that the laundry greywater in India contained between 10.7 and 79 mg L^{-1} of TN compared to 4.7–11.4 mg L^{-1} in kitchen greywater. This might be due to the presence of urine which comes from houses with children. Another source of ammonium nitrogen in greywater is the cationic surfactants of fabric softeners and laundry disinfectant agents.

Based on the data presented in Table 1.3, it can be noted that the TN concentrations in most reported studies are more than TP, except for some studies which indicated that the TP was more than TN. For example, Al-Mughalles et al. (2012) found that the TP concentrations in greywater from Yemen were higher than that of TN. A similar result has been reported in a previous study conducted in Australia, where the TN was 0.3 mg L^{-1} while the TP was 40.8 mg L^{-1} (Chin et al. 2009).

In a review of chemical parameters of greywater from developed and developing countries (Table 1.3), the results revealed that the greywater from developing countries has high concentrations of COD and BOD_5 in comparison to that of developed countries. For instance, the COD in greywater generated from household activities in Yemen ranged from 1200 to 2000 mg L^{-1}. In Uganda, it was between 4699 and 8427 mg L^{-1}, in Jordan, it was on average 16.88 mg L^{-1} and in Malaysia, the COD was between 180 and 621 mg L^{-1} (Al-Mughalles et al. 2012; Ammari et al. 2014 Katukiza et al. 2014; Mohamed et al. 2013, 2014). In contrast, the average COD was 399 mg L^{-1} in France (Chaillou et al. 2011). In Denmark, the COD ranged from 77 to 240 mg L^{-1} (Eriksson et al. 2002). Similar results were reported for BOD_5, among different countries. The maximum concentration was noted in greywater from Jordan (600–1710 mg L^{-1}) and Uganda (929–1861 mg L^{-1}) (Ammari et al. 2014; Katukiza et al. 2014). The maximum concentration of BOD_5 of the greywater from developed countries was recorded in Sweden (495–682 mg L^{-1}) (Palmquist and Hanæus 2005).

Little information is available about the concentrations of TN and TP as most studies focus on main parameters which include COD, BOD and pH. The results presented in Table 1.3 from different studies indicate the absence of an association between TN and TP concentrations and development levels for the country. Moreover, the highest concentrations of TN were noted in greywater from Jordan (54.30–155.03 mg L^{-1}) while the lowest concentration was reported in Egypt (7.5–9.2 mg L^{-1}) and Yemen (7.84-11.28 mg L^{-1}) (Al-Mughalles et al. 2012; Abdel-Shafy et al. 2014). However, these differences were not related to the development level of the countries, Jordan, Egypt and Yemen are developing countries and located in the same geographical area. So far, the studies conducted in Jordan which are presented in Table 1.3 indicate that the TN of different types of greywater is high. This might be due to the low quantities of greywater generated compared to the quantities in other countries. The high amount of water usage might lead to the dilution of the main parameters of greywater. A similar observation was also noted for TP where the maximum concentration was recorded in greywater from South Africa (40–69 mg L^{-1}) (Rodda et al. 2011), while the minimum concentration was noted in Greece (0.1–1 mg L^{-1}).This may be due to limited use of phosphorus-containing detergents in baths and washing machines (Fountoulakis et al. 2016).

Based on the above discussion, it can be concluded that the main parameters of greywater vary in their concentrations. These findings were also noted by previous studies which indicated that greywater has a wide range in the most common constituents compared to wastewater. Moreover, the differences in the characteristics of greywater among developed and developing countries might be related to the differences in tap water quality and experimental protocols used for greywater sampling as well as the small sample size in terms of participants surveyed (Eriksson et al. 2002).

1.4.3.1 Heavy Metals

Heavy metals is a term which refers to a wide range of elements in the periodic table. These metals are classified into groups based on toxicity and biological roles. Some heavy metals such as Cu, Zn and Ni play an important role as cofactors at low concentrations for biological reactions during the metabolic and anabolic pathways of cells. These metals are called trace elements or microelements, but a high concentration of these metals is toxic to the cells. In contrast, other metals such as Pb, Cd, Hg, Ag, Cd and Co are toxic even at low concentrations and have no function in biological processes. Heavy metals in greywater originate mainly from the detergents and chemical products used for bathing and clothes washing as well as dishwashing (Leal et al. 2007).

Others metals including Na, K, Ca and Mg which have no toxicity are called macro-elements (Lim et al. 2010). However, the main concern with heavy metal lies in their accumulation in the plant tissue and their transmission into humans and animals since heavy metals are non-biodegradable. Hence, their accumulation in the organs can cause several diseases in high doses (Epstein, 2002; Banana et al. 2016). The list of the most common metals in greywater is presented in Table 1.4. These heavy metals are present in concentrations less than the standard, while macro-elements are available in greywater in high concentrations.

Palmquist and Hanæus (2005) detected 22 metals ions in the greywater generated from ordinary Swedish households including Al, Ba, Bi, Ca, Cd, Co, Cr, Cu, Fe, K, Mg, Mn, Na, Ni, Pb, Pt, S, Sb, Sn, Te and Zn.

1.4.4 Microbiological Characteristics

Authors assume that greywater is free from harmful bacteria due to the absence of solid faeces (Finley et al. 2009). Many infectious agents in greywater have been reported in literature. Among them, *Staphylococcus aureus* and *Pseudomonas aeruginosa*, total coliforms (TC), faecal coliforms (FC), *Escherichia coli*, Enterococci, *Klebsiella pneumoniae* and *Salmonella* spp. as well as *Cryptosporidium parvum* and *Giardia lamblia* (Rangel-Martínez et al. 2015; Al-Gheethi et al. 2016a). Total coliform (TC) is a bacterial group which includes four genera which are *Citrobacter* sp. *Enterobacter* sp. *Escherichia* sp. and *Klebsiella* sp. These genera are Gram-negative bacteria with rod shapes and have the ability to ferment lactose and produce gas at 37 °C within 24–48 h. Faecal coliform is a subgroup of TC and includes *E. coli* and some strains of *K. pneumonia* which have the ability to ferment lactose and produce gas at 44.5 °C within 24 h.

The diversity of infectious agents in greywater depends on public health and the number of occupants in a house. These variations occur not only geographically but also seasonally on the same site. In houses with a high number of occupants, high concentrations and a diversity of microbial species are present with low levels of infectious agents even in the presence of some infected residents. In contrast, in

Table 1.4 Elements found in greywater and related sources

References	Elements concentrations (mg L^{-1})								
	Al	As	B	Ba	Ca	Cd	Cl	Cr	Cu
Matos et al. (2014)	5.8–5.1	0.01	0.2	0.02	9.0–12.0	0.07	72.0–83.0	0.1	0.16
Leal et al. (2007)	0.49–7.35		0.42–0.87		60.79–65.53				0.08–0.12
Palmquist and Hanaeus (2005)	1.48 3.39			0.02	31–38			0.002–0.005	0.047 0.07
Eriksson and Donner (2009)						0.00006 0.0002			
Eriksson et al. (2002)						0.0001–0.0002			
Surendran and Wheatley (1998)						0.52–0.63			
Jefferson et al. (2000)	0.003				47.9	0.056–2.5			0.006
Ammari et al. (2014)					52–80		126–216		
Turner et al. (2016)	0.05–1.5	0.0001–0.012	0.005–0.52			0.0002–0.002		0.0003–0.003	0.008–0.390
Finley et al. (2009)					30–44				
Rodda et al. (2011)									0.1 ± 0.1
Saumya et al. (2015)					4.654				

(continued)

Table 1.4 (continued)

References	Elements concentrations (mg L^{-1})								
	Fe	K	Mg	Mn	Na	Ni	Pb	Hg	Zn
Matos et al. (2014)	0.48–0.63		6.0–7.0	0.1	170–200	0.1	0.1		0.11–0.1
Leal et al. (2007)	0.11–1.28		6.15–30.55		86.35–159.75				
Palmquist and Hanæus (2005)	0.18–0.57	7.69–8.85	5.30–6.22	0.01.014	61.4 92.4	0.04 0.03	0.002–0.003	0.00002	0.06 0.08
Eriksson and Donner (2009)						0.005.1–0.027	0.0049–0.010	0.0006–0.036	
Eriksson et al. (2002)						<25	3–5	0.29	
Surendran and Wheatley (1998)						1.3–28	0.61–6.9	0.022–0.26	
Jefferson et al. (2000)	0.017	5.79	5.29	0.04					0.03
Ammari et al. (2014)			14.3–64.4						
Turner et al. (2016)	0.002–0.56			0.002–0.06		0.0003–0.01	0.0002–0.01		0.017–0.31
Finley et al. (2009)	0.09	2.2–2.5	8.0–9.9		20–27				0.04–0.42
Rodda et al. (2011)		31 ± 2.7	7.5 ± 1.7		188 ± 27				0.24 ± 0.4
Saumya et al. (2015)	0.203	3.286	25.23		3.768		1.72		

Otherelements reported in literature include Ag, Sn, Sb, Co and Bi houses with few occupants, the concentration and diversity of microorganisms may be low but the percentage of infectious agents might be high if some members are infected.

Pathogen diversity in greywater depends on the source of waste. A high concentration of microorganisms in greywater is related to the presence of nutrients. Suspended and dissolved solids provide a suitable medium for microbial growth in greywater. For example, kitchen sink and dishwater greywater are often highly contaminated due to the presence of food and grease particles. Showers, hand basins and washing machines might also be a source for pathogenic bacteria in greywater. However, pathogenic microorganisms in the greywater from showers are different from that of kitchen greywater. Moreover, the main hazard of greywater is associated with faecal contamination which comes from contaminated faecal washing clothes, child care and washing raw meats. Marjoram (2014) revealed that the TC and FC are high in bathroom and laundry greywater in houses with children. The greywater generated from the houses with children are associated with high concentrations of FC. This is because faeces and urine represent the main source for FC. The pathogenic bacteria diversity with their concentration in greywater from developed and developing countries can be found in Table 1.5. The most common pathogenic bacteria in greywater include *E. coli*, Enterococci, *P. aeruginosa, S. aureus, K. pneumoniae, C. perfringens* and *Salmonella* spp. The researchers in developed and developing countries are looking for different bacterial species. However, TC and FC can be detected by all of them. Furthermore, in developed countries such as the UK, Sweden and Israel, researchers focus on the detection of *P. aeruginosa, S. aureus and C. perfringens* while they focus only on TC and FC in developing countries. This would be due to the absence of advanced technologies in developing countries which is required to isolate and identify pathogenic bacteria.

The most common pathogenic bacteria in the greywater generated from bathing and laundry are *P. aeruginosa* and *S. aureus* because these bacteria are part of the normal flora which colonizes the body surface as well as the mouth, nose and ears. In contrast, the content of *Salmonella* spp. and *Campylobacter* spp. might be high in kitchen greywater. Moreover, inclusion of bathing and laundry greywater with kitchen greywater might improve the bacterial multiplication and growth due to the presence of high nutrient content (Rose et al. 2002; Ottoson and Stenström 2003; Al-Gheethi et al. 2016a). The pathogenic bacteria in greywater are available in very low concentrations. Therefore, the presence or absence of specific bacteria in greywater would depend on the quantity and number of greywater samples analysed as well as the efficiency of the analysis technique. Hence, in many countries, the evaluation of microbiological quality for greywater depends mainly on the concept of indicator bacteria (Sect. 1.4.4.1).

Besides the most common pathogenic bacteria illustrated in Table 1.5, there are other species found in greywater as reported in the literature including *Aeromonas* spp. (Albrechtsen 2002), *Legionella pneumophilia* (Birks et al. 2004), *Mycobacterium* spp. (Albrechtsen 2002) and *Campylobacter* sp. (Albrechtsen 2002). These pathogens were also listed by a study conducted by Winward (2007). Ukwubile

Table 1.5 Concentration of pathogenic bacteria in greywater

Country	Concentration of bacteria (cell 100 mL^{-1})									References
	TC	FC	E. coli	Enterococci	P. aeruginosa	S. aureus	K. pneumoniae	C. perfringens	Salmonella spp.	
USA	8.03×10^7	5.63×10^5		2.38×10^2	1.99×10^4					Casanova et al. (2001)
UK	5.4×10^5		2.8×10^2	2.8×10^2	4.4×10^4	3.4×10^3				Winward et al. (2008c)
Uganda	8.72×10^7		3.7×10^6						2.73×10^4	Katukiza et al. (2014)
Jordan	3.1×10^5	4.4×10^5								Bani-Melhem et al. (2015)
Israel			1.4×10^5		7.9×10^2	5×10^3				Maimon et al. (2014)
Portugal	$4.8\text{--}13 \times 10^7$	$3.7\text{--}43 \times 10^4$								Matos et al. (2014)
UK					4.4×10^4	3.37×10^3				Winward et al. (2008c)
Jordan	1×10^7	3×10^5	2×10^5							Halalsheh et al. (2008)
Jordan	7.1×10^7 to 1.27×10^8		$1.4\text{--}1.9 \times 10^6$							Ghunmi et al.(2010)
Israel			$4.1\text{--}6.3 \times 10^3$		9.4×10^1 to 3.1×10^4	1.2×10^2 to 4.1×10^3	4.1×10^2			Benami et al. (2016)

(continued)

Table 1.5 (continued)

| Country | Concentration of bacteria (cell 100 mL^{-1}) | | | | | | | | | References |
	TC	FC	E. coli	Enterococci	P. aeruginosa	S. aureus	K. pneumoniae	C. perfringens	Salmonella spp.	
Jordan	6.2×10^4–3.89×10^6		3×10^4 to 4.1×10^5							Ammari et al. (2014)
Malaysia	8.5–53.2×10^4		2.5×10^4 to 6.1×10^5							Oh et al. (2016)
Canada	4.7×10^4 to 8.3×10^5			110–3.8×10^5						Finley et al. (2009)
Sweden	5.5×10^5 to 8.7×10^8		4.3×10^4 to 6.8×10^6	3×10^3 to 5.1×10^5				2.3×10^2 to 4.8×10^4		Ottoson and Stenström (2003)
Israel		3.8×10^4			3.3×10^3	9.9×10^3		4.6×10^4		Gilboa and Friedler (2008)
Bangladesh		7.33×10^2 to 22.6×10^4								Abedin and Rakib(2013)
Yemen		18.5–19×10^6								Al-Mughalles et al. (2012)

(2014) revealed the presence of *S. typhi, Vibrio cholera, E. coli, C. jejuni, S. aureus* and *Shigella dysenteriea* in greywater in Nigeria.

On the other hand, the most common parasites reported in greywater include *Cryptosporidium* sp. and *Giardia* sp. with more than 10^7 cell/100 L (Birks et al. 2004; Birks and Hills 2007). Parasites in greywater might be transmitted from faeces and urine but they can also be present in kitchen greywater which are transmitted from vegetables and fruits irrigated with contaminated water since both *Cryptosporidium* sp. and *Giardia* sp. have high potential to form cysts which have a long period of survival in the environment.

1.4.4.1 Indicator Bacteria

Indicator bacteria are a bacterial group or species used as a model to indicate the presence or absence of pathogens in greywater. The indicator bacteria concept has become more popular since 1870 because the direct detection of different types of pathogenic bacteria, viruses, cysts of protozoan parasites and helminths in greywater is an unpredictable and time-consuming procedure.

The indicator bacteria are present in a concentration higher than that of pathogens. The data presented in Table 1.5 show that the concentrations of TC ranged from 10^5 to 10^6 cell 100 mL^{-1}, FC ranged between 10^4 and 10^6 cell 100 mL^{-1}, *E. coli* was present between 10^2 and 10^6 cell 100 mL^{-1} and Enterococci ranged between 10^2 and 10^5 cells 100 mL^{-1}. In contrast, *P. aeruginosa* was between 10^1 and 10^4 cell 100 mL^{-1}, *S. aureus* was present ranging between 10^2 and 10^3 cell 100 mL^{-1}, *Salmonella* spp. and *C. perfringens* were on average 10^3 cell 100 mL^{-1}.

There are several bacterial species suggested as an indicator of the presence of pathogens. Nevertheless, the historical and traditional indicator bacteria are TC and FC, since they exhibited correlation with pathogens such as viruses and parasites. Moreover, some recent studies indicated the absence of correlation between TC, FC and *E. coli*, and some pathogens such as *Salmonella* spp. and *Campylobacter* spp. (Polo et al. 1999; Hörman et al. 2004). Besides, the main criteria for indicator bacteria are that they should not increase and multiply in the environment but Byappanahalli and Fujioka (1998) revealed that FC and *E. coli* have grown in tropical soil environments. Therefore, researchers are looking for alternative indicators with more correlation to the pathogens and at the same time possess resistance towards environmental conditions without growth. In biosolids and sewage effluents, FC and *E. coli* are preferred as indicators because these wastes contain faeces and urine which represent the main source of coliform bacteria. In contrast, in medical waste, *P. aeruginosa* and *S. aureus* were suggested by STAATT (2005) as indicators because these bacteria are opportunistic organisms and thus are associated with clinical waste. *S. aureus* was proposed as an indicator of hospital hygiene for microbiological standards (Dancer 2004). Jin et al. (2012) used *S. aureus* as an indicator to evaluate the hydrothermal treatment process to achieve hygienic safety for food waste. Celico et al. (2004) claimed that Enterococci is a more reliable indicator than TC and FC because they have high potential to survive in the environment in comparison to FC. Enterococci are used frequently as a reference microorganism for pasteurized foods.

The selection of the indicator depends on the type of the waste as it has to reflect the real microbial diversity and loads of the waste (Al-Gheethi et al. 2016b). Therefore, with reference to the microbial diversity in greywater revealed by previous studies, it can be noticed that they include TC, FC, *E. coli*, Enterococci, *P. aeruginosa, S. aureus, K. pneumoniae, C. perfringens* and *Salmonella* spp. Among these organisms, TC, FC, *E. coli* and *K. Pneumoniae* as well as Enterococci originated from greywater with faecal contamination with some exception for *K. pneumoniae* as well as Enterococci which have non-faecal sources. *K. pneumoniae* was the first indicator bacteria suggested by Von Fritsch (Geldreich 1978). However, further research studies indicated that this bacteria has low correlation with pathogenic bacteria compared to *E. coli* (Al-Gheethi et al. 2013). FC and especially *E. coli* may be used as indicators but it failed to reflect the actual load of pathogens due to the absence of faecal source in many types of greywater such as kitchen and sink greywater. *Salmonella* spp. was suggested in some references as an indicator for the evaluation of the treatment process of biosolids since it is typically present in higher densities than other bacterial pathogens and has the ability to survive for a long time in the environment (USEPA 2003). So far, *Salmonella* spp. might be not suitable as an indicator for greywater because it might be present in kitchen greywater resulting from the washing of vegetables and chicken meat, but not from showers and laundry greywater. *C. perfringens* was also suggested in the literature as an indicator for wastewater (Rouch et al. 2011). However, other authors stated that the hardy spores of this bacterium make it too resistant to be useful as an indicator organism (Vierheilig et al. 2013). Besides, some pathogenic bacteria such as *Salmonella* spp. and *C. perfringens* have been suggested as indicators which have a narrow range to represent the total microbial load in greywater. These bacterial species might be more suitable as models to determine the effectiveness of the treatment process of wastewater but are not to be used as indicators to evaluate the microbiological characteristics of different types of wastewater including greywater.

Based on previous studies conducted on the microbiological characteristics of greywater especially that related to microbial loads, *S. aureus* and *P. aeruginosa* might be considered as an indicator for the pathogenic bacteria in greywater due to its importance as an opportunistic pathogen. Besides, these bacteria were reported in different types of greywater such as those generated from showers, laundry and kitchen activities. Both bacterial species are not related to faecal contamination but they can be used as indicators for pathogenic bacteria regardless of its sources. *P. aeruginosa* was suggested for the first time as an indicator in freshwater streams by Wheater et al. (1980). Before, Kenner and Clark (1974) revealed that *P. aeruginosa* has better potential as an indicator than FC in bathing and recreational water. These suggestions are in agreement with the study performed by Warrington (2001), in which *P. aeruginosa* exhibited more efficiency than *E. coli* as an indicator for marine water. Moreover, recently, *P. aeruginosa* was proposed as a new indicator for determining the microbiological quality of sewage effluent (Coronel-Olivares et al. 2011). The authors in literature have revealed that *P. aeruginosa* and *S. aureus* were isolated from wastewater in which *E. coli* was less than the detection limit (Garland et al. 2000; Gross et al. 2007a, b).

1.5 Conclusions

This qualitative review has attempted to give a comprehensive idea about the quantities and qualitative characteristics of greywater in developed and developing countries so that the reader can get an idea of the characteristics of these wastes and the differences in their composition based on the source of generation. Therefore, this aspect needs to be investigated further in order to facilitate the selection of treatment processes and best understand their adverse effects on the environment. Moreover, greywater with a high content of nutrients and pathogenic organisms need to be disposed with care or reused for irrigation purposes which is the common practice in the developing countries.

Acknowledgements The authors wish to thank the Ministry of Higher Education (MOHE) for supporting this research under FRGS vot 1574 and also the Research Management Centre (RMC) UTHM for providing grant IGSP U682 for this research.

References

Abdel-Shafy HI, Al-Sulaiman AM, Mansour MS (2014) Greywater treatment via hybrid integrated systems for unrestricted reuse in Egypt. J Water Process Eng 1:101–107

Abedin SB, Rakib ZB (2013) Generation and quality analysis of greywater at Dhaka City. Environ Res Eng Manage 64(2):29–41

Adendorff J, Stimie C (2005) Food from used water-making the previously impossible happen. The Water Wheel. South African Water Research Commission (WRC), pp 26–29

Albrechtsen HJ (2002) Microbiological investigations of rainwater and graywater collected for toilet flushing. Water Sci Technol 46(6–7):311–316

Alderlieste MC, Langeveld JG (2005) Wastewater planning in; Djenné, Mali. A pilot project for the local infiltration of domestic wastewater. Water Sci Technol 51(2):57–64

Al-Gheethi AA, Ismail N, Lalung J, Talib A, Kadir MOA (2013) Reduction of faecal indicators and elimination of pathogens from sewage treated effluents by heat treatment. Caspian J App Sci Res 2(2)

Al-Gheethi AA, Mohamed RM, Efaq AN, Hashim MA (2016a) Reduction of microbial risk associated with greywater by disinfection processes for irrigation. J Water Health 14(3):379–398

Al-Gheethi AA, Mohamed RMS, Efaq AN, Norli I, Hashim MA, Ab Kadir MO (2016b) Bioaugmentation process of sewage effluents for the reduction of pathogens, heavy metals and antibiotics. J Water Health 14(5):780–795

Al-Hamaiedeh H, Bino M (2010) Effect of treated grey water reuse in irrigation on soil and plants. Desalination 256(1):115–119

Al-Mughalles MH, Rahman RA, Suja FB, Mahmud M, Jalil NA (2012) Household greywater quantity and quality in Sana'a, Yemen. EJGE 17:1025–1034

Al-Saud M (2010) Water sector of Saudi Arabia. The 2nd Japan-Arab Economic Forum, Tunisia

Ammari TG, Al-Zubi Y, Al-Balawneh A, Tahhan R, Al-Dabbas M, Ta'any RA, Abu-Harb R (2014) An evaluation of the re-circulated vertical flow bioreactor to recycle rural greywater for irrigation under arid Mediterranean bioclimate. Ecol Eng 70:16–24

Antonopoulou G, Kirkou A, Stasinakis AS (2013) Quantitative and qualitative greywater characterization in Greek households and investigation of their treatment using physicochemical methods. Sci Total Environ 454:426–432

Atiku A, Mohamed RMSR, Al-Gheethi AA, Wurochekke AA, Kassim AH (2016) Harvesting microalgae biomass from the phycoremediation process of greywater. Environ Sci Pollut Res 23(24):24624–24641

Banana AS, Radin Maya Saphira RM, Al-Gheethi AA (2016) Mercury pollution for marine environment at Farwa Island, Libya. J Environ Health Sci Eng 14:5

Bani-Melhem K, Al-Qodah Z, Al-Shannag M, Qasaimeh A, Qtaishat M, Alkasrawi M (2015) On the performance of real grey water treatment using a submerged membrane bioreactor system. J Membrane Sci 476:40–49

Benami M, Busgang A, Gillor O, Gross A (2016) Quantification and risks associated with bacterial aerosols near domestic greywater-treatment systems. Sci The Total Environ 562:344–352

Birks R, Hills S (2007) Characterization of indicator organisms and pathogens in domestic greywater for recycling. Environ Monit Assess 129(1–3):61–69

Birks R, Colbourne J, Hills S, Hobson R (2004) Microbiological water quality in a large in-building, water recycling facility. Water Sci Technol 50(2):165–172

Bodnar I, Szabolcsik A, Baranyai E, Uveges A, Boros N (2014) Qualitative characterization of household greywater in the northern great plain region of Hungary. Environ Eng Manage J 13(11):2717–2724

Boyjoo Y, Pareek VK, Ang M (2013) A review of greywater characteristics and treatment processes. Water Sci Technol 67(7):1403–1424

Busser S, Nga PT, Morel A, Anh NV (2006) Characteristic and quantities of domestic wastewater in urban and peri-urban households in Hanoi. In: Proceedings of the environmental science and technology for sustainability of Asia, The 6th general seminar of the core university program, Kumamoto

Byappanahalli MN, Fujioka RS (1998) Evidence that tropical soil environment can support the growth of *Escherichia coli*. Water Sci Technol 38(12):171–174

Carden K, Armitage N, Sichone O, Winter K (2007) The use and disposal of greywater in the non-sewered areas of South Africa. WRC Report No. 1524/1/07, Water Research Commission, Pretoria, South Africa

Casanova LM, Gerba CP, Karpiscak M (2001) Chemical and microbial characterization of household graywater. J Environ Sci Health, Part A 36(4):395–401

Celico F, Varcamonti M, Guida M, Naclerio G (2004) Influence of precipitation and soil on transport of fecal enterococci in fractured limestone aquifers. Appl Environ Microbiol 70(5):2843–2847

Chaillou K, Gérente C, Andrès Y, Wolbert D (2011) Bathroom greywater characterization and potential treatments for reuse. Water Air Soil Pollut 215(1–4):31–42

Chin WH, Roddick FA, Harris JL (2009) Greywater treatment by UVC/H_2O_2. Water Res 43(16):3940–3947

Christova-Boal D, Eden RE, McFarlane S (1996) An investigation into greywater reuse for urban residential properties. Desalination 106(1):391–397

Coronel-Olivares C, Reyes-Gómez LM, Hernández-Muñoz A, Martínez-Falcón AP, Vázquez-Rodríguez G, Iturbe U (2011) Chlorine disinfection of *Pseudomonas aeruginosa*, total coliforms, *Escherichia coli* and *Enterococcus faecalis*: revisiting reclaimed water regulations. Water Sci Technol 64(11):2151–2157

Dancer SJ (2004) How do we assess hospital cleaning? A proposal for microbiological standards for surface hygiene in hospitals. J Hosp Infect 56(1):10–15

do Couto EDA, Calijuri ML, Assemany PP, da Fonseca Santiago A, de Castro Carvalho I (2013) Greywater production in airports: qualitative and quantitative assessment. Res Conserv Recycl 77:44–51

Dubey A, Goyal D, Mishra A (2013) Zeolites in wastewater treatment. Green Mater Sustain Water Rem Treat 23:82

Efaq AN, Saeed AA, Mohamed RMSR (2016) Current status of greywater in Middle East countries, a glance at the world. Waste Manage 49:1–5

Epstein E (2002) Land application of sewage sludge and biosolids. CRC Press

Eriksson E, Donner E (2009) Metals in greywater: sources, presence and removal efficiencies. Des 248(1):271–278

Eriksson E, Auffarth K, Henze M, Ledin A (2002) Characteristics of grey wastewater. Urban Water 4:85–104

Faruqui N, Al-Jayyousi O (2002) Greywater reuse in urban agriculture for poverty alleviation: a case study in Jordan. Water Int 27(3):387–394

Finley S, Barrington S, Lyew D (2009) Reuse of domestic greywater for the irrigation of food crops. Water Air Soil Pollut 199(1–4):235–245

Fountoulakis MS, Markakis N, Petousi I, Manios T (2016) Single house on-site grey water treatment using a submerged membrane bioreactor for toilet flushing. Sci Total Environ 551:706–711

Friedler E (2004) Quality of individual domestic greywater streams and its implication for on-site treatment and reuse possibilities. Environ Technol 25(9):997–1008

Garland JL, Levine LH, Yorio NC, Adams JL, Cook KL (2000) Greywater processing in recirculating hydroponic systems: phytotoxicity, surfactant degradation, and bacterial dynamics. Water Res 34(12):3075–3086

Geldreich EE (1978) Bacterial populations and indicator concepts in feces, sewage, stormwater and solid wastes. Galveston Bay Bibliography, GBIC Materials Available at Jack K. Williams Library

Ghaitidak DM, Yadav KD (2013) Characteristics and treatment of greywater—a review. Environ Sci Poll Res 20(5):2795–2809

Ghunmi LA, Zeeman G, Fayyad M, van Lier JB (2010) Grey water treatment in a series anaerobic–aerobic system for irrigation. Bioresour Technol 101(1):41–50

Gilboa Y, Friedler E (2008) UV disinfection of RBC-treated light greywater effluent: kinetics, survival and regrowth of selected microorganisms. Water Res 42(4):1043–1050

Gokulan R, Sathish N, Kumar RP (2013) Treatment of grey water using hydrocarbon producing Botryococcusbraunii. Int J Chem Tech Res 5(3):1390–1392

Grčić I, Vrsaljko D, Katančić Z, Papić S (2015) Purification of household greywater loaded with hair colorants by solar photocatalysis using TiO_2-coated textile fibers coupled flocculation with chitosan. J Water Process Eng 5:15–27

Gregory AW, Hansen BE (1996) Practitioners corner: tests for cointegration in models with regime and trend shifts. Oxford Bull Econ Stat 58(3):555–560

Gross A, Kaplan D, Baker K (2007a) Removal of chemical and microbiological contaminants from domestic greywater using a recycled vertical flow bioreactor (RVFB). Ecol Eng 31(2):107–114

Gross A, Shmueli O, Ronen Z, Raveh E (2007b) Recycled vertical flow constructed wetland (RVFCW)—a novel method of recycling greywater for irrigation in small communities and households. Chemosphere 66(5):916–923

Günther F (2000) Wastewater treatment by greywater separation: outline for a biologically based greywater purification plant in Sweden. Ecol Eng 15(1):139–146

Halalsheh M, Dalahmeh S, Sayed M, Suleiman W, Shareef M, Mansour M, Safi M (2008) Grey water characteristics and treatment options for rural areas in Jordan. Bioresour Technol 99(14):6635–6641

Harikumar PS, Mol B (2012) A synoptic study on the preparation of a liquid waste management plan for Kerala State, India. Environ Natural Res Res 2(2):74

Helvetas (2005) Water consumption in Switzerland (in German: Wasserverbrauch in derSchweiz), Helvetas, Schweizer Gesellschaft für Zusammenarbeit

Hörman A, Rimhanen-Finne R, Maunula L, von Bonsdorff CH, Torvela N, Heikinheimo A, Hänninen ML (2004) Campylobacter spp., Giardia spp., Cryptosporidium spp., noroviruses, and indicator organisms in surface water in southwestern Finland, 2000–2001. Appl Environ Microbiol 70(1):87–95

Howard G, Bartram J, Water S (2003) Domestic water quantity, service level and health. World Health Organization, Geneva

Jais NM, Mohamed RM, Al-Gheethi AA, Hashim MA (2016) The dual roles of phycoremediation of wet market wastewater for nutrients and heavy metals removal and microalgae biomass production. Clean Technol Environ Policy 1–16

Jamrah A, Al-Omari A, Al-Qasem L, Ghani A (2006) Assessment of availability and characteristics of greywater in Amman. Water Int 31(2):210–220

Jamrah A, Al-Futaisi A, Prathapar S, Al Harrasi A (2008) Evaluating greywater reuse potential for sustainable water resources management in Oman. Environ Monit Assess 137, 315e327

Jefferson B, Laine A, Parsons S, Stephenson T, Judd S (1999) Technologies for domestic wastewater recycling. Urban Water 1:285–292

Jefferson B, Palmer A, Jeffrey P, Stuetz R, Judd S (2004) Grey water characterization and its impact on the selection and operation of technologies for urban reuse. Water Sci Technol 50(2):157–164

Jeppesen B, Solley D (1994) Domestic greywater reuse: overseas practice and its applicability to Australia. Urban Water Research Association of Australia

Jin Y, Chen T, Li H (2012) Hydrothermal treatment for inactivating some hygienic microbial indicators from food waste-amended animal feed. J Air Waste Manage Assoc 62(7):810–816

Jong J, Lee J, Kim J, Hyun K, Hwang T, Park J, Choung Y (2010) The study of pathogenic microbial communities in graywater using membrane bioreactor. Des 250(2):568–572

Karabelnik K, Kõiv M, Kasak K, Jenssen PD, Mander Ü (2012) High-strength greywater treatment in compact hybrid filter systems with alternative substrates. Ecol Eng 49:84–92

Katukiza AY, Ronteltap M, Niwagaba C, Kansiime F, Lens PN (2014) A two-step crushed lava rock filter unit for grey water treatment at household level in an urban slum. J Environ Manage 133:258–267

Katukiza AY, Ronteltap M, Niwagaba CB, Kansiime F, Lens P (2015) Grey water characterisation and pollutant loads in an urban slum. Int J Environ Sci Technol 12(2):423–436

Kenner BA, Clark H (1974) Detection and enumeration of *Salmonella* and *Pseudomonas aeruginosa*. J Water Poll Control Fed 2163–2171

Klammer I (2013) Policy of onsite and small-scale wastewater treatment options in Finland. Environmental Engineering, Tampere University of Technology, Finland

Laghari A, Ali Z, Haq MU, Channa A, Tunio M (2015) An economically viable method by indigenous material for decontamination of greywater. Sindh Univ Rese J-SURJ (Sci Ser) 47(3)

Leal LH, Zeeman G, Temmink H, Buisman C (2007) Characterization and biological treatment of greywater. Water Sci Technol 56(5):193–200

Ledin A, Eriksson E, Henze M (2001) Aspects of groundwater recharge using grey wastewater. Chapter 18. In: Decentralized sanitation and reuse. IWA Publishing

Lim S, Chu W, Phang S (2010) Use of *Chlorella vulgaris* for bioremediation of textile wastewater. J Bioresour Technol 101:7314–7322

Maimon A, Friedler E, Gross A (2014) Parameters affecting greywater quality and its safety for reuse. Sci Total Environ 487:20–25

Mara DD, Cairncross S (1989) Guidelines for the safe use of wastewater and excreta in agriculture and aquaculture: measures for public health protection. World Health Organization, Geneva

Marjoram C (2014) Graywater research findings at the residential level. Department of Civil and Environmental Engineering, Colorado State University Fort Collins, Colorado

Matos C, Pereira S, Amorim EV, Bentes I, Briga-Sá A (2014) Wastewater and greywater reuse on irrigation in centralized and decentralized systems—an integrated approach on water quality, energy consumption and CO_2 emissions. Sci Total Environ 493:463–471

Mohamed RM, Kassim AH, Anda M, Dallas SA (2013) Monitoring of environmental effects from household greywater reuse for garden irrigation. Environ Monit Assess 185(10):8473–8488

Mohamed RM, Chan CM, Senin H, Kassim AH (2014) Feasibility of the direct filtration over peat filter media for bathroom greywater treatment. J Mater Environ Sci 5(6):2021–2029

Mohamed, RMSR, Al-Gheethi AA, Kassim AHM (2016a) Reuse of ablution water to improve peat soil characteristics for ornamental landscape plants cultivation. In: 2nd international conference on Sustainable Environment and Water Research (ICSEWR 2016), 5–6 December 2016, Novotel Melaka, Malaysia

Mohamed RM, Al-Gheethi AA, Jackson AM, Amir HK (2016b) Multi natural filter for domestic greywater treatment in village houses. J Am Water Works Assoc (AWWA)

Morel A, Diener S (2006) Greywater management in low and middle-income countries, review of different treatment systems for households or neighbourhoods. Swiss Federal Institute of Aquatic Science and Technology (Eawag), Dubendorf

Mourad KA, Berndtsson JC, Berndtsson R (2011) Potential fresh water saving using greywater in toilet flushing in Syria. J Environ Manage 92(10):2447–2453

Nnaji CC, Mama CN, Ekwueme A, Utsev T (2013) Feasibility of a filtration-adsorption grey water treatment system for developing countries. Hydrol Curr Res

Oh KS, Poh PE, Chong MN, Chan ES, Lau EV, Saint CP (2016) Bathroom greywater recycling using polyelectrolyte-complex bilayer membrane: advanced study of membrane structure and treatment efficiency. Carbohyd Polym 148:161–170

Ottoson J, Stenström TA (2003) Faecal contamination of greywater and associated microbial risks. Water Res 37(3):645–655

Palmquist H, Hanæus J (2005) Hazardous substances in separately collected grey-and blackwater from ordinary Swedish households. Sci Total Environ 348(1):151–163

Patil YM, Munavalli GR (2016) Performance evaluation of an integrated on-site greywater treatment system in a tropical region. Ecol Eng 95:492–500

Polo F, Figueras MJ, Inza I, Sala J, Fleisher JM, Guarro J (1999) Prevalence of Salmonella serotypes in environmental waters and their relationships with indicator organisms. Antonie Van Leeuwenhoek 75(4):285–292

Prathapar SA, Jamrah A, Ahmed M, Al Adawi S, Al Sidairi S, Al Harassi A (2005) Overcoming constraints in treated greywater reuse in Oman. Desalination 186(1):177–186

Rangel-Martínez C, Jiménez-González DE, Martínez-Ocaña J, Romero-Valdovinos M, Castillo-Rojas G, Espinosa-García AC, Maravilla P (2015) Identification of opportunistic parasites and helminth ova in concentrated water samples using a hollow-fibre ultrafiltration system. Urban Water J 12(5):440–444

Rodda N, Salukazana L, Jackson SA, Smith MT (2011) Use of domestic greywater for small-scale irrigation of food crops: effects on plants and soil. Phys Chem Earth Parts A/B/C 36(14):1051–1062

Rose BE, Hill WE, Umholtz R, Ransom GM, James WO (2002) Testing for Salmonella in raw meat and poultry products collected at federally inspected establishments in the United States, 1998 through 2000. J Food Prot 65(6):937–947

Rouch DA, Mondal T, Pai S, Glauche F, Fleming VA, Thurbon N, Deighton M (2011) Microbial safety of air-dried and rewetted biosolids. J Water Health 9(2):403–414

Santamasas C, Rovira M, Clarens F, Valderrama C (2013) Grey water reclamation by decentralized MBR prototype. Res Conserv Recycl 72:102–107

Saumya S, Akansha S, Rinaldo J, Yayasri MA, Suthindhiran K (2015) Construction and evaluation of prototype subsurface flow wetland planted with Heliconia angusta or the treatment of synthetic greywater. J Cleaner Prod 91:235–240

Schäfer AI, Nghiem LD, Óschmann N (2006) Bisphenol A retention in the direct ultrafiltration of greywater. J Membr Sci 283(1):233–243

STAATT (2005) Technical assistance manual: state regulatory oversight of medical waste treatment technology. Report of the state and territorial association on alternative treatment technologies (STAATT)

Surendran S, Wheatley AD (1998) Grey-water reclamation for non-potable re-use. Water Environ J 12(6):406–413

Teh XY, Poh PE, Gouwanda D, Chong MN (2015) Decentralized light greywater treatment using aerobic digestion and hydrogen peroxide disinfection for non-potable reuse. J Cleaner Prod 99:305–311

Turner RD, Warne MSJ, Dawes LA, Vardy S, Will GD (2016) Irrigated greywater in an urban sub-division as a potential source of metals to soil, groundwater and surface water. J Environ Manage 183:806–817

Ukwubile CA (2014) Microbial analysis of greywater from local bathrooms and its health implications in bali local government area Taraba state Nigeria. J Adv Biotechnol 2(1):48–57

USEPA (2003) Ultraviolet disinfection guidance manual. Office of Water; 2003. U.S. Environmental Protection Agency, Washington, D.C

Vakil KA, Sharma MK, Bhatia A, Kazmi AA, Sarkar S (2014) Characterization of greywater in an Indian middle-class household and investigation of physicochemical treatment using electrocoagulation. Sep Purif Technol 130:160–166

Vierheilig J, Frick C, Mayer RE, Kirschner AKT, Reischer GH, Derx J, Farnleitner AH (2013) *Clostridium perfringens* is not suitable for the indication of fecal pollution from ruminant wildlife but is associated with excreta from nonherbivorous animals and human sewage. Appl Environ Microbiol 79(16):5089–5092

Warrington PD (2001) Water quality criteria for microbiological indicators overview report. Resource Quality Section Water Management Branch, Ministry of Water, Land and Air Protection, U.K

Wheater DWF, Mara DD, Jawad L, Oragui J (1980) *Pseudomonas aeruginosa* and *Escherichia coli* in sewage and fresh water. Water Res 14(7):713–721

Wilderer PA (2004) Applying sustainable water management concepts in rural and urban areas: some thoughts about reasons, means and needs. Water Sci Technol 49(7):7–16

Winward GP (2007) Disinfection of Greywater. Thesis submitted in partial fulfilment of the requirements for the degree of Doctor of Philosophy, Cranfield University Centre for Water Sciences, Department of Sustainable systems, School of Applied Sciences, p 189

Winward GP, Avery L, Stephenson T, Jefferson B (2008a) Chlorine disinfection of grey water for reuse: effect of organics and particles. Water Res 42, 483e491

Winward GP, Avery LM, Frazer-Williams R, Pidoua M, Jeffrey P, Stephenson T, Jefferson B (2008b) A study of the microbial quality of grey water and an evaluation of treatment technologies for reuse. Ecol Eng 32:187–197

Winward GP, Avery LM, Stephenson T, Jefferson B (2008c) Chlorine disinfection of grey water for reuse: effect of organics and particles. Water Res 42:483–491

Wurochekke AA, Mohamed RMS, Al-Gheethi AA, Amir HM, Matias-Peralta HM (2016) Household greywater treatment methods using natural materials and their hybrid system. J Water Health 14(6):914–928

Chapter 2
Consequences of the Improper Disposal of Greywater

**Efaq Ali Noman, Adel Ali Saeed Al-Gheethi,
Radin Maya Saphira Radin Mohamed, Balkis A. Talip, H. Nagao,
Amir Hashim Mohd Kassim and Siti Asmah Bakar**

Abstract Discharge of greywater into the environment and natural water bodies is the main challenge in the management of greywater. The increase of greywater disposed into the environment has drawn serious attention from the society and the government who endeavour to find a safe alternative way for the disposal of these wastes. The implication for the improper disposal of greywater is associated with infectious agents. This is because the organisms are able to multiply in the environment and might reach the infective dose which causes several diseases in human and animals. In this chapter, the health risks and effects posed by pathogens and heavy metals in disposed greywater to the environment and humans are reviewed. The chapter discusses the level of risk for each component in greywater. It has appeared that eutrophication and water bloom are associated with the discharge of greywater into the natural water due to the high level of nutrients.

Keywords Discharge · Health risk · Heavy metals · HABs · Regulations

E. A. Noman
Faculty of Applied Sciences and Technology (FAST), Universiti Tun Hussein Onn Malaysia
(UTHM), Pagoh, Johor, Malaysia

E. A. Noman
Department of Applied Microbiology, Faculty Applied Sciences, Taiz University, Taiz, Yemen

A. A. S. Al-Gheethi (✉) · R. M. S. Radin Mohamed · A. H. Mohd Kassim · S. A. Bakar
Micro-Pollutant Research Centre (MPRC), Department of Water and Environmental Engineering,
Faculty of Civil and Environmental Engineering, Universiti Tun Hussein Onn Malaysia (UTHM),
86400 Parit Raja, Batu Pahat, Johor, Malaysia
e-mail: adel@uthm.edu.my

R. M. S. Radin Mohamed
e-mail: maya@uthm.edu.my

B. A. Talip
Faculty of Applied Sciences and Technology (FAST), Universiti Tun Hussein Onn Malaysia
(UTHM), 84000 KM11, Jalan Panchor, Pagoh Muar, Johor, Malaysia

H. Nagao
School of Biological Sciences, Universiti Sains Malaysia (USM), 11800 George Town, Penang,
Malaysia

© Springer International Publishing AG, part of Springer Nature 2019   33
R. M. S. Radin Mohamed et al. (eds.), *Management of Greywater in Developing Countries*,
Water Science and Technology Library 87,
https://doi.org/10.1007/978-3-319-90269-2_2

2.1 Introduction

The disposal of greywater into the environment is common in developing countries, which have several water resources, while countries which have deficiency in water resources normally reuse greywater for the purpose of irrigation. In the coastal cities, the greywater is discharged into the sea. In contrast, the disposal of these wastes into the valley (*wadi*) is common in mountain cities (Al-Gheethi et al. 2015a). In village regions, the discharge of greywater into the drainage is performed due to the absence of sewerage network systems and centralized wastewater treatment plants (Fig. 2.1).

The absence of these facilities in the rural regions is due to economic considerations. The separation of greywater from black water and then the discharge of greywater into the environment mainly aim to reduce the heavy load on the septic tank. Nonetheless, the discharge process of these wastes directly into drains or rivers without treatment represents a real hazard for the environment and humans (Atiku et al. 2016). The adverse effects of the discharge of greywater into the environment and natural water bodies include the contamination of water resources, transmission of infectious agents, reduction the biological diversity in nature as well as eutrophication of freshwater and water bloom (Wurochekke et al. 2016).

Fig. 2.1 Greywater discharge into the drain without any treatment. *Photo* was taken at 20/11/2016

Some of these effects occur for a short time while others have long-term effects. For wastewater, several countries have adopted strict regulations for the disposal process. However, the regulations for the discharge of greywater are still not available for most developing countries. The health risks associated with the discharge of greywater represent a real problem in developing countries due to the absence of the service and sanitation (Rodda et al. 2011). Most studies focused on the chemical contamination of water bodies which receive released greywater from various sources due to their potential to persist for a long period in the environment. Moreover, microbial contamination should be highlighted due to their ability to survive and propagate. Therefore, microbes can multiply to reach the infective dose and become more hazardous for humans and animals. Nevertheless, chemical risk might have more effect on plant growths while microbial risk has less effects on plant growth due to the absence of virulence factors to infect plants as these pathogens are specialized to infect humans and animals at an optimal temperature of 37 °C (Al-Gheethi et al. 2013). Therefore, they can be transmitted through plants to animals and humans where it becomes more effective. Besides, microorganisms have a more complicated mechanism to survive in the environment compared to chemical compounds which might exhibit resistance towards degradation. Moreover, the presence of chemical compounds in the environment even at low concentrations as in the case of antibiotics might induce the microbial resistance of toxic compounds and survival for long periods. The discharge of greywater is a common practice in countries with several water resources such as Malaysia. In contrast, the reuse of greywater for irrigation purposes is most common in countries that have water shortage problems. The chemical composition of the greywater such as high chemical oxygen demand (COD) and total suspended solids (TSS) as well as pH might change natural water characteristics and induce the eutrophication phenomenon. This chapter focuses on the health risks associated with the chemical and microbial contents of discharged greywater while the reuse of greywater for irrigation is reviewed in Chap. 4.

2.2 Microbial Risks

Information regarding infectious agents in greywater might highlight their potential risk to the environment and humans. This information is very important and must be considered by the regulations adopted by countries. The evaluation of the risks associated with greywater represents one of the main goals of the management process, as the reduction of these pathogens may reduce the biohazard risk of greywater and its safe discharge into the environment. Moreover, the critical step in the management process of greywater lies in the reduction of infectious agents in greywater as they might have the ability to multiply during the storage process which precedes its discharge into the environment. This is because environmental factors of greywater such as the presence of carbon, nitrogen and phosphorus source might improve the multiplication and growth of pathogens and then reduce dissolved oxygen (DO) which would add more challenges to the safe disposal of greywater into the water bodies.

Microbial load of greywater depends on its source and composition. Pathogenic bacteria generated from bathing and showering (from traces of urine and faeces), laundry (from soiled nappies, clothing or bedclothes) and kitchen greywater (from washing of food) are suspected to be associated with the contamination in greywater (Rodda et al. 2011). *Bacillus* spp., *Staphylococcus* spp., *Klebsiella* spp., *Enterococcus* sp., *Salmonella* spp., *Escherichia coli*, *Pseudomonas aeruginosa* and *Campylobacter* sp. *Clostridium perfringens, Cryptosporidium parvum* and *Giardia lamblia* have been detected as the most common pathogenic bacteria in greywater (Winward et al. 2008; Casanova et al. 2001; Katukiza et al. 2014; Maimon et al. 2014; Rangel-Martıneza et al. 2015; Bani-Melhem et al. 2015; Al-Gheethi et al. 2016a).

Pathogens in discharged greywater might cause a wide range of infections depending on the nature of contact with these wastes. The direct contact of greywater with the skin might lead to infections of the eyes and infection of the skin as the ingestion of pathogens may cause gastroenteritis (Winward 2007). The basic hygienic risk with wastewater includes greywater as the occurrence of pathogens represents the starting point for understanding the epidemiological reflections and necessary precautions (Strauch 1991).

There are three types of health risks related to the presence of pathogenic bacteria in greywater which have to be considered during the disposal process. These risks include occupational health risks, risks concerning product safety and environmental risks (Strauch 1998). Factors which affect the potential of microorganisms to represent a real risk for humans or animals include the ability to survive under hard environmental conditions and the production of toxins and spores which differ among microorganism species. In this section, the potential survival and transmission of the most common pathogens in discharged greywater are reviewed.

The main concern of pathogens in greywater is associated with the bacterial species because the bacterial cell can multiply and regrow in the environment without the need for an intermediate host as in viruses and parasites. In contrast, parasites such as *Cryptosporidium parvum* and *G. lamblia* have high potential to produce cysts which represent the dormant state for these pathogens and can survive for long periods in the environment (Efaq et al. 2015; Al-Gheethi et al. 2016a).

Pathogenic bacteria in greywater can be divided into two groups: plant pathogenic bacteria and human pathogenic bacteria. Plant pathogenic bacteria are those coming from kitchen greywater. These pathogens grow at an ambient environmental temperature. Therefore, the treatment processes performed at mesophilic temperature might be enough to reduce their concentrations. In contrast, human pathogens, which include those resulting from laundry and shower greywater, have an optimal temperature growth of 37 °C. However, they might survive in an ambient environment. In terms of reduction, the inactivation of these pathogens needs a treatment process with a temperature more than 37 °C as thermophilic, but the mesophilic treatment might also have a significant effect for reducing these pathogens based on the competition with indigenous microorganisms (Al-Gheethi et al. 2013).

In terms of human pathogenic bacteria, wide ranges of pathogens and opportunistic pathogens, which have the potential to cause primary or secondary infection in humans and animals, have been reported in greywater. It has been estimated that the infections caused by pathogens from water affect more than 250 million people annually lead to 10–20 million deaths (Anonymous 1996). Pathogenic bacteria in greywater represent real hazards for humans due to virulence factors as well as the ability to resist a range of antibiotics. Several bacterial species with the resistance for antibiotics have been isolated from the environment. The source of these bacteria might be greywater or other types of wastewater but bacteria with high potential to resistant antibiotics indicate the health risks associated with the presence of these pathogens in the environment. Al-Gheethi et al. (2015b) have recovered a resistant bacterial species belonging to *Staphylococcus aureus, Enterococcus faecalis, Klebsiella pneumonia and E. coli* from discharged restaurant greywater. These bacterial species exhibited multiresistance for amoxicillin, ampicillin, tetracycline, ciprofloxacin and erythromycin. The restaurant greywater might not represent the main source for antibiotics residues in waste but the resistance among bacterial species might have resulted from gene transmission between bacterial strains when they are living in close proximity.

Among 2000 serotypes of *Salmonella* spp., two serotypes of *Salmonella, S. typhi* and *S. paratyphi* (A, B, C), are most dangerous to people. The high pathogenicity of *Salmonella* spp. is due to their ability to infect nearly all living vectors from insects to mammals as well as resistance towards a wide range of antibiotics. It has revealed that 39% of *S. typhimurium* exhibited high resistance towards one or more antibiotics, while 23% had multiresistance towards five antibiotics (Strauch 1991). *S. typhi* and *S. paratyphi* are the main causes for typhoid and paratyphoid fever (Bumann et al. 2000). *Salmonella* spp. in greywater might come from kitchen wastewater. The main sources of these bacteria are chickens which are considered the main reservoir host for *Salmonella* spp. while humans are the primary host for these bacteria.

E. coli poses less risk for humans except for three serotypes including Enteropathogenic *E.coli* (EPEC), Enterotoxigenic *E. coli* (ETEC) and Enteroinvasive *E. coli* which have high potential to cause gastrointestinal disorders such as cramping, abdominal pain and bloody diarrhoea (Buzrul 2009). *E. coli* is one of the main causes of meningitis and pneumonia infection in new babies (Dobrindt and Hacker 2008). *E. coli* is a Gram-negative bacterium where 80% of the strains are motile with fimbriates type K which are used for hemagglutinins to attach on the epithelial cells. *E. coli* has three types of antigens including antigen O (somatic), K (capsular) and H (flagellar). These antigens are used for identifying the pathogenic strains of *E. coli* strains (Fijalkowski et al. 2014)

K. pneumonia is a Gram-negative bacterium with rod shapes. They are classified as the main cause of bronchopneumonia and bronchitis. The difficulty in the treatment of *K. Pneumonia* infection lies in their ability to resist different types of antibiotics and the production of β-lactamases which degrade antibiotics belonging to β-lactam group due to their ability to produce extended-spectrum β-lactamases (Podschun and

Ullmann 1998; Renois et al. 2011). *P. aeruginosa* in greywater has been reported by many authors in the literature and has been suggested as indicator bacteria to evaluate the quality of greywater (Al-Gheethi et al. 2016a). *P. aeruginosa* is a rod bacterium with Gram-negative reaction. It has a potential pathogenicity to cause secondary infection among patients infected with autoimmune diseases or HIV. *P. aeruginosa* differs from other Pseudomonas species in its ability to cause disease in humans and other mammals (Kiil et al. 2008).

E. *faecalis* has also been reported as a major nosocomial pathogen (Facklam 1991). *E. faecal* is a type of Gram-positive bacteria and is usually spherical in shape. It has been reported as being multiresistant for β-lactam and aminoglycosides antibiotics (Al-Gheethi et al. 2015b). It can cause several diseases as a hospital infection (Freitas et al. 2011). *Staphylococcus* spp. are defined as normal flora bacteria colonizing human skin and nose. However, it has a high potential to become more pathogenic for humans if it has entered the blood system. There are many reasons that induce this bacterium to convert from normal flora to opportunistic bacteria. Among those factors are the deficiencies in the opsonization in the human body, phagocytosis and chemotaxis as well as immunodeficiency. *S. aureus* with coagulase test positive represents the most medical strains due to the ability of *S. aureus* to produce enterotoxins (class A, B, C, D and E) and its high resistance for numerous antibiotics. It is also known as the first penicillinase-producing bacteria (recently known as β-lactamase enzyme) which leads to the inactivation of penicillin antibiotics. Methicillin-resistant *S. aureus* (MRSA) is a *S. aureus* strain which has the ability to resist most antibiotics in medical care (Vickery 1993). *S. aureus* has been suggested as indicator bacteria for greywater and medical wastes. Despite that, *S. aureus* has the ability to grow in salt medium. They are not common in seafood, therefore the source of this bacterium in the kitchen wastewater might come from other food or vegetables or from humans because these bacteria can be found in concentrations of 10–100 cells on the skin. One of the recent diseases caused by *S. aureus* is called shock syndrome toxin-1. *S. aureus* also produces epidermolytic toxins which lead to the damage of mucoid polysaccharide matrix of the skin.

In order to assess the microbial risk of wastewater before reuse or disposal, it is necessary to determine the microbial pathogen species present as well as its relative numbers (El-Lathy et al. 2009). Several factors affect the potential of pathogenic bacteria from discharged greywater to cause infection in humans including the ability of bacteria to survive in the environment, infective dose and pathogenicity. Many authors have documented the survival of pathogenic microorganisms in landfills. Winfield and Groisman (2003) indicated that *Salmonella* spp. have the potential to survive for different periods based on the environmental conditions. For example, it has been found to survive for weeks in aquatic environments and survived for more than one month in sediments and soils (Moore et al. 2003). Al-Gheethi et al. (2014) found that *Salmonella* spp. survived and recovered from wastewater stored for 28 days at room temperature. *E. coli* has a survival period ranging between 3 months to a year (AWWA 2006). *E. coli* can survive in non-sterile mineral water up to 70 days, while it persisted for 7 weeks in sterile mineral water and 3 weeks in sterile distilled deionised water (Kerr et al. 1999). This might be due to the absence of nutrients and trace elements in sterilized mineral water since the bacterial cells reduced quickly in sterile distilled deionised water. However, WHO (2003) indicated that the pathogen cells survive better in sterile water than in natural mineral water due to the absence of competition with the indigenous bacterial flora.

Vibrio cholera has short to long survival periods (1–4 weeks) in water based on the presence or absence chlorine and trace elements. In de-chlorinated tap water, it can persist for 10 days, while the presence of Fe element might increase the survival ability to 2 weeks (Patel et al. 1995; Joseph and Bhat 2000). These bacteria are available in concentrations of more than 10^2 cells in raw seafood. They normally live on the outer surface of fish and other types of seafood. Therefore, they can be transmitted into kitchen wastewater during the washing process of these foods. *Vibrio cholera* has no pathogenicity in nature because they do not have the ability to produce toxins. On the other hand, the induction for the production of toxicity takes place inside the human host.

Clostridium sp. is a type of Gram-positive bacteria with a cylindrical shape with heat stable and subterminal endospores. This bacterium has been suggested as an indicator for testing the efficiency of wastewater treatment processes due to their ability to produce spores and survive for a long time in the environment (Fujioka et al. 1997; Rouch et al. 2011). High diversity has been identified among the *Clostridium* spp. species. Most of their species are strictly anaerobic with exceptions for some species such as *Clostridium histolyticum* and *Clostridium tertium* which appear to occur as aero-tolerant bacteria. This bacterium has a Gram-positive reaction. However, *Clostridium ramosum* and *Clostridium clostridioforme* are Gram-negative bacteria. Some species of this bacterium such as *C. perfringens* are nonmotile cells and might be found in kitchen wastewater because it is recognized as the most common cause of food poisoning. *C. perfringens* has a high potential to produce enterotoxin that causes diarrheal diseases and cytotoxin which causes necrosis disease. *Clostridium difficile* is another species of *Clostridium* spp. which is classified as the third bacteria causing diarrhoea after *Salmonella* spp. and *S. aureus*. The infection for this bacterium is associated with the consumption of antibiotics (antibiotic-associated diarrhoea). *Shigella* spp. is very sensitive to environmental conditions such as high temperature and the presence of salts. It can survive for a maximum of 5–7 days in fresh water but dies within 2 days in salt water (FDA 2012).

Greywater has less organic content in comparison to domestic wastes. However, the amount of organic matter and nutrients in greywater is enough to support bacterial growth or at least prolong its survival period. Therefore, the concerns of health risks related to greywater lie in the potential of infectious agents for regrowth and the transmission to humans and animals through the food chain (Rose et al. 1999). So far, the ability of bacterial cells to survive does not mean the ability of these cells to cause infection in humans or animals. Survival is known as the potential of the pathogen cell to propagate indefinitely when placed in a suitable environment (Davis et al. 1968). Survivability is a critical stage for infectivity; however, microorganism cells might survive in the environment but have no ability to cause infection (Cox 1987).

The survival of pathogenic microorganisms in the environment depends on the type of microorganism and the nature of pathogens. For instance, Gram-positive bacteria such as *Bacillus* spp. with thick cell walls and the formation of endospores have a higher chance to survive compared to Gram-negative bacteria with thin cell walls. The temperature is a critical factor for organisms to survive. Bacterial cells from showers and laundry might survive with a temperature of 37 °C which is the optimal temperature for their growth. In contrast, the bacterial species from kitchen greywater might survive better at ambient temperatures between 25 and 30 °C. Moreover, the increase of the surrounding temperature to more than that required for optimal growth may lead to the destruction of metabolic enzymes. Bacterial cells can survive longer in warm-temperate zones than subtropical zones (Wang et al. 2014). Others factors which have a direct influence on the survival of bacteria in the environment include pH and competition between pathogenic bacteria from greywater and indigenous microorganisms in the environment and natural water system (Guan and Holley 2003). Most bacteria grow well at a neutral pH between 6 and 8. Therefore, the discharge of greywater into acidic soil might accelerate the death of bacterial cells. In contrast, the bacterial cells have a high potential to survive in neutral or alkaline soils (Wang et al. 2014).

Both *G. lamblia* and *Cryptosporidium oocysts* have a high ability to form cysts which enable them to resist the disinfection process of greywater and survive better in the environment (Iacovski et al. 2004). *Giardia lamblia* cysts can survive for two months in natural water (DeRegnier et al. 1989). In contrast, *C. oocysts* have the ability to persist for 6–12 months in the environment (King et al. 2005). *G. lamblia* and *C. oocysts* are among the parasites which cause gastrointestinal illness in humans (Coupe et al. 2006). The main source for these pathogens is contaminated drinking water (Kumar et al. 2016).

In terms of the competition between microorganisms in the environment that receive greywater, the bacterial cells from showers and laundry greywater might be more competitive than those generated from kitchen greywater. Faecal bacteria might be transmitted from tier primary environment which is the human body to a secondary environment. Franz et al. (2011) revealed that *E. coli* strains recovered from human specimens have been found to survive for 211 days, while those strains obtained from samples taken from animals survived for 70 days. These findings demonstrate that the source of bacteria plays an important role in their ability to face harsh environmental conditions. In order to best assess the microbial risk of greywater, the nature of competitive interactions between infectious agents in greywater and the indigenous organisms in the environment and water systems that received these wastes must be understood. The ability of pathogens from greywater to compete with normal flora in the natural ecosystem shows the ability of these pathogens to remain active during the transmission to humans via drinking water and thereby causing diseases (Konopka 2009).

Ducluzeau et al. (1976) first reported the influence of autochthonous bacteria on indicator bacteria in mineral water. The study revealed that *E. coli* reduced by 90% within 90 days, while the complete loss of viability occurred within 35 and 55 days in the presence of autochthonous mineral flora. Furthermore, studies indicated that

Salmonella spp. and *E. coli* are hardy pathogens which possess moderate to high potential to survive in water but are not highly competitive microorganisms in water ecosystems (Merikanto et al. 2012).

Gram-negative bacteria, such as *Proteus mirabilis, P. aeruginosa, E. coli, Salmonella* spp.*, K. pneumonia*, and *Campylobacter* spp. have no mechanism to produce endospores as in some Gram-positive bacteria, but their ability to survive in the environment has been reported by many authors in the literature. This potential belongs to the characteristics of Gram-negative bacteria to be available but non-culturable (VBNC). Therefore, the absence of these pathogens in the greywater or natural water which receive discharged greywater does not mean that these systems have no pathogens. Instead, it indicates that the efficiency of the isolation technique used to recover these pathogens is not accurate or sufficient to detect their presence or absence. In order to overcome these challenges, the authors suggested the use of molecular analysis methods.

The potential of pathogens from environment received discharged wastewater to cause infections for human or animals depend mainly on the high exposure to these pathogens via water or foods which might overwhelm the immune system (Epstein and Kanwisher 1998; NIOSH 2002).

In terms of pathogenicity of the infectious agents, the main virulence factor among several microorganism species is the production of toxins. Gram-positive bacteria such as *Clostridium* spp. has a high potential to produce exotoxins, which might lead to poisoning of food and water. In Gram-negative bacteria, two types of toxins are produced including exotoxins and endotoxins. Exotoxins are toxic substances produced outside the cells while endotoxins represent a part of the outer cell membrane of Gram-negative bacteria and consist of lipopolysaccharide (LPS). Therefore, Gram-negative bacteria are more dangerous than Gram-positive ones since they have the potential to produce exotoxin as a live cell and their structure of LPS is toxic even if the cells are dead.

Pathogens have different infective doses (ID) which range from less than ten infectious particles or cysts for most enteric viruses and protozoa to more than 10^3 cells for bacteria (Molleda et al. 2008). Besides, the infectious dose of pathogens might also rely on the health status of the host as well as his age, time of infection and nutritional condition (Leggett et al. 2012). Moreover, the pathogenicity of microbial cells depends on their ability to survive in the environment. For example, *Shigella* sp. has low ID (10–100 cells) but it has no potential to survive in the environment. Therefore, infections caused by *Shigella* sp. might take place when humans are exposed to freshly contaminated wastewater. In contrast, both *E. coli* and *Salmonella* spp. require high ID (ranged from 10^3–10^6 cells) to cause the infection and these bacteria are more dangerous for humans and animals because it can multiply in the environment to reach the ID and are then transmitted to humans (Gordon et al. 2002). *Vibrio cholera* has high health significance with 10^3–10^8 of infective dose cells, but their ability to resist thermal treatment is low (Vinnerås et al. 2003). Most viruses such as Hepatitis A virus (HAV), Rotavirus Virus, Astrovirus and Enterovirus have high health significance with more than 100 infectious particles. These viruses cause hepatitis, fever, gastrointestinal infection, meningitis and in some cases the infection might lead to death (Rao et al. 1984; Clark and Graz 2010; Nocker and Gebra 2010). Parasites such as *G. lamblia* and *Cryptosporidium* spp. have high health significance and low dosage with 10 cysts for *G. lamblia* and 15–100 cysts for *Cryptosporidium* spp. (Robertson et al. 1992; Robertson and Nocker 2010).

The pathogenicity of pathogens depends on the mechanism of infection which relies on the infective dose. Schmid-Hempel and Frank (2007) revealed two infection mechanisms for pathogens which are direct and indirect mechanisms. In the direct mechanism, the pathogen cells attach to the host cells as in the case of parasites and viruses. In this mechanism, specific cells act as a receptor for the pathogen cells. In the indirect mechanism, the pathogens cells produce toxins which consist of protein or lipo-poly-saccharides which lead to the destruction of the host cell membrane and then allow for the pathogens to enter the host cell.

2.3 Chemical Risks

The chemical risk associated with the presence of chemical pollutants in greywater ranges from direct to indirect effects on humans and the environment. The high concentrations of total suspended solids in discharged greywater might reduce the light penetration of water bodies, and the growth of aquatic life and thereby decrease the biodiversity in water. The concentrations of suspended solids (SS) in greywater are used as a general indicator of the quality of treated greywater and for assessing compliance with discharge regulations (APHA 2005). The direct discharge of greywater into natural water bodies without the reduction of SS might increase the level of sediments. The high concentrations of TSS in the wastewater discharged into the water lead to sedimentation which negatively affect fish growth as it can clog their gills (Jais et al. 2017). Oil and grease in greywater are available in low concentrations but they might create a layer on the surface of the water and reduce the level of oxygen required for aquaculture growth. Besides that, it can also increase the level of odour, which leads to an increase in insect attraction, and subsequently more transmission of diseases via insect vectors (Jameel and Olanrewaju 2011). The high levels of BOD in greywater and the water which receive the waste lead to a lowering of the pH value, oxygen levels and accelerate bacteria growth (Al-Gheethi et al. 2015c). The presence of high nutrients in natural water is associated with eutrophication and overgrowth of algae as discussed below.

2.3.1 *Eutrophication and Harmful Algae Blooms (HABs)*

Eutrophication is a natural phenomenon related to microalgae growth in freshwater due to high levels of nutrient content. Eutrophication is correlated with harmful algae blooms (HABs) that are associated with seafood poisoning and negatively affect biodiversity in these waters (Anderson et al. 2002). HABs can be transported by the ballast of ships in water, travelling of shellfish from one place to another, nutrient enrichment and climatic shift which are common in developing countries in the Middle East and Africa. Those are some of the factors which contribute to the distribution of HABs (Hallegraeff 1993; Anderson et al. 2002).

The occurrence of eutrophication depends mainly on phosphorus (P) and nitrogen (N) because P is needed during the photosynthesis process of algae, while N represents the building unit of amino acids in an organism (Anderson et al. 2002). Both elements are available in moderate to high concentrations in greywater which means that the discharge of greywater into natural waters might contribute to the occurrence of eutrophication and thus HABs. Among several species of microalgae which are classified as HABs, the cyanobacterial species is of concern due to their ability to produce different types of toxins which are called cyanobacterial toxins. There are two ways where HABs can cause infections in humans including the inhalation of airborne toxins and the consumption of seafood contaminated with toxins (Glibert 2007). Among different types of HABs, neurotoxins, hepatoxins and dermatotoxins are more dangerous especially for children. HABs can also infect humans who consume seafood such as fish. The toxins released by HABs might cause vomiting, diarrhoea, skin rashes, eye irritation and respiratory symptoms which are used for the diagnosis of the infections by these toxins, since there are no specific diagnostic procedures for these toxins in human blood. Moreover, the presence of toxicity in natural water which are generated by HABs is determined using the KDHE procedure which was developed by researchers at the Kansas Department of Health and Environment in USA in 2015. This method depends on the determination of microcystin toxicity (<4 to >2000 µg/L) or cell concentration of microalgae in the water sample (<80,000 to >10,000,000 cells/mL). The high levels of microcystin toxicity or cells count are associated with the presence of high risk in the water body resulting from HABs. In this method, the water sample is collected from under the surface of the target water into a screw test tube. The test tube should be kept in a cold refrigerator overnight. On the next day, the occurrence of sedimentation of algae in the bottle of the test tube indicates the absence of HABs while the formation of a green ring surface indicates the presence of HABs. The presence of HABs should be followed by the determination of microcystin toxicity or cell count to detect the risk level (KDHE 2015).

2.3.2   Heavy Metals

The microbiological contents of greywater are insufficient for assessing the consequences of the improper disposal of greywater. Therefore, the risks associated with heavy metals in greywater are discussed in this section. Before, in the review for the health risks associated with the heavy metals in greywater, it was mentioned that the concentration of these metals in greywater is low and in most cases, it is within the standard limits recommended for the disposal process. Nonetheless, the main concern with the discharge of greywater containing heavy metals into the environment and water bodies lies in their ability to be accumulated in organism tissues to reach a toxic dose (Al-Gheethi et al. 2015d). Besides, the presence of these metals in the environment even in low concentrations might develop antimicrobial resistance among indigenous and pathogenic bacteria in the environment. Some reports indicated that there is a significant correlation between microbial resistance for heavy metals and antibiotics (Al-Gheethi et al. 2015d). In this section, the health risks associated with heavy metals in greywater are discussed, while the positive and negative effects for these metals on plants would be discussed in Chap. 5.

Heavy metals are transition elements. Some of these metal ions such as Cu, Zn and Ni play an important role in low concentrations as trace elements or microelements in sophisticated biological reaction. However, they are toxic at high concentrations because their ions are active and are able to form unspecific complex compounds in the cell which lead to the destruction of metabolic pathways in the cells and then death. Hg, Cd and Ag are very active and are capable of forming strong toxic complexes which make them too dangerous for any physiological function (Nies 1999; Al-Gheethi et al. 2016b). The disposal of greywater with heavy metals into the sea negatively affects biodiversity. Among several heavy metals, Hg^{2+} is the most toxic element for organisms (Banana et al. 2016). Hg ions are one of the most toxic elements (Banana et al. 2016). It has revealed that the accumulation of Hg in the oceans is correlated with the rising tide of mercury pollution.

The mechanism of heavy metal toxicity is explained based on the reaction between metals ions and the sulfhydryl group (-SH) in the protein structure of the cell. This reaction forms bridges between the protein groups and then binds protein molecules together. As a result, cellular metabolism is disrupted and the microorganism dies because the proteins represent the backbone of the enzyme structure (Smith 1996). Several studies have confirmed that heavy metals affect human behaviour directly by impairing mental and neurological functions, influencing neurotransmitter production and utilization, and altering numerous metabolic body processes. Heavy metals ions can induce impairment and dysfunction in the blood, urinary, immune, gastrointestinal, nervous, reproductive and cardiovascular systems as well as detoxification pathways which can cause negative changes in the liver, colon and kidneys. They also have adverse effects on the harmonic functions, pathways of energy production and enzymatic reactions (Martin and Griswold 2009; Singh et al. 2011; Al-Gheethi et al. 2015d).

The accumulation of heavy metals in the human body causes a mutation in cell genomes and competes with trace metals required for metabolic and anabolic pathways of the cells. Besides, the heavy metals have antimicrobial agents against the normal bacterial flora in the digestion system which play an important role in digestive health. Some heavy metals such as Zn and Cu are cofactors for several enzymes in living cells. However, high levels of oxidative free radicals of these metals, which are generated from antioxidant deficiencies often lead to tissue damage (Haley et al. 1990).

In the blood system, the presence of heavy metals is associated with the decreasing pH and high acidity. The human body overcomes this problem by drawing calcium from the bones but this process leads to the hardening of artery walls and progressive blockage of the arteries. The low concentrations of Ca in the bones may cause osteoporosis. Finally, the health risks posed by heavy metals towards humans depend on age, nutritional status, metabolic rate and the time and mode of metal exposure. Children and the elderly, whose immune systems are either underdeveloped or age-compromised, are more vulnerable to toxicity.

The negative effects of heavy metals depend also on the transportation process of metals across the cell membrane. The heavy metals ions are transported into the cells by unspecific processes which are driven by the chemiosmotic gradient as well as by specific processes which rely on adenine Triphosphate (ATP) hydrolysis (Nies 1999). Moreover, in the presence of high concentration of heavy metals in the surrounding environment, they are transported into the cell by an unspecific system and this is the explanation for the toxicity of heavy metals (Nies 1999; Al-Gheethi et al. 2015d).

2.4   Regulations for Greywater Disposal

The determination of a safe and effective management of greywater is one of the main objectives in many countries which aim to protect the surrounding environments of humans and animals. The regulations for the proper management of greywater have been adopted by many developed countries. However, there are no regulations for the discharge of greywater into the environment in developing countries. These are associated with the absence of studies which assess the adverse effects of the disposal of greywater on soil composition and ecosystems. In Malaysia, greywater is not used for irrigation due to the presence of many water resources and hence greywater is directly disposed into the environment. However, the country has no regulation for the disposal process of greywater. Moreover, the regulations for sewage and industrial effluents have started in 1989. These regulations are described in more detail in a new version of Environmental Quality (Scheduled Wastes) Regulations (2009). There are two standards regarding the discharge of wastewater into the environment (DOE 2010) (Table 2.1). Standard A is regulated for the discharge of effluents for the upstream of water intake points, while Standard B is for downstream discharge of water intake points. The absence of specific regulations for the discharge of greywater in many countries resulted in the use of standards for the discharge of general wastewater. Table 2.1 shows the national and international standards of wastewater discharge for effluent discharge. The international environmental law for the discharge of wastewater is expressed based on ECOLEX which has been adopted by the International Union for the Conservation of Nature and Natural Resources (IUCN), the United Nations Environment Programme (UNEP) and Food and Agriculture Organization (FAO). In India, wastewater should meet the standards regulated by the National Environmental Protection Act (1986) before it is discharged into the sea. On the other hand, one of the main challenges in the adoption of a regulation for XOCs in greywater lies in the selection of chemicals which should be included in the standards list due to the presence of these compounds in greywater or storm water (Baun et al. 2006). Moreover, the presence of many XOCs in the list which

Table 2.1 Wastewater discharge standards

Parameter (mg/L)	Standard limits			
	Malaysia		Standard (India)	ECOLEX
	Standard A (DOE)	Standard B (DOE)		
pH	6.0–9.0	5.5–9.0	6.0–9.0	5.0–9.0
BOD$_5$ at 20 °C	20	50	100	40
COD	50	100	250	120
Suspended solids	50	100	100	35
Oil and grease	1.0	10.0	10.0	10.0
Sulphide	0.5	0.50	5.0	0.002
Cadmium	0.0	0.0	0.05	0.01
Chromium	0.2	1.0	2.0	NA
Copper	0.2	1.0	3.0	0.5
Zinc	1.0	1.0	15.0	2.0
Lead	0.10	0.5	0.1	0.05
Manganese	0.20	1.0	2.0	0.2
Nickel	0.20	1.0	3.0	0.1
Aluminium	10	15	NA	5
Iron (Fe)	1.0	5.0	3.0	2.0

should be tested in the greywater before disposal creates more difficulties to apply the regulation due to the high cost of the assessment process.

2.5 Conclusions

It can be concluded that greywater has two main risks which are associated with the presence of pathogenic microorganisms and toxic heavy metals. However, the microbial risk represents the main change due to their ability to reproduce in the environment as well as their ability to survive and be transmitted to humans directly via the contact with greywater or indirectly via insect vectors. Heavy metals pose a long-term risk for humans and animals because it is available in low concentrations in greywater, while eutrophication leads to algae bloom in fresh water with more threats towards biodiversity in these systems.

Acknowledgements The authors wish to thank the Ministry of Higher Education (MOHE) for supporting this research under FRGS vot 1574 and also the Research Management Centre (RMC) UTHM for providing grant IGSP U682 for this research.

References

Al-Gheethi AA, Norli I, Lalung J, Azieda T, Ab Kadir MO (2013) Reduction of faecal indicators and elimination of pathogens from sewage treated effluents by heat treatment. Caspian J Appl Sci Res 2(2):29–45

Al-Gheethi AA, Abdul-Monem MO, Al-Zubeiry AH, Al-Amery R, Efaq AN, Shamar AM (2014) Effectiveness of selected wastewater treatment plants in Yemen for reduction of faecal indicators and pathogenic bacteria in secondary effluents and sludge. Water Practice Technol 9(3):293–306

Al-Gheethi AA, Norli I, Efaq AN, Bala JD, Al-Amery A (2015a) Solar disinfection and lime treatment processes for reduction of pathogenic bacteria in sewage treated effluents and biosolids before reuse for agriculture in Yemen. Water Reuse Des 5(3):419–429

Al-Gheethi AA, Aisyah M, Bala JD, Efaq AN, Norli I (2015b) Prevalence of antimicrobial resistance bacteria in non-clinical environment 4th International Conference on Environmental Research and Technology (ICERT 2015) on 27–29 May 2015 at Parkroyal Resort, Penang, Malaysia

Al-Gheethi AA, Mohamed RM, Rahman AA, Mas Rahayu J, Amir HK (2015c) Treatment of wastewater from car washes using natural coagulation and filtration system, International Conference On Sustainable Environment & Water Research (ICSEWR2015), 25–26 Oct 2015, Johor Baru, Malaysia

Al-Gheethi AA, Lalung J, Efaq AN, Bala JD, NorliI (2015d) Removal of heavy metals and β-lactam antibiotics from sewage treated effluent by bacteria. Clean Technol Environ Policy 17(8):2101–2123

Al-Gheethi AA, Mohamed RM, Efaq AN, Amir HK (2016a) Reduction of microbial risk associated with greywater utilized for irrigation. Water Health J 14(3):379–398

Al-Gheethi AA, Mohamed RMS, Efaq AN, Norli I, Hashim A, Ab Kadir MO (2016b) Bioaugmentation process of sewage effluents for the reduction of pathogens, heavy metals and antibiotics. J Water Health 14(5):780–795

American Water Works Association (AWWA) (2006) Waterborne Pathogens. AWWA Manual M48, 2nd edn. American Water Works Association

Anderson DM, Glibert PM, Burkholder JM (2002) Harmful algal blooms and eutrophication: nutrient sources, composition, and consequences. Estuaries 25(4):704–726

Anonymous (1996) Waterborne pathogens kill 10 M–20 M people/year. World Water Environ Eng

APHA (2005) Standard methods for the examination of water and wastewater, 21st edn. American Public Health Association (APHA), Washington, D.C

Atiku A, Mohamed RMSR, Al-Gheethi AA, Wurochekke AA, Kassim Amir H (2016) Harvesting microalgae biomass from the phycoremediation process of greywater. Environ Sci Pollut Res 23(24):24624–24641

Banana AS, Mohamed RM, Al-Gheethi AA (2016) Mercury pollution for marine environment at Farwa Island, Libya. J Enviro Health Sci Eng 14:5

Bani-Melhem K, Al-Qodah Z, Al-Shannag M, Qasaimeh A, Qtaishat MR, Alkasrawi M (2015) On the performance of real grey water treatment using a submerged membrane bioreactor system. J. Membr Sci 476:40–49

Baun A, Eriksson E, Ledin A, Mikkelsen PS (2006) A methodology for ranking and hazard identification of xenobiotic organic compounds in urban stormwater. Sci Total Environ 370(1):29–38

Bumann D, Hueck C, Aebischer T, Meyer TF (2000) Recombinant live *Salmonella* spp. for human vaccination against heterologous pathogens. FEMS Immunol Med Microbiol 27(4):357–364

Buzrul S (2009) Modeling and predicting inactivation of *Escherichia coli* under isobaric and dynamic high hydrostatic pressure. Innovative Food Sci Emerging Technol 10(4):391–395

Casanova LM, Gerba CP, Karpiscak M (2001) Chemical and microbial characterization of household greywater. J Environ Sci Health Part A Toxic/Hazard Subst Environ Eng 36(4):395–401

Clark S, Graz M (2010) Rotavirus. Retrieved from http://waterbornepathogens.susana.org/menuviruses/rotaviruses on 12 Oct 2016

Coupe S, Delabre K, Pouillot R, Houdart S, Santillana-Hayat M, Derouin F (2006) Detection of *Cryptosporidium*, *Giardia* and *Enterocytozoon bieneusi* in surface water, including recreational areas: a one-year prospective study. FEMS Immunol Med Microbiol 47(3):351–359

Cox CS (1987) The aerobiological pathway of microorganisms. Wiley, New York, NY

Davis BD, Dulbecco R, Eisen HN, Ginsberg HS, Wood WE (1968) Principles of microbiology and immunology, New York, NY; Harper and Row Publishers, Inc., Denmark. APMIS 106(6):606–622

DeRegnier DP, Cole L, Schupp DG, Erlandsen SL (1989) Viability of *Giardia* cysts suspended in lake, river, and tap water. Appl Environ Microbiol 55:1223–1229

Dobrindt U, Hacker J (2008) Targeting virulence traits: potential strategies to combat extraintestinal pathogenic *E. coli* infections. Curr Opin Microbiol 11(5):409–413

DOE (2010) Environmental requirements: a guide for investor, Appendix K1 & K2: acceptable condition of sewage discharge of Standard A and B. Department of Environment Malaysia, KL

Ducluzeau R, Hudault S, Galpin JV (1976) Inoculation of the digestive tract of axenic mice with the autochthonous bacteria of mineral water. Eur J Appl. Microbiol 2(2):127–134

Efaq AN, Ab Rahman NNN, Nagao H, Al-Gheethi AA, Shahadat Md, Ab Kadir MO (2015) Supercritical carbon dioxide as non-thermal alternative technology for safe handling of clinical wastes. J Environ Proces 2(4):797–822

El-Lathy AM, El-Taweel GE, El-Sonosy M, Samhan FA, Moussa TA (2009) Determination of pathogenic bacteria in wastewater using conventional and PCR techniques. Environ Biotechnol 5:73–80

Epstein R, Kanwisher N (1998) A cortical representation of the local visual environment. Nature 392(6676):598–601

Facklam RR (1991) *Streptococcus* and related catalase-negative gram-positive cocci. Manual Clin Microbiol 238–257

Fijalkowski KL, Kacprzak MJ, Rorat A (2014) Occurrence changes of *Escherichia coli* (including O157:H7 serotype) in wastewater and sewage sludge by quantitation method of (EMA) real time-PCR. Des Water Treat 52:19–21

Food and Drug Administration (FDA) (2012) Bad bug book, foodborne pathogenic microorganisms and natural toxins, 2nd edn. Retrieved from http://www.fda.gov/downloads/Food/FoodborneIllnessContaminants/UCM297627.pdf on 12 Oct 2016

Franz E, van Hoek AH, El BouwAarts HJM (2011) Variability of *Escherichia coli* O157 strain survival in manure-amended soil in relation to strain origin, virulence profile, and carbon nutrition profile. Appl Environ Microbiol 77(22):8088–8096

Freitas AR, Coque TM, Novais C, Hammerum AM, Lester CH, Zervos MJ, Peixe L (2011) Human and swine hosts share vancomycin-resistant *Enterococcus faecium* CC17 and CC5 and *Enterococcus faecalis* CC2 clonal clusters harboring Tn1546 on indistinguishable plasmids. J Clin Microbiol 49(3):925–931

Fujioka RS, Hurst CJ, Knudsen GR, McInerney MJ, Stezenbach LD, Walter MV (eds) (1997) Indicators of marine recreational water quality. In: Manual of environmental microbiology. ASM Press, Washington, D.C, pp 176–183

Glibert PM (2007) Eutrophication and harmful algal blooms: a complex global issue, examples from the Arabian seas including Kuwait Bay, and an Introduction to the Global Ecology and Oceanography of Harmful Algal Blooms (GEOHAB) programme. Int J Oceans Oceanogr 2:157–169

Gordon DM, Bauer S, Johnson JR (2002) The genetic structure of *E. coli* populations in primary and secondary habitats. J Microbiol 148(5):1513–1522

Guan TY, Holley RA (2003) Pathogen survival in swine manure environments and transmission of human enteric illness—a review. J Environ Qual 32(2):383–392

Haley PJ, Finch GL, Hoover MD, Cuddihy R (1990) The acute toxicity of inhaled beryllium metal in rats. Fund Appl Toxicol 15(4):767–778

Hallegraeff GM (1993) A review of harmful algal blooms and their apparent global increase. Phycologia 32(2):79–99

Iacovski RB, Barardi CRM, Simões CO (2004) Detection and enumeration of *Cryptosporidium* sp. oocysts in sewage sludge samples from the city of Florianópolis (Brazil) by using immuno-magnetic separation combined with indirect immunofluorescence assay. Waste Manage Res 22(3):171–176

Jais NM, Mohamed RMSR, Al-Gheethi AA, Hashim A (2017) Dual role of phycoremediation of wet market wastewater for nutrients and heavy metals removal and microalgae biomass production. Clean Technol Environ Policy. https://doi.org/10.1007/s10098-016-1235-7

Jameel AT, Olanrewaju A (2011) Aerobic biodegradation of oil and grease in palm oil mill effluent using consortium of microorganisms. In: Alam MDZ, Jameel AT, Amid A (eds) Current research and development in biotechnology engineering at International Islamic University Malaysia (IIUM), vol III. IIUM Press, Kuala Lumpur, pp 43–51. ISBN 9789674181444

Joseph S, Bhat KG (2000) Effect of iron on the survival of *Vibrio cholerae* in water. Ind J Med Res 111:115–117

Katukiza AY, Ronteltap M, Niwagaba CB, Kansiime F, Lens PNL (2014) Grey water treatment in urban slums by a filtration system: optimization of the filtration medium. J Environ Manage 146:131–141

KDHE (2015) Harmful Algae Blooms: A TOOL KIT for Health Departments. Kansas Department of Health and Environment Bureau of Epidemiology and Public Health Informatics, 1000 SW Jackson Street, Suite 330. Topeka, KS. 66612-1365. www.kdheks.gov/algae-illness/index.htm

Kerr M, Fitzgerald M, Sheridan J, McDowell DA, Blair IS (1999) Survival of *Escherichia coli* O157: H7 in bottled natural mineral water. J. Appl Microbial. 87(6):833–841

Kiil K, Binnewies TT, Willenbrock H, Hansen SK, Yang L, Jelsbak L, Ussery DW, Friis C (2008) Chapter 1 comparative genomics of pseudomonas. In: Rehm BHA (ed) Pseudomonas model organism, pathogen, cell factory. Wiley-VCH Verlag GmbH & Co. KGaA, Weinheim

King BJ, Keegan RA, Monis PT, Saint CP (2005) Environmental temperature controls *Cryptosporidium Oocyst* metabolic rate and associated retention of infectivity. Appl Environ Microbiol 71:3848–3857

Konopka A (2009) What is microbial community ecology & quest. ISME J 3(11):1223–1230

Kumar T, Majid MA, Onichandran S, Jaturas N, Andiappan H, Salibay CC, Phiriyasamith S (2016) Presence of *Cryptosporidium parvum* and *Giardia lamblia* in water samples from Southeast Asia: towards an integrated water detection system. Infect Dis Poverty 5(1):1

Leggett HC, Cornwallis CK, West S (2012) Mechanism of pathogenesis, infective dose and virulence in human parasites. PLoS Pathog 8(2):e1002512. https://doi.org/10.1371/journal.ppat.1002512

Maimon A, Friedler E, Gross A (2014) Parameters affecting greywater quality and its safety for reuse. Sci Total Environ 487:20–25

Martin S, Griswold W (2009) Human health effects of heavy metals. Environ Sci Technol Briefs Citizens 15:1–6

Merikanto I, Laakso J, Kaitala V (2012) Outside-host growth of pathogens attenuates epidemiological outbreaks. PLoS ONE 7(11):50–58

Molleda P, Blanco I, Ansola G, de Luis E (2008) Removal of wastewater pathogen indicators in a constructed wetland in Leon, Spain. Ecol Eng 33:252–257

Moore BC, Martinez E, Gay JM, Rice DH (2003) Survival of *Salmonella enterica* in freshwater and sediments and transmission by the aquatic midge *Chironomus tentans* (Chironomidae: Diptera). Appl Environ Microbiol 69(8):4556–4560

Nies DH (1999) Microbial heavy metals resistance. Appl Microbiol Biotechnol 51(6):730–750

NIOSH (2002) Violence: occupational hazards in hospitals, Centers for Disease Control and Prevention (2002-101)

Nocker A, Gerba C (2010) Enterovirus. http://waterbornepathogens.susana.org/menuviruses/enterovirus on 12 Oct 2016

Patel M, Isaacson M, Gouws E (1995) Effect of iron and pH on the survival of *Vibrio cholerae* in water. Trans R Soc Trop Med Hyg 89(2):175–177

Podschun R, Ullmann U (1998) *Klebsiella* spp. as nosocomial pathogens: epidemiology, taxonomy, typing methods, and pathogenicity factors. Clin Microbiol Rev 11(4):589–603

Rangel-Martıneza C, Jime'nez-Gonza'lezb DE, Martı'nez-Ocanaa J, Romero-Valdovinosa M, Castillo-Rojasc G, Espinosa-Garcıad AC, Lo'pez-Vidalc Y, Mazari-Hiriartd M, Maravillaa P (2015) Identification of opportunistic parasites and helminth ova in concentrated water samples using a hollow-fibre ultrafiltration system. Urban Water J 12:440–444

Rao V, Seidel KM, Goyal SM, Metcalf TG, Melnick JL (1984) Isolation of enteroviruses from water, suspended solids, and sediments from Galveston Bay: survival of poliovirus and rotavirus adsorbed to sediments. Appl Environ Microbiol 48:404–409

Renois F, Jacques J, Guillard T, Moret H, Pluot M, Andreoletti L, de Champs C (2011) Preliminary investigation of a mice model of *Klebsiella pneumoniae* subsp. ozaenae induced pneumonia. Microbes Infect 13(12):1045–1051

Robertson L, Nocker A (2010) *Giardia*. Retrieved from http://waterbornepathogens.susana.org/menuprotozoa/giardia on 12 Oct 2016

Robertson LJ, Campbell AT, Smith H (1992) Survival of oocysts of *Cryptosporidium parvum* under various environmental pressures. Appl Environ Microbiol 58:3494–3500

Rodda N, Salukazana L, Jackson SAF, Smith MT (2011) Use of domestic greywater for small-scale irrigation of food crops: Effects on plants and soil. Phys Chem Earth Parts A/B/C 36(14):1051–1062

Rose JB, Atlas RM, Gerba CP, Gilchrist MJR, Le-Chevallier MW, Sobsey MD, Yates MV, Cassell GH, Tiedje JM (1999) Microbial pollutants in our nation's waters: environmental and public health issues. Am Soc Microbiol

Rouch DA, Mondal T, Pai S, Glauche G, Fleming VA, Thurbon N, Blackbeard J, Smith SR, Deighton M (2011) Microbial safety of air-dried and rewetted biosolids. J Water Health 9(2):403–414

Schmid-Hempel P, Frank SA (2007) Pathogenesis, virulence, and infective dose. PLoSPathog 3:1372–1373. https://doi.org/10.1371/journal.ppat.0030147

Singh R, Gautam N, Mishra A, Gupta R (2011) Heavy metals and living systems: an overview. Ind J Pharmacol 43(3):246

Smith EH (1996) Uptake of heavy metals in batch systems by a recycled iron-bearing material. Water Res 30(10):2424–2434

Strauch D (1991) Survival of pathogenic micro-organisms and parasites in excreta, manure and sewage sludge. Revue Scientifiqueet Technique (Int Off Epizootics) 10(3):813–846

Strauch D (1998) Pathogenic micro-organisms in sludge. Anaerobic digestion and disinfection methods to make sludge usable as fertiliser. Eur Water Manage 1(2):12–26

Vickery AM (1993) Strains of methicillin-resistant *Staphylococcus aureus* isolated in Australian hospitals from 1986 to 1990. J Hosp Infect 24(2):139–151

Vinnerås B, Björklund A, Jönsson H (2003) Thermal composting of faecal matter as treatment and possible disinfection method—laboratory-scale and pilot-scale studies. Bioresour Technol 88(1):47–54

Wang H, Ibekwe AM, Ma J, Wu L, Lou J, Wu Z, Liu R, Xu J, Yates SR (2014) A glimpse of *Escherichia coli* O157:H7 survival in soils from Eastern China. Sci Total Environ 476–477:49–56

WHO (2003) Heterotrophic plate counts and drinking-water safety. In: Bartram J, Cotruvo J, Exner M, Fricker C, Glasmacher A (eds) IWA Publishing, London, UK. ISBN:1 84339 025 6

Winfield MD, Groisman EA (2003) Role of nonhost environments in the lifestyles of *Salmonella* and *Escherichia coli*. Appl Environ Microbiol 69(7):3687–3694

Winward GP (2007) Disinfection of Grey water. PhD thesis, Centre for Water Sciences, Department of Sustainable Systems, School of Applied Sciences, Cranfield University

Winward GP, Avery LM, Frazer-Williams R, Pidoua M, Jeffrey P, Stephenson T, Jefferson B (2008) A study of the microbial quality of grey water and an evaluation of treatment technologies for reuse. Ecol Eng 32:187–197

Wurochekke AA, Mohamed RMS, Al-Gheethi AA, Amir HM, Matias-Peralta HM (2016) Household greywater treatment methods using natural materials and their hybrid system. J Water Health 14(6):914–928

Chapter 3
Determination of Pathogens in Greywater

Adel Ali Saeed Al-Gheethi, Efaq Ali Noman,
Radin Maya Saphira Radin Mohamed and Amir Hashim Mohd Kassim

Abstract There are many methods used for isolation and the enumeration of pathogenic organisms. The direct methods depend on the culture medium and microscopic examination. However, these techniques are not effective for all pathogenic organisms in the environment since many organisms require a specific condition to grow on the culture medium. Therefore, the using of enrichment methods might exhibit more efficiency in the determination of pathogens from the wastewater samples. The main challenge in the microbiological assessment of greywater lies in finding the most effective method to detect the presence or absence of pathogens which are available in low concentrations. In this chapter, traditional methods including direct culture and enrichment methods are reviewed.

Keywords Culture methods · Greywater · Pathogens · Antibiotics resistant bacteria

A. A. S. Al-Gheethi (✉) · R. M. S. Radin Mohamed · A. H. Mohd Kassim
Department of Water and Environmental Engineering, Faculty of Civil and Environmental
Engineering, Micro-Pollutant Research Centre (MPRC), Universiti Tun Hussein Onn Malaysia
(UTHM), 86400 Parit Raja, Batu Pahat, Johor, Malaysia
e-mail: adel@uthm.edu.my

R. M. S. Radin Mohamed
e-mail: maya@uthm.edu.my

E. A. Noman
Faculty of Applied Sciences and Technology (FAST), Universiti Tun Hussein Onn Malaysia
(UTHM), Pagoh, Johor, Malaysia

E. A. Noman
Department of Applied Microbiology, Faculty Applied Sciences, Taiz University, Taiz, Yemen

© Springer International Publishing AG, part of Springer Nature 2019   51
R. M. S. Radin Mohamed et al. (eds.), *Management of Greywater in Developing Countries*,
Water Science and Technology Library 87,
https://doi.org/10.1007/978-3-319-90269-2_3

3.1   Introduction

Determination and isolation of pathogenic organisms from greywater and other environments require critical techniques due to the nature of greywater which has different contaminated compounds that can negatively affect the detection of these pathogens. There are two techniques which are used for the detection of microorganisms in greywater including culture-based methods and non-culture-based methods (molecular-based methods). The culture-based method depends on the detection of microorganisms based on their growth on selective and enrichment media. This method does not need highly qualified technicians but the technicians should have a very good background in isolation and identification methods using biochemical tests. In the non-culture-based method, the determination of microorganism merely on the utilization of advanced procedures such as molecular techniques which identify the microorganism based on the sequences in the 16S rRNA and require well-trained technicians. The selection of detection methods depends on the type of wastewater such as before or after the treatment process. The non-culture-based technique is efficient in the detection of microorganisms in both raw and treated wastewater. In contrast, the culture-based method is not as efficient for the detection of microorganisms in treated and disinfected wastewater due to the failure of some inactivated pathogens to grow on the culture medium while living based on its metabolic activity and energy status.

A comparison was done on the viability of advanced technology for the determination of pathogenic microorganisms between developed and developing countries. The absence of advanced cutting protein techniques in developing countries represents the main challenge in the evaluation of risk associated with the reuse of greywater for irrigation or discharge into the environment. Therefore, this chapter aims at presenting both traditional and advanced processes used for the detection of infectious pathogens in order to give the researchers from the developing countries more ideas about the techniques which can be used to give a clear indication of the risk levels of greywater, since the developing countries have strict regulations for wastewater (but not greywater). However, the main limitation lies in the absence of qualified technicians and advanced techniques which are necessary to apply those regulations.

Pathogenic bacteria and fungi in greywater are the most important among several pathogens due to their ability to multiply in the environment without the need for a host as in the case of viruses and parasites (Al-Gheethi et al. 2016). Moreover, most standard regulations for greywater require tests to gauge the concentration of pathogenic bacteria. The ability of bacteria to convert from being active to inactivate during the disinfection process of greywater or due to the presence of different chemical compounds such as xenobiotic organic compounds (XOCs) make the isolation of pathogens more difficult. Therefore, this chapter focuses on the techniques used for isolation and purification as well as the identification of bacteria and fungi from greywater or any type of wastewater. This chapter provides technicians with the basic skills required for the recovery of pathogens from greywater by using traditional methods with the minimum requirements especially in developing countries where there are no more facilities to utilize advanced techniques. Advanced technologies have also been highlighted.

The greywater samples are collected using sterilized glass bottles (1 L or more). In order to get a more accurate evaluation of the microbiological characteristics of greywater, the samples should be collected in triplicates at different times. Some references recommend the collection of seven samples at a rate of one sample per week (U.S. EPA 2003). The obtained samples are then transported to a microbiological laboratory within an ice box and subject to analysis within 24 h.

3.2 Culture-Based Methods

The culture-based method includes procedures which depend on the observation of microbial growth in synthetic media (selective and enrichment medium) after the incubation period which ranges from 24 to 72 h at an appropriate temperature (30, 37 and 44.5 °C) and can be conducted by direct isolation on agar plates or in broth test tubes which is known as most probable numbers (MPN) method. This method might also be conducted by using membrane filters which are used for filtration of wastewater. The membrane filter will be placed on appropriate selective media to show microbial growth after the incubation period. The standard methods for isolation and determination of pathogenic organisms in wastewater are described in detail by APHA (1999). Moreover, in this section, some notes will be highlighted in order to obtain more accurate results.

3.2.1 Standard Plate Method

This technique depends on the preparation of serial dilutions of greywater samples based on the density of microorganisms which are expected to be available in the greywater. The method is appropriate for the isolation of pathogens from raw wastewaters, which are present in great numbers, since one millilitre of greywater sample may contain between 10^6 and 10^9 cells depending on the source of these wastes and microorganism species as well as other factors mentioned in Chap. 1.

The use of the standard plate method can be conducted by spreading or pouring procedure depending on the bacterial species. Pouring technique is used for the isolation of facultative anaerobic bacteria while spreading technique is used for the isolation of both obligate aerobic and facultative anaerobic bacteria. In brief; the procedure for the spreading technique is performed by preparing a culture medium one day before the isolation process. The medium is sterilized by autoclave or heating based on the manufacturing instructions for each medium. After pouring the agar, it is supposed to be left at room temperature overnight to ensure the absence of contamination. The colonies growing on the medium are separated from each other to distinguish their characteristics and ease the purification process.

The most common indicator bacteria which are counted in greywater include FC, TC, *Enterococcus faecalis*, *Escherichia coli*, *Staphylococcus aureus* and *Pseudomonas aeruginosa*. These bacterial species can be counted by direct isolation from the appropriate selective medium while pathogenic bacteria such as *Salmonella* spp. need to be enriched by using an enrichment medium before detecting their presence or absence. The use of an enrichment medium is discussed more in Sect. 3.2.5. *Clostridium* spp. is an aerobic bacterium which has to be isolated using specific procedures using an anaerobic cabinet. The fungi most commonly detected in greywater especially that generated from the kitchen include *Aspergillus* spp. and *Penicillium* spp. The culture medium used for isolation and enumeration of indicator bacteria and pathogenic bacteria and fungi are described in APHA (1999) standard guidelines.

The dilution of greywater should be carried out by using sterilized saline solution immediately before the isolation process. Leaving the diluted sample for a long time (more than 10–15 min) before the isolation process might negatively affect the viability of the bacterial cells, and thus reduce the opportunity to determine the actual number of pathogens in the sample. The volume of the sample used to prepare the first dilution should be appropriate to represent the sample and obtain a more accurate bacteria count. For example, the dilution of 30 mL of greywater using 270 mL of sterilized saline solution or distilled water might be better than diluting 1 mL of greywater in 9 mL of sterilized saline solution. The selection of a dilution solution depends on the source of the samples. Wastewater may be diluted with distilled (or deionized) water. Moreover, the use of physiologic saline (0.85% NaCl) might be the best option to prevent any possible cell lysis due to osmotic stress. The diluted sample should be shaken well to physically separate the microorganisms to manageable levels (30 times to the front and back by hand or by shaking at 100 rpm for 5 min) but not vigorously to prevent the destruction of cells (Prescott 2002).

The usual amount of pipetted sample used for the isolation is around 0.1 mL (100 μL). The micropipette with sterilized yellow or blue tips is used to transfer the inoculums sample. The pipette is filled and emptied with the inoculums three or more times before withdrawal of the diluting solution to homogenize the bacterial cells in the solution. Thereafter, the remaining amounts of the inoculums are added to the diluting solution and mixed well. A fixed amount (1 mL) of the diluted solution is transferred to the next tube containing 9 mL of the physiologic saline solution to prepare the next dilution. This procedure is the same for the preparation of all dilutions required (Collee et al. 1989).

The spreading processes of the inoculums are carried out by using flamed-sterilized stainless-steel spreader (L-shape) or glass spreader onto the surface of the selected medium. The spreading process should be performed carefully to ensure the best distribution of the inoculums on the whole surface of the medium. For primary technicians, large Petri dishes (150 mm × 20 mm) might be used for the isolation, while the medium-sized Petri dishes (80 mm × 15 mm) are used for the purification process. The media plates are left in the laminar flow for 10 min to allow the medium to absorb the inoculum water.

The pouring method is conducted with a similar procedure, but in this technique, the media are prepared and kept at 45 °C. After the transfer of 1 mL of the greywater sample or diluted solution into the Petri dish, 20 mL of the medium is poured and then homogenized slowly by movement of the plate to the right and left and to front and back. Isolation of filamentous fungi is performed only by using the spreading method due to the nature of fungi as aerobic organisms.

The culture plates are incubated at the optimal temperature required for each organism species. Media plates used for heterotrophic bacterial counts (HBC), TC, *E. faecalis*, *S. aureus* and *P. aeruginosa* are incubated at 35–37 °C, while plates used for counting FC and *E. coli* are incubated at 44.5 °C. The media used for the isolation of fungi are incubated at 28 °C. In order to prevent the drying of the media, the plate cultures are sealed with parafilm before incubation. In regions with dry weather, a wet cotton might be placed inside the plastic page together with the culture medium to provide the moisture required for microbial growth and to prevent the drying of the medium.

The incubation period might range from 24 to 72 h for bacteria and between 5 and 7 days for fungi. The incubation period between 24 and 36 h is enough for bacteria growth and the development of the colony. However, in order to get more morphological characteristics of the colony such as colours and pigments, the medium is suggested to be incubated for more than 72 h. This is because the production of pigments take place during the late phase of bacterial growth. The production of pigments by bacteria is a response towards unfavourable conditions and to protect the cells from external factors such as low temperature. Therefore, keeping the culture medium in a fridge might accelerate the resealing of these pigments and give more distinguishing characteristics to the colony. For instance, the laboratory observations found that *S. aureus* and *Staphylococcus epidermidis* have similar colony morphology with the yellow colour on the mannitol salt agar (MSA) after 24 h. However, an extended incubation period of 48 h or keeping the culture plates at 4 °C for next 24 h improved the colony morphology for both species. *S. aureus* changed to gold with a gold colour medium while *S. epidermidis* colony appeared in yellow pink. The description of the morphological characteristics of bacterial colonies are presented in detail by Prescott (2002), Morello et al. (2003) and Benson (2005).

After the incubation period, plates with 30–300 colonies are counted. In some references, the counting colonies ranged from 20 to 200 colonies (Götz et al. 2006). The selection of the counting range is related to statistical considerations which aimed to reduce the standard division values (Collee et al. 1989). In the fungi plates, APHA (1999) recommended to count plates with 15–150 colonies. The viable numbers of the bacterial and fungal cells are expressed as CFU mL^{-1} according to Eq. (3.1).

$$\text{Number of bacteria or fungi in the sample (CFU/mL)}$$
$$= \text{Number of colonies grown on the culture medium} \times \frac{1}{\text{Dilution factor}} \quad (3.1)$$

The standard plate method was certified by the US. EPA as a standard procedure for the environmental monitoring of wastewater plants in 1979. These standards have been developed with more guidelines in a new version. The method has also been used for the counting of pathogenic bacteria from different wastewater samples by several authors in literature (Al-Gheethi et al. 2013, 2014). It exhibited high efficiency for the counting of bacteria in highly contaminated samples, but it was unsuccessful with the low density of bacteria in wastewater samples. The detection limit of this method is around 10 CFU/mL, where the minimum concentration might be available in 1 mL of the sample is one cell. The determination of pathogenic bacteria in greywater with a concentration of less than 10 CFU/mL requires more efficient methods such as the membrane filter method or MPN.

Götz et al. (2006) indicated that direct surface plating procedures are more accurate than MPN for detecting *S. aureus* in raw or unprocessed foods. Moreover, the sensitivity of this method might be increased with larger volumes of the tested samples (>1 ml). In contrast, the best results with a desired number of colonies might be achieved by using three replicate plates for two or more decimal dilutions.

One of the methods used for the direct isolation of pathogenic bacteria on the selective media is by using the spiral plate technique. This method is conducted by using a machine in which a known volume of the sample is disturbing a rotating agar medium. Thereafter, the total number of bacteria is counted by selecting an appropriate area of the plate. The efficiency of this method is better than the pour plate method in terms of the total number of bacteria recovered. Besides, this technique needs less time compared to the traditional direct plating procedure (Gilchrist et al. 1973). This technique has been proposed recently by a few authors who suggest it to be used for the direct isolation of *Salmonella* spp. from different sources on the selective medium (Brichta-Harhay et al. 2007).

3.2.2 Purification of Bacterial and Fungal Isolates

In order to identify the bacteria and fungi isolated from the greywater, it has to be purified as well. The morphological characteristics of the colonies grown on the culture plates might provide the primary identification of the grown microorganisms. However, the identification process should be confirmed either by phenotypic methods or molecular techniques. Moreover, the purification of the isolates represents the critical step to obtain the correct name for the pathogenic microorganism. In this section, the purification procedure for bacteria and fungi are presented.

3.2.2.1 Purification of Bacterial Isolates

The methods used for the purification of bacterial isolates have been reviewed by many manual books. However, this section provides some key notes which might contribute in obtaining a pure bacterial isolate. The streak procedure is the most common method for purifying the bacterial colony. The method is conducted by picking up a small portion of the grown colony on the isolation media and then streaking it on the surface of a new medium. The notes discussed in this procedure include the type of medium used for the purification process and the direct streak of the picked colony on the new medium.

In terms of media used for purification, the use of nutrient agar (NA) or brain heart infusion medium (BHI) might affect the purification process. This is because the selective medium used for the isolation of pathogenic bacteria from wastewater has inhibitory substances which inhibit the growth of floral bacteria available together with the pathogenic bacteria in the wastewater sample. These inhibited bacteria on the selective medium might grow on non-selective agar such as NA or BHI. Therefore, in order to obtain pure isolates of the selected pathogenic bacteria, it should be purified on the same selective medium used for the first isolation for two times or more and then subject to purification on a non-selective medium. Second, the bacterial cells which have capsules or produce slim layers and have a mucus colony texture on the isolation medium as in the case of *Klebsiella* spp. are attached to other bacteria cells. Therefore, purification of these bacteria is quite difficult using the direct subculture on a new medium by the streak technique. Therefore, in order to overcome this problem Benson (2005) described a technique which depends on the preparation of a serial dilution for the picked colony using a nutrient broth and then subjected to the subculture on a new media. This procedure is effective for getting a pure culture of selected bacteria. However, it might not be enough to separate the bacterial cells which have very mucoid capsules or slim layers. Theatrically, the capsule and slim layer consist of polysaccharides which are more soluble in water. Therefore, the use of sterilized distilled water might be more efficient to obtain separate cells and a pure culture (Al-Gheethi et al. 2013). The purification method which depends on the use of sterilized distilled water is carried out as the following; one loopful of bacterial colony grown on an isolated medium is transported to 10 mL of sterilized distilled water to prepare bacterial suspension. The suspension is mixed well and immediately one loopful is rubbed onto the agar surface of the new medium using a sterilized glass spreader which is made in order to distribute and separate the bacterial cells as much as possible. After 24–48 h of the incubation period at 35–37 °C, a single colony is transferred to another 10 mL of sterilized distilled water, mixed well and then one loopful is transferred to the agar surface of the same medium in a similar manner. One of the developed bacteria colonies after 24 h is used for checking the purity by using Gram stain (Park et al. 2006).

3.2.2.2 Purification of Fungal Isolates (Single Spore Isolation)

In order to identify fungal isolates more successfully, the fungal isolates should be purified by using very accurate techniques. Fungal growth is quite different from that of bacteria because the production of external spores by the fungi makes their purification more difficult since the spores spread quickly and might cause contamination. There are many methods for purifying fungi. The single spore isolation technique represents the best method in terms of simple procedure and the absence of contamination yeast, bacteria and mites. The purification process by using single spore isolation is conducted under a microscope (4–10×). The fungal colony grown on the culture medium is raised from only one spore. Hence, this method is more suitable for obtaining a pure fungi culture and also for the identification using phenotypic methods. This technique was described by Choi et al. (1999). Moreover, Noman et al. (2016) have described a modification of this method which has better performance and named it the Dr. Nagao method. In brief; the culture medium which has fungal growth from the first isolation is placed under a light microscope with a magnification of 4×. One colony is selected and picked up using a sterilized scalpel into a new culture medium. The medium should be dried at 28 °C for 24 h before use to ensure that there is no contamination. In order to get separate colonies on the culture medium, 0.1 mL of sterilized distilled water (SDW) is spread aseptically on the surface of agar by using stainless-steel or glass spreaders (L-shape). The plate is left to dry at room temperature for 10 min and then incubated at 28 °C for 16–18 h. At the end of the incubation period, the germinated spores are observed under a light microscope (4×). A small piece (0.2 × 0.2 cm) of culture medium with one germ spore is cut out aseptically by using a scalpel, transferred onto a new culture medium and then incubated at 28 °C for 5–7 days.

In order to prevent the contamination of pure cultures by mites, the stage of light microscope, equipment, laminar flow, containers, incubator, as well as the lab wares used in the purification process, should be sterilized with ethanol (70%). All plate cultures should be sealed with parafilm before the incubation, while fungal cultures should be stored in a refrigerator at 4 °C. The description of fungal culture characteristics is performed according to manual references such as Promputtha et al. (2005) and Benson (2005).

The culture characteristics of fungal colonies subcultured by using the single spore technique occurred more clearly than those subcultured directly using the needle. It might be due to the fact that in the single spore technique, the fungal colony was raised from only one spore. Therefore, the colonies grow on the medium separately. In the direct culture technique, several spores on the needle are transferred onto the culture medium, and thus the colonies imbricate. It is difficult to distinguish their characteristics. Besides that, the measurement of the colony diameter for the fungal culture purified by single spore isolation is easier than that purified by direct isolation. The fungal colony in single spore isolation possesses a circular shape while they are irregular in shape when they are cultured by direct isolation. These observations were made during work at the laboratory.

3.2.3 Identification of Bacterial and Fungal Isolates

The identification of bacteria and fungi is conducted based on phenotypic character-istics and molecular analysis of the cell genome. Phenotypic identification relies on culture characteristics of the grown colonies and cell morphology under the micro-scope. These characteristics differ among fungal species such as *Penicillium* sp., *Aspergillus* sp., *Trichoderma* sp., *Rhizopus* sp. and *Curvularia* sp., which have different shapes and sizes in their conidiophores and spores that play an important role in the identification of the species level (Emine et al. 2010). Further, the use of scanning electronic microscopy (SEM), flow cytometry as well as fluorescent microscopy provides an accurate recognition of fine structure in the conidiophore and spores (Guarro et al. 1999). The uses of SEM with high magnification reveal the slight differences in the surface spore which are quite different from one fungal species to another.

In contrast, bacterial cells have no high diversity in their morphology (cocci, bacilli and rod) and the response towards Gram staining (Gram-positive and Gram-negative). Therefore, phenotypic identification depends on biochemical tests which detect the ability of bacterial cells to produce specific enzymes such as oxidase, urease and catalase enzymes. The biochemical tests are conducted using analytical profile index (API) System which includes the biochemical API 20 test system, API 20NE and RapiD 20E that are used for different bacterial species. The API system was developed by Analytab Products of Plainview, New York. This system depends on the enzymatic reaction with the substrates available in a plastic strip with 20 separate compartments. The number of biochemical tests is different from one class to other. The bacteria suspension is prepared with physiological saline provided with these kits and then a small portion of each bacterial suspension is inoculated for each tube containing a specific medium or substrate depending on the test. The kits are incubated for 18–24 h and the results are read based on the changes in colour. In some tests, the reagents should be used, and then the identification process is performed by using computerized software which provides more than 99% accuracy (Benson 2005).

API systems might also be conducted by using VITEK® 2 Densi-CHEK Plus, which is used to determine the optical density of bacterial suspension. The instrument provides the optical density values in terms of McFarland units. This system is more commonly used in a vitro diagnostic process of pathogenic bacteria in medial units. The preparation of bacteria suspension is performed by inoculating a pure bacterial colony grown on the culture medium for 18–24 h into a test tube containing 3 mL of physiological solution. The suspension is prepared in a density of 0.5–0.63 McFarland units. APi kits inoculated with bacterial suspension are incubated in VITEK® 2 Densi-CHEK Plus is also used to read the results. The results of the API system might be read manually and the identification process is conducted by using the API software program. The recent identification for bacteria and fungi are conducted using the molecular technique which has several advantages in comparison to the traditional method in terms of accuracy and time.

3.2.4 Membrane Filtration (MF) Methods

Membrane filtration is a culture-based method with high efficiency to recover pathogenic bacteria from water and wastewater due to the high quantity of water samples filtered using a membrane filter 0.45 μm in diameter. Therefore, membrane filtration is more applicable to the recovery and assessment of the quality of drinking water or highly treated wastewater. In some cases, the wastewater needs to be diluted before the filtration process in order to prevent clogging of the membrane pores. This method is described as a direct-plating method, where the membrane filters are transferred to a Petri dish containing selective isolation agar or an absorbent pad saturated with selective broth. Membrane filtration is more effective than the standard plate method with high limitation level (1 cell per 100 mL) depending on the level of the dilution process. In many cases, it is used as an alternative to the standard plate method if the concentration of pathogenic bacteria in polluted wastewater is less than the detection limits of the standard plate method. The protocol details for using this method are described by APHA (1999) and U.S. EPA (2006a, b).

The incubation conditions depend on the target bacterial species as in the standard plate method. However, the range of colony counted is between 20 and 80 colonies. The number of bacteria grown is reported as colony-forming units (CFU) per 100 mL which is determined according to Eq. (3.2).

$$\text{CFU } 100 \text{ mL}^{-1} = \frac{(\text{Number of colonies on agar medium})}{\text{volume (mL) of undiluted sample}} \times 100 \qquad (3.2)$$

This technique has been used by many researchers for the isolation of pathogenic bacteria from wastewater before and after the disinfection process. Friedler et al. (2005) used this method for evaluating the microbiological quality of greywater onsite. In the study, mTEC agar was used for the isolation of FC, M-PA-C agar for *P. aeruginosa* and Staphylococcus Medium 110 agar for *S. aureus*. Plates of FC were incubated at 44.5 °C for 22–24 h and the colonies grown were confirmed as FC based on the occurrence of yellow to yellowish brown colour after placing the filter on a pad saturated with urea for 15 min. The plates containing *P. aeruginosa* were incubated at 41.5 °C for 48–72 h. The confirmation test of this bacteria included the hydrolyses casein and production of yellowish to green diffusible pigment on Milk Agar. *S. aureus* plates were incubated at 35 °C for 18–48 h. The coagulase-positive results confirmed the presence of *S. aureus*. Membrane filtration is the national standard method for enumerating *S. aureus* from the water samples regulated by the Health Protection Agency (HPA 2004) in the UK.

3.2.5   Enrichment Methods

The use of enrichment methods aims to recover pathogenic bacteria which are available in low concentrations in the tested samples or which are available in less concentrations than the detection limits of surface direct plating method (10 CFU/mL) or membrane filtrations (1 CFU/mL). Besides, the high amount of pollutants in the samples with floral bacteria makes the isolation of relevant bacteria quite difficult. Therefore, the enrichment method is necessary for the recovery of pathogenic bacteria available among all the other bacteria species. There are two types of enrichment media used, which are non-selective (pre-enrichment medium) and selective (enrichment medium). Zhang (2005) indicated that BHI broth is the best for the pre-enrichment process while brilliant green broth (BG) is most efficient for enriching *Salmonella* from milk products. The selection of an enrichment medium plays an important role in the detection of pathogenic bacteria, regardless of the methods used. Stone et al. (1994) revealed that the Rappaport–Vassiliadis and tetrathionate broth media have inhibitory effects on the detection of *Salmonella* spp. by PCR, while BHI broth and selenite cystine broths were the best for detecting this bacterium by using the PCR procedure.

3.3   Comparison Between Direct Surface Plate Method on Different Selective Media and Enrichment Methods

A comparison between the direct surface plate method on selective medium and enrichment methods is discussed here based on the type of wastewater sample (treated or untreated). In the non-treated samples, the direct surface plate method is more effective than MPN and membrane filtrations while in the treated or disinfected samples, the efficiency of this method is negligible and the enrichment process of pathogenic bacteria is necessary before plating on the selective media. In this section, the comparison between using direct plating on different selective media as well as using enrichment methods for *Salmonella* spp. and *Shigella* spp. is done to best understand the best methods for recovering both bacteria from different sources.

Shaban and El-Taweel (1999) examined the detection of *Listeria monocytogenes* from different wastewater samples using direct surface plate counts and the MPN method. The study revealed that the direct surface plate technique detected a higher count of *L. monocytogenes* than the selective enrichment MPN technique. However, in the disinfected samples by chlorination, the variations were less than that recorded in the raw samples which were between 1 and 2 log. The study indicated that the surface plate technique is the most practical method for the determination and enumeration of *L. monocytogenes* in the wastewater samples. El-Lathy et al. (2009) studied the detection of *Salmonella* spp. in raw and disinfected wastewater by using PCR, MPN and surface plate techniques. The results revealed that *Salmonella* spp. was found to be positive in all samples when PCR was used. In contrast, it was detected in 93% of the samples using the direct surface plate technique and in 90% of the samples using the MPN technique. These findings indicated that the molecular technique is more efficient than the culture method in the isolation of bacteria from wastewater.

Nonetheless, the efficiency of surface plating depends mainly on the type of isolation medium, the use of an enrichment process and the samples tested. Below are some of the studies which investigated the isolation of pathogenic bacteria from different waste samples. King and Metzger (1968) evaluated the efficiency of Hektoen enteric agar (HE), *Salmonella* and *Shigella* agar (SS agar) and EMB Agar for the isolation of *Salmonella* spp. and *Shigella* spp. from 2855 stool specimens by direct and indirect methods. Twice the amount of *Shigella* spp. was recovered on HE Agar than on SS Agar through both methods. Among 98 isolates, 97 isolates were obtained on HE agar, 74 on EMB agar and only 40 isolates on SS agar through the direct method. Moreover, HE agar exhibited higher efficiency than SS agar and EMB in recovering *Salmonella* spp. by direct or indirect methods. Bhat and Rajan (1975) compared the efficiency between desoxycholate citrate agar (DCA) medium and xylose lysine desoxycholate (XLD) medium in isolating *Shigella* spp. from stool specimens. The results revealed that XLD exhibited higher efficiency in recovering *Shigella* sp. than DCA, where the morphological characteristics of the colonies on XLD were clearly visible after 18–24 h. Clear characteristics on DCA appeared after 48 h. Warren (2006) reported that the DCA and XLD media are more useful as intermediate selective media for isolating *Shigella* spp. from food.

Kelly et al. (1999) estimated the cost-effective methods for the isolation of *Salmonella enterica* from clinical specimens. The study was conducted using 8717 faecal specimens. The results revealed that the primary isolation of *S. enterica* on XLD agar has enhanced the speed of the diagnostics process but with less sensitivity than that achieved with Selenite enrichment. However, the enrichment process by using selenite broth and then plating on both brilliant green and XLD agar offered no further advantage than the direct plating onto XLD without enrichment.

Nye et al. (2002) evaluated of the efficiency of XLD, mannitol lysine crystal violet brilliant green (MLCB), DCA and α-β chromogenic ABC agars for the isolation of *S. enterica* from 2409 human faeces samples by direct plating method with enrichment and non-enrichment processes. The results revealed that the 46 of the 60 possible isolates of *Salmonella* sp. were recovered by direct plating after the enrichment process via selenite enrichment while no isolates were obtained by the direct plating method without the enrichment process. MLCB recorded the high recovery rate individually (84.8%) with less amounts of competing flora (CF) which have no effect on the recognition of *Salmonella* spp. colonies. ABC was the best in terms of specificity, but CF has affected the sensitivity of this medium. DCA recognizes only 9.01% of picked positive bacteria. The study concluded that XLD and MLCB are the best media for recovering *Salmonella* spp. from faeces samples while XLD is the most effective for routine diagnostic process. Pant and Mittal (2008) claimed that the direct plating method needs less time in comparison to the conventional method based on the enrichment process. The study detected *Salmonella* sp. and *Shigella* sp. in the wastewater samples for six months and found that the isolation of both bacteria by direct methods improved by 105 and 276%, respectively, in comparison to the conventional method.

Létourneau et al. (2010) investigated the presence of TC, FC, *Enterococcus* sp., *E. coli*, *Y. enterocolitica* and *C. perfringens* from feed, water and manure samples of hog finishing houses using the direct plating method. The counting of bacteria in feed and manure was conducted using a dilution process where sterile sodium metaphosphate buffer (2 g/L) was used while bacterial loads in the water samples were enumerated via the membrane filtration method. The bacterial colonies grown on *m*Endo-LES agar with a metallic green sheen appearing after 18–20 h at 37 °C were counted as TC, while those grown on *m*FC agar at 44.5 °C with a distinctive indigo blue colour were counted as FC. *m*Enterococcus agar was used for the isolation of *Enterococcus* spp. which showed a burgundy colour after 48 h at 37 °C. *E. coli* appeared as blue indicative colonies on *m*FC basal medium after 18–24 h at 44.5 °C. *E. coli* colonies exhibited yellow colour which indicated β-glucuronidase activity in the medium due to the presence of 3-bromo-4-chloro-5-indolyl-β-D-glucuronide (100 mg/L) in the medium components. *C. perfringens* have yellow colonies surrounded by a yellow halo on *m*CP agar. The exposure of the colonies to ammonium hydroxide fumes changed it to magenta. The stormy fermentation of skim milk broth inoculated with these bacteria after 24 h confirmed it as *C. perfringens*. Cefsulodin-Irgasan-Novobiocin agar was used for enumerating *Y. enterocolitica* which grew after 18 h at 30 °C as small colonies measuring 2 mm in diameter. Lysine arginine iron agar slants (LAIA) were used to confirm the identity of this bacteria which appeared as alkaline slant and acid butt without the production of gas and H_2S.

Brichta-Harhay et al. (2008) added antibiotics into the isolation medium in order to isolate *Salmonella* spp. and *E. coli* O157:H7 from hides and carcasses of culled cattle in the United States by the direct plating method, while the identification process was performed using serology tests and PCR analysis. The enumeration of *Salmonella* spp. was performed by using a spiral process for 50 μl aliquot of each 20-ml sponge sample on XLD_{tnc} medium supplied with 4.6 mL/L of tergitol (niaproof), 15 mg/L of novobiocin, and 10 mg/L of cefsulodin. The isolation media was incubated for 18–20 h at 37 °C, thereafter the plates were incubated at room temperature (23–25 °C) for 18–20 h. The most probable bacteria grown on the medium were identified as *Salmonella* spp. by using PCR. *E. coli* O157:H7. They were isolated on ntChrom-O157 agar and incubated for 18–20 h at 42 °C. The identification of *E. coli* O157:H7 was conducted by using the DrySpot agglutination test kit. The colonies with positive results for this test were subcultured inctSMac and then subjected to PCR analysis.

Nagvenkar and Ramaiah (2009) studied the abundance of TC, *E. coli*, *Vibrio cholera*, *Salmonella* spp. and *E. faecalis* from sea water contaminated with sewage. TC was isolated on McConkey Agar, *E. coli* on Thiosulphate citrate bile salts sucrose (TCBS) Agar, *Salmonella* spp. on Hi-Crome *Salmonella* Agar and *E. faecalis* on *Enterococcus* confirmatory Agar. Seawater was added to each medium to provide a salinity of 15 PSU, except for TCBS which was prepared using deionized water. The isolation procedure was conducted by using the direct plating method. The study revealed that direct isolation on the selective media exhibited efficiency in recovering pathogenic bacteria from polluted sea water.

Ongeng et al. (2011) used the surface spread plate method on selective media for the isolation of *E. coli* O157:H7 and *S. enterica* serovar *Typhimurium* from manure and manure-amended soil. The enrichment procedure was used when the concentrations of both bacteria were less than the detection limit of 2 log CFU g^{-1} to detect the presence of residual viable cells. The authors in this study used highly selective isolation media which exhibited high efficiency. These media included XLT4-Rif100-Ny50-Cy50 for *S. typhimurium* and CT-SMAC-Rif100-Ny50-Cy50 for *E. coli* O157:H7. EC broth medium containing novobiocin antibiotics was used as an enrichment media for the recovery of *E. coli*. Thompson et al. (2013) isolated *S. aureus* from hospital wastewater and sewage treatment plants using the direct method. The isolation of bacteria from raw wastewater was carried out using the direct surface plate method, while the membrane filtration method was used for recovering the bacteria from treated samples. The isolation process was conducted on Vogel-Johnson agar and MSA containing cefoxitin antibiotic according to Broekema et al. (2009).

In some bacteria such as *Shigella* spp., it has been reported that the direct isolation on selective media is better compared to direct isolation carried out on enriched media. Dunn and Martin (1971) claimed that the best method for the isolation of *Shigella* spp. from faecal specimens is by using the direct surface plate method on the selective media. Iveson (1973) reported that the direct surface plating of *Shigella* spp. on DCA was superior to other enrichment methods. Morris et al. (1970) conducted a study to compare the effectiveness of XLD, MacConkey and SS agar for the direct isolation and transport of *Shigella* spp. from faecal specimens. The study revealed that XLD agar was better than MacConkey and SS agar for isolating *Shigella sonnei*, while SS agar and XLD were better than MacConkey agar for recovering *Shigella flexneri*. The direct plating method exhibited high efficiency for the isolation of *Shigella* spp. from faecal specimens in comparison to the transport medium or enrichment broth. The study also indicated that the use of buffered glycerol saline has improved the recovery of *Shigella* spp. by 83% within 48 h in comparison to direct plating. Finally, the study recommended XLD agar and SS agar for the direct isolation of *Shigella* spp. without the need for an enrichment process.

Pollock and Dahlgren(1974) investigated the use of the direct plating method on selective media (XLD, HE agar, MacConkey agar and SS agar) for isolating *Shigella* spp. and compared them to enrichment methods. The study was conducted for a period of one year in which 455 stool specimens were used in this period. The total number of *Shigella* spp. recovered was 53 isolates. Among them, 56% was identified as *S. sonnei* while 13% was identified as *S. flexneri*. The results showed that 90% of the isolates were recovered on XLD, 87% on HE agar, 80% on MacConkey, while only 28% was isolated on SS agar. In contrast, only 0.5% of *Shigella* spp. was recovered after the enrichment process using Selenite-F enrichment medium.

In terms of using enrichment cultures, Muniesa et al. (2005) revealed that the presence of bacteriophages was associated with *Salmonella* spp. bacteria in the enrichment culture medium. Therefore, it might negatively affect pathogenic diversity and density during the enrichment process. The study compared filtrated samples (0.45 microns) and non-filtrated samples. The results showed that the *Salmonella* biotypes recovered from the filtrated samples were changed in comparison to the non-filtrated samples.

Cassar and Cuschieri (2003) studied the isolation of *Salmonella* spp. on *Salmonella* chromogenic medium (SCM) in comparison to DCLS from 500 stool specimens by the direct plating method and the selenite enrichment method. The results found that among the 44 Salmonella-positive stool samples, the specifications for the direct plating method and the enrichment method were 82.5% versus 72.8%, respectively. For DCLS agar and SCM, the specifications were 98.5 versus 95.8%, respectively. Castillo et al. (2006) investigated the isolation of *Salmonella* spp. and *Shigella* spp. from different sources including food and clothes by direct plating, pre-enrichment process only, and pre-enrichment process followed by enrichment-only methods. The selective media used were MacConkey agar, XLD agar, SS agar, brilliant green sulfadiazine agar (BGS) and bismuth sulfite agar (BSA) while the enrichment medium was tetrathionate broth, GN pre-enrichment broth, selenite cystine and tetrathionate broths. The direct plating on XLD, SS agar and MacConkey agar media was used for isolating the bacteria from the samples and then incubated for 24 h at 35 °C. Direct enrichment process was carried out with tetrathionate broth which was incubated for 24 h at 43 °C and then plated onto BGS agar and BSA and incubated again for 24 h at 35 °C. The pre-enrichment method was conducted using GN broth and incubated for 18 h at 35 °C. Thereafter, one loopful was streaked onto XLD, SS agar and MacConkey agar. The enrichment process was performed with selenite cystine and tetrathionate broths. After the incubation period for 24 h at 43 °C, a loopful was streaked onto BGS and BSA and incubated for 24–48 h at 35 °C. The study found that the number of positive samples increased by using the direct plating method and not by the pre-enrichment method. Moreover, the direct plating method was the best for the recovery of *Shigella* spp. in comparison to the enrichment medium.

Maddocks et al. (2002) investigated *Salmonella* spp. growth on a new formulation of CHROM agar Salmonella (CAS) medium in comparison with XLD, SS agar and HE agar. The results revealed that the efficiency of direct plating on the CAS medium for the detection of *Salmonella* spp. was 83%, while the efficiency was 55% for both XLD and SS agar. Those findings confirmed the sensitivity and specificity of the CAS medium for the direct isolation of *Salmonella* spp. Even though *Candida* spp. and *Pseudomonas* spp. have also grown on the CAS medium, their colony morphology was quite different than that of *Salmonella* spp. colonies. Based on this study, it can be indicated that the direct isolation of *Salmonella* spp. requires using a very sensitive and specific medium.

Examples of studies which have used the direct culture method for the isolation of pathogenic bacteria from different sources are presented in Table 3.1. Based on the studies reviewed above, it can be concluded that the main criteria used for the selection of the appropriate isolation method is the source of samples. Wastewater is highly polluted and contains a high concentration of pathogenic bacteria. Therefore, the direct method might be effective for the isolation and enumeration of these pathogens. In contrast, the isolation of pathogenic bacteria from food products and drinking water has to be enriched first due to the presence of chemical substances such as preservations in the foods and chlorine or ozone residues in the water which inhibit or inactivate the bacterial cells.

3.4 Antibiotic Resistance Among Pathogenic Bacteria in Greywater

Antimicrobial resistance among pathogenic bacteria in greywater needs to be assessed because bacterial resistance towards antibiotics is one of the main factors

Table 3.1 Direct plating method and membrane filter method for the isolation and enumeration of pathogenic bacteria from different sources

Bacteria	Source	Method	Reference
Shigella sp.	Faecal specimens	Direct inoculation on selective media	Morris and Fulton (1970)
Shigella sp.	Faecal specimens	Direct inoculation on selective media	Dunn and Martin (1971)
Shigella sp.	Faeces	direct inoculation on selective media	Iveson (1973)
Shigella sp.	Stool specimens	Direct plating on selective media and compared to enrichment media	Pollock and Dahlgren (1974)
TC, FC, faecal streptococci, *Staphylococcus* sp., *Klebsiella* sp., *Pseudomonas* sp.	Sewage	Membrane filtration, spread-plating directly onto appropriate selective media	Dudley et al. (1980)
Shigella isolates	Stool samples	Direct plating on selective media	Vila et al. (1994)
L. monocytogenes	wastewater	Surface plate counting with the selective media	Shaban and El-Taweel (1999)
Salmonella enterica	Faecal specimens	Direct plating method	Kelly et al. (1999)
Salmonella group	Wastewater	Direct Surface Plating Procedures	El-Taweel and Ali (2000)
Salmonella spp. *Campylobacter* spp.	Raw, whole chickens	Spread-plating	Jørgensen et al. (2002)
Salmonella enterica	faeces	Direct plating media	Nye et al. (2002)
Salmonella sp.	Freshwater and Seawater	Direct plate counts	Sugumar and Mariappan (2003)
Salmonella sp.	Stool Specimens	Direct plating	Cassar and Cuschieri (2003)
E.coli O157:H7, *S. enteric Serovar Enteritidis*	Chicken manure	Direct plating method	Erickson et al. (2004)
S. aureus	Water	Membrane filtration	HPA (2004)
FC, *P. aeruginosa, S. aureus*	Wastewater	Membrane Filter Method	Friedler et al. (2005)
Staphylococcus sp.	Water	Surface plate technique	Götz et al. (2006)
Shigella sp.	Freshly Squeezed Orange Juice, Fresh Oranges, and Wiping Cloths	Direct plating method	Castillo et al. (2006)
Salmonella E. coli	Compost material with cow dung	Direct counting method	Chun-ming (2007)
Salmonella sp., *E. coli* O157:H7	Ground beef, cattle carcass, hide and faecal samples	Direct plating method	Brichta-Harhay et al. (2007)
Salmonella sp.	Poultry carcass rinses	Direct plating method	
S. aureus, E.coli, P. auroginosa, Proteus sp., *Bacillus cerues, Streptococcus* sp.	Urine and swap	Direct plating method	Ogbulie et al. (2008)
Salmonella sp., *Shigella* sp.	Sewage	Direct method	Pant and Mittal (2008)

(continued)

Table 3.1 (continued)

Bacteria	Source	Method	Reference
Salmonella sp., *E. coli* O157:H7	Hides and carcasses of cull cattle	Direct plating method	Brichta-Harhay et al. (2008)
Salmonella spp., *Vibrio* spp., *Listeria* spp.	Wastewater	Surface plate technique	El-Lathy et al. (2009)
S. aureus	Milk	Direct plating method	Synnott et al. (2009)
TC, FC, Enterococci, *E. coli*, *Salmonella* spp., *Vibrio cholerae*, *Vibrio parahaemolyticus*	Water and sediment samples	Direct plating method	Nagvenkar and Ramaiah (2009)
TC, FC, *E. coli*, Enterococci, *Salmonella* sp., *Shigella* sp., *Proteus* sp., *Klebsiella* sp., *Vibrio cholera*, *P. aeruginosa*	Coastal water	Direct surface plate	Patra et al. (2009)
TC, FC, *Enterococcus* sp, *E. coli*, *Y. enterocolitica*	Manure from hog finishing houses	Membrane filtration	Létourneau et al. (2010)
E. coli O157:H7, *Salmonella enterica* serovar Typhimurium	Manure and manure-amended soil	Surface spread plate	Ongeng et al. (2011)
TC, FC, *E. coli*, *Shigella* like organisms, *Vibriocholera* like organisms, *Salmonella* like organisms	Sewage	Direct plating method	Rodrigues et al. (2011)
Salmonella sp.	Organic and conventional broiler feed	Direct plating method	Petkar et al. (2011)
S. aureus	Hospital wastewaters and sewage	Membrane filtration	Thompson et al. (2013)
S. viridins, *Staphylococcus* sp., *Corynebacterium* sp., *Enterobacter* sp., *Heamophilus* sp.,	Pus specimens	Direct plating on selective medium	Chunduri et al. (2012)

which play an important role in bacterial pathogenicity to humans. However, studies on bacterial resistance towards antibiotics in non-clinical environments should be conducted in different ways and compared to those performed in hospitals or clinical environments.

The difference between the concept of antimicrobial resistance in clinical and non-clinical environments has occurred recently (Al-Gheethi et al. 2013). Walsh (2013) explained more about the differences between antimicrobial resistance tests in clinical and non-clinical environments. This term is used to define natural antibiotic resistance. Clinically, the classification of bacteria as susceptible or resistant depends on whether an infection with the bacterium responds to therapy. In contrast, in the environment, resistant microorganisms from a microbiological point of view are those that possess any kind of resistance mechanism or resistance gene.

Al-Gheethi et al. (2015a, b) investigated microbial resistance among pathogenic bacteria obtained from surface water and treated sewage effluents. The study used two different methods. The disk diffusion technique with the minimum inhibition concentrations (MICs) of antibiotics was carried out to determine the antibiotic resistance among pathogenic bacteria from treated sewage effluents. However, the results were read as resistant, moderate and sensitive based on non-clinical applications in which the bacteria was considered resistant at 0.7–1 cm in diameter of the inhibition zone, moderately susceptible at 1.0–1.2 cm diameter of the inhibition zone and susceptible when a diameter of more than 1.2 cm of the inhibition zone was observed (Audra and Andrew 2003). This method is different from the Clinical and Laboratory Standards Institute (CLSI) guidelines which require different readings for antibiotic sensitivity based on antibiotics class and bacterial species. In this study, however, the authors used a constant range for all pathogenic bacteria and antibiotics.

In contrast, the antibiotic resistance among bacteria obtained from surface water was investigated using the culture-based technique, in which high concentrations of antibiotics (100 mg L^{-1}) were added into the culture medium after autoclaving and then the medium was used for isolating the bacteria from the surface water. The percentage of the resistance was calculated in comparison with the number of bacterial cells (CFU) in the culture medium without antibiotics (control). The authors explained the use of two different methods based on the survival period of the bacteria in the environment and the natural adaptation for antibiotic resistance. Antibiotic resistance can be due to the transmission of antibiotic gene resistance or the availability of antibiotics in the environment which leads to the development of a resistance mechanism in the bacterial cell. The authors claimed that the pathogenic bacteria isolated immediately from domestic sewage generated from households have not developed their resistance mechanism against a wide range of antibiotics and still reflect their potential to resist antibiotics which have been exposed to them in the human body. Furthermore, a constant range to classify these bacteria as either resistant, moderate or sensitive was applied because these pathogens would develop their resistance mechanism in an environment where they will be discharged along with wastewater. In time, they will be acclimatized to resist several antibiotics during their survival in the wastewater. Therefore, the antimicrobial resistance among pathogenic bacteria obtained from surface water is supposed to be tested using a culture media containing antibiotics where the bacterial cells will be exposed to high concentrations of antibiotics.

Nevertheless, the concentrations of antibiotics in surface water remain low, but the use of high concentrations aim to detect bacterial tolerance towards antibiotics which is developed during its long survival period in the environment. Bacterial resistance and tolerance are not the same. More discussion on the differences between bacterial resistance and tolerance are available in the work published by Lewis (2008) and Al-Gheethi et al. (2015b).

The most common media used for antibiotic sensitivity test is Muller Hinton agar (MHA). However, Niederstebruch and Sixt (2013) indicated that standard nutrient agar 1 (StNA1) might be the best alternative for MHA for antibiograms in developing countries. Most international method standards and guidelines for the microbiological assessment of wastewater are not applicable in developing countries due to the lack of experience among technicians as well as the absence of equipment and facilities to meet the standard requirements. Nevertheless, MHA might be common for antibiotic susceptibility tests of pathogenic bacteria from clinical environments but not for that obtained from non-clinical environments. Wiggins (1996) used Trypticase soy agar plates to determine the antimicrobial sensitivity among faecal Streptococci isolates from natural water. Tao et al. (2007) used nutrient agar plates to study the antibiotic resistance of *Sphingomonas* sp. against cefuroxime, erythromycin, chloramphenicol, amoxicillin and tetracycline.

3.5 Conclusion

It can be concluded that different methods can be used for the determination of pathogens in greywater and the environment. The selection of specific techniques depends mainly on the aim of the isolation. The enumeration of pathogens in the raw greywater might be conducted via the culture-based method. However, pathogens such as *Salmonella* sp. need to be enriched first in order to facilitate their isolation. The enrichment method is used to detect the presence or absence the pathogenic bacteria available in a concentration less than the detection limits of the direct count plate but it cannot be used for counting their cells. This is because the pre-enrichment process would increase their number and would not reflect the actual number of the greywater samples. Moreover, in the treated samples, the use of non-culture methods might be the best alternative to detect the validity and potential of the inactivated bacteria to regrow. Molecular techniques can be used for counting the pathogenic bacteria in greywater which failed to grow on the isolation medium. However, this technique also has limitations in terms of its detection limits of bacterial cell number.

Acknowledgements The authors wish to thank the Ministry of Higher Education (MOHE) for supporting this research under FRGS vot 1574 and also the Research Management Centre (RMC) UTHM for providing grant IGSP U682 for this research.

References

Al-Gheethi AA, Norli I, Lalung J, Azieda T, Ab Kadir MO (2013) Reduction of faecal indicators and elimination of pathogens from sewage treated effluents by heat treatment. Caspian J Appl Sci Res 2(2):29–45

Al-Gheethi AA, Abdul-Monem MO, AL-Zubeiry AHS, Al-Amery R, Efaq AN, Shamar AM (2014) Effectiveness of selected wastewater treatment plants in Yemen for reduction of faecal indicators and pathogenic bacteria in secondary effluents and sludge. Water Pract Technol 9(3):293–306

Al-Gheethi AA, Aisyah M, Bala JD, Efaq AN, Norli I (2015a) Prevalence of antimicrobial resistance bacteria in non-clinical environment 4th International Conference on Environmental Research and Technology (ICERT 2015) on 27–29 May 2015 at Parkroyal Resort, Penang, Malaysia

Al-Gheethi AA, Lalung J, Efaq AN, Bala JD, Norli I (2015b) Removal of heavy metals and β-lactam antibiotics from sewage treated effluent by bacteria. Clean Technol Environ Policy 17(8):2101–2123

Al-Gheethi AA, Mohamed RM, Efaq AN, Amir HK (2016) Reduction of microbial risk associated with greywater utilized for irrigation. Water Health J 14(3):379–398

APHA (1999) Coliforms-total, faecal and *E. coli*, Method 8074, m-Endo. Standard methods for the examination of water and wastewater, American Public Health Association, American Water Works Association, Water Environment Federation. 9221 B, 9222 B, 9225B, 9230C. Adopted by U.S. EPA. DOC316.53.001224

Audra M, Andrew J (2003). Fate of a representative pharmaceutical in the environment. Final Report Submitted to Texas Water Resources Institute

Benson JH (2005) Microbiological applications: laboratory manual in general microbiology, 7th edn. WCB/McGraw-Hill Co., New York, USA

Bhat P, Rajan D (1975) Comparative evaluation of desoxycholate citrate medium and xylose lysine desoxycholate medium in the isolation of *Shigellae*. Am J Clin Pathol 64(3):399–404

Brichta-Harhay DM, Arthur TM, Bosilevac JM, Guerini MN, Kalchayanand N, Koohmaraie M (2007) Enumeration of *Salmonella* and *Escherichia coli* O157: H7 in ground beef, cattle carcass, hide and faecal samples using direct plating methods. J Appl Microbiol 103(5):1657–1668

Brichta-Harhay DM, Guerini MN, Arthur TM, Bosilevac JM, Kalchayanand N, Shackelford SD, Koohmaraie M (2008) *Salmonella* and *Escherichia coli* O157: H7 contamination on hides and carcasses of cull cattle presented for slaughter in the United States: an evaluation of prevalence and bacterial loads by immunomagnetic separation and direct plating methods. Appl Environ Microbiol 74(20):6289–6297

Broekema NM, Van TT, Monson TA, Marshall SA, Warshauer DM (2009) Comparison of cefoxitin and oxacillin disk diffusion methods for detection of *mec*A mediated resistance in *Staphylococcus aureus* in a large scale study. J ClinMicrobiol 47:217–219

Cassar R, Cuschieri P (2003) Comparison of *Salmonella* chromogenic medium with DCLS agar for isolation of *Salmonella* species from stool specimens. J ClinMicrobiol 41(7):3229–3232

Castillo A, Villarruel-López A, Navarro-Hidalgo V, Martínez-González NE, Torres-Vitela MR (2006) *Salmonella* and *Shigella* in freshly squeezed orange juice, fresh oranges, and wiping cloths collected from public markets and street booths in Guadalajara, Mexico: incidence and comparison of analytical routes. J Food Protect 69(11):2595–2599

Choi YW, Hyde KD, Ho WH (1999) Single spore isolation of fungi. Fungal Divers 3:29–38

Chunduri NS, Madasu K, Goteki VR, Karpe T, Reddy H (2012) Evaluation of bacterial spectrum of orofacial infections and their antibiotic susceptibility. Ann Maxillofac Sur 2(1):46

Collee J, Duguid JP, Fraser AG, Marmion BP (1989) Practical medical microbiology, 13th edn. Longman-FE. Ltd, London

Dudley DJ, Guentzel MN, Ibarra MJ, Moore BE, Sagik BP (1980) Enumeration of potentially pathogenic bacteria from sewage sludges. Appl Environ Microbiol 39(1):118–126

Dunn C, Martin WJ (1971) Comparison of media for isolation of *Salmonellae* and *Shigellae* from fecal specimens. ApplMicrobiol 22(1):17–22

El-Lathy AM, El-Taweel GE, El-Sonosy MW, Samhan FA, Moussa TA (2009) Determination of pathogenic bacteria in wastewater using conventional and PCR techniques. Environ Biotechnol 5:73–80

El-Taweel GE, Ali GH (2000) Evaluation of roughing and slow sand filters for water treatment. Water Air Soil Pollut 120(1):21–28

Emine S, Kambol R, Zainol N (2010) Morphological characterization of soil *Penicillium* sp. strains—potential producers of statin. In: Biotechnology Symposium IV, 1–3 Dec 2010, Universiti Malaysia Sabah, Sabah, Malaysia

Erickson MC, Islam M, Sheppard C, Liao J, Doyle MP (2004) Reduction of *Escherichia coli* O157: H7 and *Salmonella enterica* serovar enteritidis in chicken manure by larvae of the black soldier fly. J Food Protect 67(4):685–690

Friedler E, Kovalio R, Galil NI (2005) On-site greywater treatment and reuse in multi-storey buildings. Water Sci Technol 51(10):187–194

Gilchrist JE, Campbell JE, Donnelly CB, Peeler JT, Delaney JM (1973) Spiral plate method for bacterial determination. Appl Microbiol 25(2):244–252

Götz F, Bannerman T, Schleifer KH (2006) The genera *Staphylococcus* and *Micrococcus*. In: The prokaryotes. Springer, US, pp 5–75

Guarro J, Gene J, Stchigel AM (1999) Developments in fungal taxonomy. Clin Microbiol Rev 12(3):454–500

HPA (2004) Health protection agency, enumeration of *Staphylococcus aureus* by membrane filtration. National Standard Method W 10 Issue 3 (2004). http://www.hpa-standardmethods.org.uk/pdf_sops.asp

Iveson JB (1973) Enrichment procedures for the isolation of *Salmonella, Arizona, Edwardsiella* and *Shigella* from faeces. Epidemiol Infect 71(2):349–361

Jørgensen F, Bailey R, Williams S, Henderson P, Wareing DRA, Bolton FJ, Humphrey TJ (2002) Prevalence and numbers of *Salmonella* and *Campylobacter* spp. on raw, whole chickens in relation to sampling methods. Int J Food Microbiol 76(1):151–164

Kelly S, Cormican M, Parke L, Corbett-Feeney G, Flynn J (1999) Cost-effective methods for isolation of *Salmonella enterica* in the clinical laboratory. J Clin Microbiol 37(10):3369

King S, Metzger WI (1968) A new plating medium for the isolation of enteric pathogens I. Hektoen Enteric Agar. Appl Microbiol 16(4):577–578

Létourneau V, Nehmé B, Mériaux A, Massé D, Cormier Y, Duchaine C (2010) Human pathogens and tetracycline-resistant bacteria in bioaerosols of swine confinement buildings and in nasal flora of hog producers. Int J Hygiene Environ Health 213(6):444–449

Lewis K (2008) Multidrug tolerance of biofilms and persister cells. In: Bacterial Biofilms. Springer, Berlin, Heidelberg, pp 107–131

Maddocks S, Olma T, Chen S (2002) Comparison of CHROM agar *Salmonella* medium and xylose-lysine-desoxycholate and *Salmonella-Shigella* agars for isolation of *Salmonella* strains from stool samples. J Clin Microbiol 40(8):2999–3003

Morello JA, Granato PA, Mizer HE (2003) Laboratory manual and work book in microbiology application to patient care, 7th edn. The McGraw – Hill Companies

Morris RF, Fulton WC (1970) Heritability of diapause intensity in *Hyphantria cunea* and correlated fitness responses. Can Entomol 102(8):927–938

Muniesa M, Blanch AR, Lucena F, Jofre J (2005) Bacteriophages may bias outcome of bacterial enrichment cultures. Appl Environ Microbiol 71(8):4269–4275

Nagvenkar GS, Ramaiah N (2009) Abundance of sewage-pollution indicator and human pathogenic bacteria in a tropical estuarine complex. Environ Monit Assess 155(1–4):245

Niederstebruch N, Sixt D (2013) Standard nutrient agar 1 as a substitute for blood-supplemented Müller-Hinton agar for antibiograms in developing countries. Eur J Clin Microbiol Infect Dis 32(2):237–241

Noman EA, Al-Gheethi AA, Rahman NNNA, Nagao H, Kadir MA (2016) Assessment of relevant fungal species in clinical solid wastes. Environ Sci Pollut Res 23(19):19806–19824

Nye KJ, Fallon D, Frodsham D, Gee B, Graham C, Howe S, Warren RE (2002) An evaluation of the performance of XLD, DCA, MLCB, and ABC agars as direct plating media for the isolation of *Salmonella enterica* from faeces. J ClinPathol 55(4):286–288

Ogbulie JN, Adieze IE, Nwankwo NC (2008) Susceptibility pattern of some clinical bacterial isolates to selected antibiotics and disinfectants. Polish J Microbiol 57:199–204

Ongeng D, Muyanja C, Geeraerd AH, Springael D, Ryckeboer J (2011) Survival of *Escherichia coli* O157: H7 and *Salmonella enterica* serovar Typhimurium in manure and manure-amended soil under tropical climatic conditions in Sub-Saharan Africa. J Appl Microbiol 110(4):1007–1022

Pant A, Mittal AK (2008) New protocol for the enumeration of *Salmonella* and *Shigella* from wastewater. J Environ Eng 134(3):222–226

Park JE, Ahn TS, Lee HJ, Lee YO (2006) Comparison of total and faecal coliforms as faecal indicator in eutrophicated surface water. Water Sci Technol 54(3):185–190

Patra AK, Acharya BC, Mohapatra A (2009) Occurrence and distribution of bacterial indicators and pathogens in coastal waters of Orissa. Indian J Mar Sci 38(4):474–480

Petkar A, Alali WQ, Harrison MA, Beuchat LR (2011) Survival of *Salmonella* in organic and conventional broiler feed as affected by temperature and water activity. Agric Food Anal Bacteriol 1:175–185

Pollock HM, Dahlgren BJ (1974) Clinical evaluation of enteric media in the primary isolation of *Salmonella* and *Shigella*. Appl Microbiol 27(1):197–201

Prescott H (2002). Laboratory exercises in microbiology, 5th edn. The McGraw-Hill Companies

Promputtha I, Jeewon R, Lumyong S, McKenzie EHC, Hyde KD (2005) Ribosomal DNA finger-printing in the identification of non sporulating endophytes from *Magnolia liliifera* (Magnoliaceae). Fungal Divers 20:167–186

Rodrigues V, Ramaiah N, Kakti S, Samant D (2011) Long-term variations in abundance and distribution of sewage pollution indicator and human pathogenic bacteria along the central west coast of India. Ecol Indic 11(2):318–327

Shaban AM, El-Taweel GE (1999) Prevaience of *Listeria* and *Listeria* monocytogenes in certain aquatic environments in Egypt. Egypt J Microbiol 34(1):67–78

Stone GG, Oberst RD, Hays MP, McVey S, Chengappa MM (1994) Detection of *Salmonella* serovars from clinical samples by enrichment broth cultivation-PCR procedure. J Clin Microbiol 32(7):1742–1749

Sugumar G, Mariappan S (2003) Survival of *Salmonella* sp. in freshwater and seawater microcosms under starvation. Asian Fish Sci 16(3/4):247–256

Synnott AJ, Kuang Y, Kurimoto M, Yamamichi K, Iwano H, Tanji Y (2009) Isolation from sewage influent and characterization of novel *Staphylococcus aureus* bacteriophages with wide host ranges and potent lytic capabilities. Appl Environ Microbiol 75(13):4483–4490

Tao XQ, Lu GN, Dang Z, Yang C, Yi XY (2007) A phenanthrene-degrading strain *Sphingomonas* sp. GY2B isolated from contaminated soils. Process Biochem 42(3):401–408

Thompson JM, Gündoğdu A, Stratton HM, Katouli M (2013) Antibiotic resistant *Staphylococcus aureus* in hospital wastewaters and sewage treatment plants with special reference to methicillin-resistant Staphylococcus aureus (MRSA). J Appl Microbiol 114(1):44–54

U.S. EPA (2003) Control of pathogens and vector attraction in sewage sludge; 40 CFR Part 503. U.S. Environmental Protection Agency, Cincinnate, OH 45268

U.S. EPA (2006a) Method 1103.1, *E. coli* in water by membrane filtration using membrane-thermotolerant Escherichia coli Agar (mTEC). U.S. Environmental Protection Agency. Office of Water (4303T) 1200 Pennsylvania Avenue, N W. Washington, USA

U.S. EPA (2006b) Method 1106.1: *Enterococci* in water by membrane filtration using membrane-Enterococcus-Esculin Iron Agar (mE-EIA). U.S. Environmental Protection Agency. Office of Water (4303T) 1200 Pennsylvania Avenue, N W. Washington, USA

Vila J, Gascon J, Abdalla S, Gomez J, Marco F, Moreno A, De Anta TJ (1994) Antimicrobial resistance of *Shigella* isolates causing traveler's diarrhea. Antimicrob Agent Chemother 38(11):2668–2670

Walsh F (2013) Investigating antibiotic resistance in non-clinical environments. Front Microbiol 4:19

Warren BR (2006) Improved sample preparation for the molecular detection of *Shigella sonnei* in foods (Doctoral dissertation, University of Florida)

Wiggins BA (1996) Discriminant analysis of antibiotic resistance patterns in *faecal streptococci*, a method to differentiate human and animal sources of faecal pollution in natural waters. Appl Environ Microbiol 62(11):3997–4002

Zhang S (2005) Development of a biosensor for the rapid detection of *Salmonella Typhimurium* in milk. PhD thesis, Auburn University, Alabama

Chapter 4
Reuse of Greywater for Irrigation Purpose

**Adel Ali Saeed Al-Gheethi, Efaq Ali Noman,
Radin Maya Saphira Radin Mohamed, Balkis A. Talip, Abd Halid Abdullah
and Amir Hashim Mohd Kassim**

Abstract Reuse of greywater for the irrigation is an alternative water source in the new water management strategy of the countries that face a severe deficiency of water resources such as the Middle East Countries. Several studies have been evaluated the effects of greywater on the soil structure and plants. Greywater with a high level of nitrogen and phosphorus as well as macro-elements induce the plant's growth. However, the reuse of these effluents at excessive rates might produce detrimental effects on soil and crops. Some of the heavy metals in the greywater are toxic to plants, while others have toxicity for human and animals. The main consideration in the reuse of greywater in the irrigation lies in the transfer of pathogenic microorganisms to humans directly or indirectly. In developed countries, the utilisation of greywater for the irrigation subject for strict regulation which lies in the method of irrigation as surface or subsurface. In contrast, the surface irrigation is the common practice in

---

A. A. S. Al-Gheethi (✉) · R. M. S. Radin Mohamed (✉) · A. H. Mohd Kassim
Micro-Pollutant Research Centre (MPRC), Department of Water and Environmental Engineering, Faculty of Civil and Environmental Engineering, Universiti Tun Hussein Onn Malaysia (UTHM), 86400 Parit Raja, Batu Pahat, Johor, Malaysia
e-mail: adel@uthm.edu.my

R. M. S. Radin Mohamed
e-mail: maya@uthm.edu.my

A. H. Abdullah
Department of Architecture and Engineering Design, Faculty of Civil and Environmental Engineering, Universiti Tun Hussein Onn Malaysia (UTHM), 86400 Parit Raja, Batu Pahat, Johor, Malaysia

E. A. Noman
Faculty of Applied Sciences and Technology (FAST), Universiti Tun Hussein Onn Malaysia (UTHM), Pagoh, Johor, Malaysia

E. A. Noman
Department of Applied Microbiology, Faculty Applied Sciences, Taiz University, Taiz, Yemen

B. A. Talip
Faculty of Applied Sciences and Technology (FAST), Universiti Tun Hussein Onn Malaysia (UTHM), 84000 KM11, Jalan Panchor, Pagoh Muar, Johor, Malaysia

© Springer International Publishing AG, part of Springer Nature 2019
R. M. S. Radin Mohamed et al. (eds.), *Management of Greywater in Developing Countries*,
Water Science and Technology Library 87,
https://doi.org/10.1007/978-3-319-90269-2_4

the developing countries. In this chapter, the benefits of the greywater for the soil and plants as well as the adverse effects are reviewed. Based on the literature review in this chapter, it can be concluded that the criteria required to reuse greywater in the irrigation include aesthetics, hygienic safety, environmental tolerance, and technical and economic feasibility.

Keywords Greywater · Microbial risk · Human health · Reuse · Irrigation

4.1 Introduction

Greywater represents very good alternative water resource among the countries with air and semi-arid zones, since these countries face many of the challenges related to food security due to the absence of water resources. Therefore, the application of greywater for the irrigation purpose is an indispensable solution for the irrigation purpose. These practices might limit the deficiency in the food supply. Besides, the composition of greywater with nutrients and microelements might improve the quality of plant growth and crops. So far, the chemical and biological aspects of greywater should be considered before applying the greywater in the irrigations. The developed countries such as USA, Japan, Germany and Australia have good experience in greywater reuse for irrigations (Ottoson and Stenström 2003). The main concepts in those counties are to reuse the treated or partially treated greywater. In contrast, many of developing countries used the untreated greywater due to the absence of facilities required for the treatment process.

On the other hands, the concerns related to the utilisation of greywater for irrigation lie in the distribution of pollutants such as chemical agents, organic micro-pollutant (OMPs) and pathogens into the soil and plant crops and then the direct or indirect transmission into the human via the food chain. Besides, the high salinity of greywater deriving from detergents represents the major concern. The level of salinity in the soil is quantified in terms of sodium adsorption ratio (SAR) index (Lazarova and Asano 2005). SAR in greywater generated from laundry might reach 12.32 mg L^{-1} which is generated from the detergents that have more 3000 mg L^{-1} of SAR (Abu-Zreig et al. 2003). The accumulation of Na in the soil due to the frequent irrigation with greywater leads to the degradation of soil permeability and composition, and SAR in the soil leads to reduce saturated hydraulic conductivity (Ksat) (Gross et al. 2008).

The overutilisation of greywater in the irrigation might cause an elevation in pH values due to the high contents of alkaline detergents (Travis et al. 2010; Sivongxay 2005). The increase in pH values in the soil is associated with the increase in soil cation exchange capacity (CEC) (Sivongxay 2005; Anwar 2011). It has revealed that the soil properties of the sandy type have changed after being irrigated with surfactant-rich laundry greywater (Anwar 2011). The present chapter focuses on the benefits of reuse greywater in the irrigation and their negative effects on the plant growth, crops and soil characteristics.

4.2   Effect of Irrigation with Greywater on Plant and Soil

4.2.1   Chemical Effects

The domestic greywater is rich with many of chemical detergents which are used in the bathing and washing process of vegetables, fruits and washing machines. The surfactant which is most common in the laundry greywater is organic compounds consisting of the hydrophilic and hydrophobic group with the long chain of alkyl C10–C20 (Anwar 2011). The agricultural surfactants which are known as "spreaders and stickers" help the fertiliser and pesticides to spread through the soil matrix and then comply with plant leaves. The reduction in the surface tension of surfactant-rich greywater may increase the hydraulic conductivity of the soil.

Surfactants (surface-active agent) represent the major xenobiotic compounds in the greywater because it used in the generation of detergents and hygiene products which are utilised extensively in the batting and clothes washing. The surfactants included the compounds generated from amphoteric, cationic, anionic and nonionic detergents. Among these classes, anionic and cationic surfactants include methyl ester sulphonate, olefin sulphonate, alkyl benzene sulphonates, alkyl ether sulphates, isotridecanol ethoxylates, benzalkonium chloride, n-hexadecyl trimethyl and ammonium chloride. The utilisation of these detergents depends on their potential to provide cleaning action, disinfection agents and low price (Jakobi and Lohr 1987; Lange 1994; Belanger et al. 2002).

Another negative effect of soil irrigated with the untreated greywater is the elevation in pH values due to the high contents of alkaline detergents (Sivongxay 2005; Travis et al. 2010). High pH in laundry greywater acts as the dispersing agent which causes soil particles to split and leads to the increase of soil cation exchange capacity (CEC) (Sivongxay 2005; Anwar 2011).

It has been revealed that the soil properties of the sandy type have changed after being irrigated with surfactant-rich laundry greywater (Anwar 2011). However, the effects discharged laundry greywater on the soil composition in developing countries have not been reported yet. This might be related to the concept of greywater and the effects associated with the disposal in these countries raised within few last years, in comparison with the developed countries. For instance, in Australia, the local governments are gravely considering the application of greywater (generated from laundries and bathrooms) as an option for irrigating household lawns and gardens, thereby reducing residents demand for filtered water (Mohamed et al. 2013). Besides, there are no regulations for discharge of greywater into the environment in the developing countries. These are associated with the absence of the studies which assessed the adverse effects of disposal of greywater on the soil composition in many of developing countries.

The toxicity of the high concentrations of Na ions in the greywater on the plants lies in the bioaccumulation in the roots and leaves. The bioaccumulation of Na ions affects negatively the uptake of water from the soil. The increasing of soil salinization and sodic conditions leads to reduce the ability of the soil to support plant growth

(Morel and Diener 2006). Rodda et al. (2011) investigated the use of household's greywater for irrigating of *Beta vulgaris* (an aboveground crop) and *Daucus carota* (a belowground crop) in comparison with tap water and a hydroponic nutrient solution. The sub-irrigation method was used for six growth cycles. The characteristics examined included growth rate of plant and crop yield as well as the concentrations of macro- and micronutrients in the yielded crops and soil composition. The results revealed that the utilisation of greywater improved the plant growth and nutrients contents with the slight differences in the plants irrigated with tap water. Moreover, the highest plant growth and crop yields were recorded with the hydroponic nutrient solution. The electrical conductivity of the soil and metal contents increased with the utilisation of greywater and correlated with the metal and sodium concentrations in the crops. The study concluded that the greywater might provide an alternative water resource for plant irrigation and some fertiliser properties. However, the precautions regarding the metal and salts accumulation have to be considered.

The dispersion of clay particles in Ca–soil column leached by Na ions and NaCl aqueous solution led to an irreversible decrease in soil hydraulic conductivity (Yaron et al. 2012). Datnoff et al. (2001) stated that the surface-induced swelling clay is the main mechanism for decreasing of Ksat in the clay. The main cause of Ksat reduction can be rationalised as a small pore clogging in the soil due to the adsorption of surfactant. Surfactant effects on water infiltration and on percolation in soils are a function of soil type and surfactant characteristics (Kuhnt 1993). Surfactants decrease capillary rise of water in soil columns when mixed with sandy or clayey soils by decreasing water surface tension (Smith and Gillham 1999). This might be caused by swelling of some clay particles, which changes the hydraulic capacity of the soil profile and resulted in reduced retention of soil water and increased the depth of infiltration (Karagunduz et al. 2001; Crites et al. 2014).

Heavy metals in the greywater are originated from the detergents. Therefore, the reuse of the greywater for irrigation might lead to transmission of the metal ions into the soil and plants. The transfer of heavy metal ions from greywater used in the irrigation to plant might lead to reduce plant's growth and affect soil microorganisms, and their activities depend on the application rate and its frequency, the soil removal mechanism and the removal capabilities of plant species (Dahdoh and El-Hassanin 1994; Logan et al. 1997; Giller et al. 1998).

Metal has important role in the metabolic and anabolic pathway as cofactors for the enzymes and as stabilisers of protein structures in the living cells. Some of these metals including Fe, Cu, Mn, Na, Zn, Co, K and Ca ions are essential nutrients as trace elements/macro-elements. In contrast, others such as Ag, Hg, Pb and Al ions have no biological role (Bruins et al. 2000; Jais et al. 2017). Moreover, the presence of macro-elements with high concentration leads to increase their toxicity for the biological functions in the cell (Al-Gheethi et al. 2016a).

4.2.2   Microbial Effects

The potential effect for reusing greywater for irrigation on the human health is associated with the presence of pathogens (Finley et al. 2009). The level of hazards depends on the plant type and pathogens, while the mode of transmission takes place directly or indirectly. In terms of plant types, the vegetables irrigated with the greywater represent the high risk for human, because they are consumed without cooking. The trees, which produce their crops far away from the soil surface, have less risk because the crops have not contacted directly with the greywater. In contrast, the plants that produce their crops on the soil surface or underground have more risk due to the direct contact with the pathogens in the greywater.

The pathogens in the greywater are originated mainly from the human bodies (shower and laundry greywater) and plants and animals (kitchen greywater). Therefore, the pathogens from shower and laundry greywater are more likely to cause the infection for human, because they have already adapted to the human temperature and acquired resistance for the human immunity. Some of those pathogens have no potential to compete with the indigenous organisms in the environment and would not multiply in the environment due to the absence of growth factors required for their growth. However, they can survive for a long time and transmit through the food chain into the human.

The pathogens from kitchen greywater especially those originated from the washing of fruits and vegetables have high levels of the pathogens which are pathogenic for the plant more than their pathogenicity for the human. They are more likely to cause the diseases in the plant, and so far affect negatively on the plant growth and crop production. These pathogens have high potential to grow on the plant's roots and leaves. The most common pathogens in those wastes included helmets and microbes, which are more available in the rhizosphere layers of the soil as well as caused the potato brown rot. However, Finley et al. (2009) indicated that there were no differences between the concentrations of faecal coliforms (FC) and faecal streptococci (FS) in the crops irrigated with treated and untreated domestic greywater as well as tap water. Moreover, the differences were noted between the bacterial species and the plant, where FC was presented with high concentrations in carrots, while FS was being highest on lettuce leaves. The study mentioned that the concentrations of these pathogens in the greywater were high in the greywater. However, their concentrations were low and did not represent a significant health risk. Similar findings were also reported by Jackson et al. (2006) who indicated that there is no significant difference in the bacterial concentrations on the plant surfaces irrigated with greywater, tap water, or hydroponic solution. So far, previous studies have found that the crop portions matured underground or near the surface of the soil irrigated with wastewater contain high concentrations of pathogenic bacteria (Armon et al. 1994).

In comparison between the health risk of human pathogens and plant pathogens, it can be indicated that most plant pathogens are sensitive to the treatment process of greywater; therefore, the primary and secondary process might be effective to reduce these pathogens to less than the detection limits. Conversely, the human pathogens exhibited more resistance to the treatment process. Besides, most of the plant pathogens are very specificity to infect the plant and have a narrow host range. Indeed, the ability of pathogens to reach the human and cause the infection depends on the survive time which is defined as the time in which the pathogenic cells have the potential to cause the infection for the human.

Regarding the ability of pathogens to survive and transmits through the food chain into the human. The irrigation method is the main factor which influences effectively and can detect the final fate of the pathogens available in the greywater. The utilisation of surface irrigation leads to the distribution of pathogens. However, the survival period of these pathogens is less due to the effect of sunlight and deficiency in the water contents in the surface soil which is required for microbial multiplication and growth. So far, the pathogens, which have the ability to form spores or cysts such as *Bacillus* spp. and helminths, might tolerate the hard environment and then transmitted to the human. The utilisation of subsurface irrigation method might be alternative to prevent the distribution of pathogens into the plant crops. However, the sanitary implications lie in the transmission of these pathogens to underground waters (Al-Gheethi et al. 2016b).

Najafi et al. (2015) examined three irrigation methods including surface, subsurface drip, and furrow irrigation for evaluating the distribution of faecal coliform bacteria from wastewater. The study revealed the usage of surface drip and furrow irrigation associated with high pollutions of the plants and soil. In contrast, the utilisation of subsurface drip irrigation reduces the distribution of faecal coliform bacteria.

Dixon et al. (1999) stated that four actors are associated with the presence of microbial health risk from greywater reuse which included populations, exposure, dose-response and delay before reuse. Based on this factor, the utilisation surface irrigation method of stored greywater for a long time with the high concentration of infectious agents has high risk. The health risk for reuse of greywater depends also on when and where the greywater will be used. The level of the risk associated with the greywater relies on the hazards level and exposure time (Fig. 4.1). It can be noted that the exposure to the hazards for long time (greywater at the source) might be more dangerous from the exposure to high dose of the hazards for only short time (greywater at the reusing point).

Indeed, the absence of epidemiology reports on the infection diseases resulted from the exposure to the hazard risks in the greywater might represent the main obstacle to assess the health risk for the greywater. But it has to be mentioned that the presence of pathogenic organism in the greywater indicates to present a risk, the level of these hazards depends on the transmission method. The availability of epidemiological studies and quantitative microbial risk assessment (QMRA) provides more details on the level of microbial and hazardous chemicals risks correlated with the reuse of greywater. QMRA consists of four steps include; Hazard Identification (Hazard ID), which aim is to identify the pathogenicity of the microbe cell and transmission routes, infective dose depends on the pathogens species and whether the pathogen need for intermediate host before the reaching to the final host and

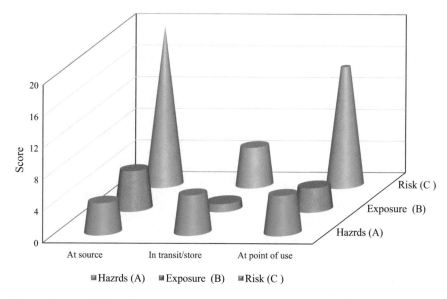

Fig. 4.1 Show the health risk associated with reuse of greywater at source, in transit and at point of use; for A and B the score is 1 (low), 3 (Intermediate), 4 (Interm-Higher) and 5 (Higher). C is calculated as A*B, the score was calculated as 5 (low), 15 (Intermediate), 20 (Interm-Higher) and 25 (higher) (Dixon et al. 1999)

the exposure assessment which describe the exposure to the pathogens (inhalation, direct contact, digesting). The risk for each pathogen depends on the hazard, infective dosage and exposure levels (Fig. 4.2).

Based on Fig. 4.2, it can be indicated that health risk level of infectious agents is a result of available three factors which contribute to the increase of the risk level. The exposure for the high dose of pathogens with high pathogenicity increases the level of health risk. An available one of the factor with the low score (1) in the presence high score of other factors reduce the health risk percentage level to 28.57%, however, presents an intermediate score for one factor lead to increase the percentage of the risk level to 57.2%.

According to World Health Organization (WHO 2005), the infectious pathogens are classified into four levels including risk group I which included the infectious agents with no or low individual and community risk. The pathogens with moderate individual risk but low community risk are classified in risk group II. The potential of pathogens to have the high individual risk and low community risk is classified as risk group III. Finally, the pathogens classified within risk group IV included those having high individual and community risk. Based on the pathogens available in the greywater which were presented in Chap. 1, *Enterococcus faecalis*, *Pseudomonas aeruginosa* and *Staphylococcus aureus* are classified within the risk group (II), while *E. coli* and *Salmonella* spp. are classified within the risk group II and III. In comparison between those pathogens, both *E. coli* and *Salmonella* spp.

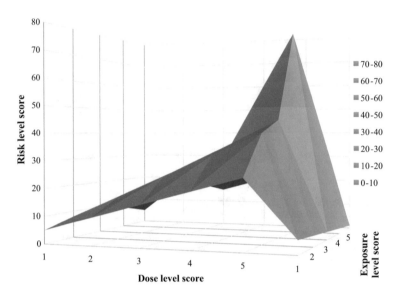

Fig. 4.2 Show the QMRA of pathogens in the greywater, the dose, exposure and hazards levels is expressed in the range (1–5); very low (1); low (2); Intermediate (3); Interm-Higher (4); Higher 95) (Higher); the risk level is divided into seven classes based on the correlation between dose, exposure and hazards levels; the high-risk level takes place above 30 scores

have high pathogenicity with the low dose (ranged from 10^3–10^6 cells). *E. coli* is presented in the greywater with concentrations ranged from 10^2 to 10^6 which is within the dose concentrations required to cause the infection. There are no more studies which have estimated the *Salmonella* spp. concentrations in the greywater because these studies focused only on the presence or absence this bacterium. However, Katukiza et al. (2014) mentioned that the concentrations of *Salmonella* spp. in greywater from Uganda were 2.73×10^4 cell/100 mL. These concentrations might be enough to cause disease if the person consumed more 100 mL of greywater with the food or with the drinking water. *P. aeruginosa* and *S. aureus* are available with concentrations between 101 and 10^4 cell/100 mL of the greywater. The infective dose of *P. aeruginosa* is 10^8–10^9 cells and for *S. aureus*, it is 10^3–10^8 cells (Sewell 1995; Rusin et al. 1997). Therefore, based on this information, it can be indicated that the pathogens in the greywater might have no high health risk for the human and animals. So far, one of the serious points associated with the pathogenic bacteria is their ability to multiply and regrow in the environment. It can be increased in their number to reach the infective dose which enables them to cause the disease for the human and animals as well as the plants (Al-Gheethi et al. 2016b).

The helminths such as *Ascaris* sp., *Necator* sp., *Ancylostoma* sp., *Strongyloides* sp., *Hymenolepis* sp., *Trichuris* sp., *Taenia* spp. and *Toxocara* sp. as well as Trematodes (*Opisthorchis* sp., *Clonorchis* sp., *Schistosoma* spp. and *Fasciola* sp.) and parasites such as *Giardia lamblia* and *Cryptosporidium* spp. have no ability to increase in their numbers in the environment because they need the intermediate host for their life cycle. Nevertheless, they have long survival periods and low infective dose (10–100 cysts) which enable them to represent a risk for the human (Robertson and Nocker 2010).

Based on QMRA, the microbial risk associated with the reusing greywater in the irrigation lies in the ability of these pathogens to survive in the environment sufficiently long to pose health risks (QMRA 2015). These pathogens are transmitted by the direct contact with these pathogens during the irrigation of the plant or harvesting process of the crops and by the consumption, the fruits and vegetables contaminated with the pathogens (Fig. 4.3). Therefore, the washing of fruits and vegetables are required for preventing the infection by the infectious agents. Some

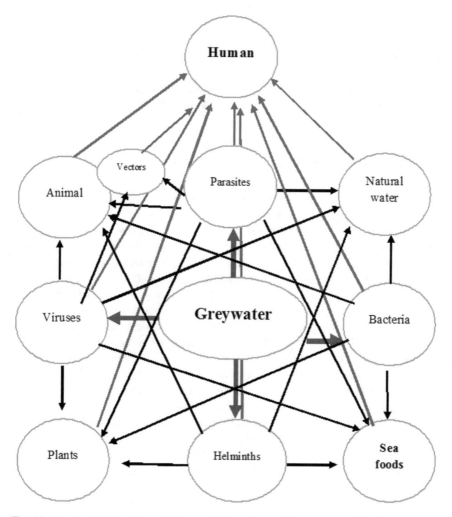

Fig. 4.3 Direct and indirect transmission routes for pathogens from greywater into human

of the food is subjected to cooking which might contribute effectively to the reduction of the pathogens and prevent their transmission in the human. Another concern related to the presence of pathogens such as dengue virus, Plasmodium spp. Japanese encephalitis virus and *Wuchereria bancrofti* lies in the transmission route of these pathogens which are transmitted to human by the vector contact (Chorus and Bartram 1999; Van der Hoek et al. 2005).

The pathogens in the greywater represent a health risk for a wide range of the people included consumers, farm workers and their families, and nearby communities (WHO 2006). The presence of these pathogens on the plant surface and the crops has been reported in the literature which is an evidence of the presence of a health risk for the plants irrigated with the greywater. However, there is no evidence confirming that the disease recorded among the consumers was originated from these pathogens. This might be due to the absence of advance technology which might detect the source of the pathogen which caused the infection (Al-Gheethi et al. 2016b). Among the farm workers and their families and nearby communities, the health risk is considered to be based on the exposure factors since the studies confirmed the presence of pathogens in the greywater and on the surface of plant leaves and crops. The farmers are more susceptible to infectious agents (Blumenthal and Peasey 2002).

4.3 Regulations for Reuse of Greywater

The effects reusing greywater in the irrigation on the soil structure and composition in many developing countries have not reported yet. This might be related to the fewer developments in the field of greywater in those countries, and the application and reuse of greywater in the developing countries located in the Middle East region such as Jordan raised within few last years. However, there are no specific regulations for the reuse of greywater in the agriculture. In contrast, in Australia, the local governments are gravely considering the application of greywater (generated from laundries and bathrooms) as an option for irrigating household lawns and gardens, thereby reducing residents demand for filtered water (Mohamed et al. 2013). Many of the developing countries adopted the WHO guidelines for microbiological regulation of the wastewater (WHO 1989) in which wastewater are classified into three classes (A–C) based on the final concentrations of FC and treatment methods (Table 4.1). The microbiological and chemical regulations for reusing greywater in the irrigation among different countries are illustrated in Tables 4.2 and 4.3.

Dixon et al. (1999) proposed a framework for regulating the reuse of greywater and some more guidelines to that adopted by WHO (1989) as follows: the storage period of the greywater should be minimised to reduce the microbial multiplication and formation of biofilm as well as the exposure. Besides, the odour is supposed to be kept at the minimum level.

Table 4.1 Guidelines for using treated wastewater in agriculture (WHO 1989)

Category	Reuse conditions	Exposed group	Intestinal nematode[a] (arithmetic mean no eggs/L)[b]	Coliform (geometric mean/100 mL)[b]	Wastewater treatment expected to achieve the required microbiological guideline
A	Irrigation of crops likely to be eaten uncooked, sports fields, public parks[c]	Workers, consumers, public	≤ 1	≤ 1000	A series of stabilisation ponds designed to achieve the microbiological quality indicated, or equivalent treatment
B	Irrigation of cereal crops, industrial crops, fodder crops, pasture and trees[d]	Workers	≤ 1	No standard recommended	Retention in stabilisation ponds for 8–10 d or equivalent helminthic and FC removal
C	Localised irrigation of crops in category B if exposure to workers and the public does not occur	None	Not applicable	Not applicable	Pre-treatment as required by irrigation technology, but not less than primary sedimentation

[a] *Ascaris* and *Trichuris* species and hookworms
[b] During the irrigation period
[c] A more stringent guideline (200 FC/100 mL) is appropriate for public lawns, such as hotel lawns, with which the public may come into direct contact
[d] In the case of fruit trees, irrigation should cease 2 weeks before the fruit is picked, and no fruit should be picked off the ground. Sprinkler irrigation should be used

Table 4.2 Microbiological regulation for greywater reused for irrigation

Country/Organization	TC	FC	*Pseudomonas aeruginosa*	References
USA, California		2.2 MPN/100 mL (23 MPN/100 mL in 30 days)		U.S. EPA (2004)
Australia		<30 CFU/100 mL^{-1}		NSW (2000)
Australia		<10 CFU/100 mL^{-1}		
Germany	<10^4 CFU/100 mL	<1000 CFU/100 mL	<100 CFU/100 mL	Nolde (1999)
WHO		Class A (<1000 CFU/100 mL) Class (B) No standard recommended		WHO (1989)
Mexico		≤2000 CFU/100 mL		Peasey (1999)
Korea		Must not be detected		Jong et al. (2010)
UK	10 CFU/100 mL	–	–	BS-8525-1 (2010, BS-8525-2 2011)
Portugal	104 CFU/100 mL	200 CFU/100 mL	–	ANQIP (2011)
Jordan		Category A (Cooked Vegetables) 100 MPN/100 mL Category B (Tree crops) 1000 MPN/100 mL		Duqqa (2002)
Japan	<50	<10		Al-Jayyousi et al. (2003)
USA		<200		
China	<10,000			
Egypt		100		Egyptian regulation (1994)

Table 4.3 Chemical regulations for reuse greywater in the irrigation

Parameters	Korea	USA	Japan	China	Germany	Egypt
BOD	<10 mg L	30	<3	<6	5	20
COD	<20 mg L					40
pH	5.8–8.5	6–9	5.8–8.6	6–9	6–9	
Turbidity	<2NTU		<5	<5	1–2	
TSS		30				20
TP				<0.5		
Reuse application	Reuse application Restricted reuses[a]	Reuse application Restricted reuses	Environmental (limited public contact)	Restricted impoundments and lakes	Toilet flushing	

[a]Irrigation of areas where public access is infrequent and controlled golf courses, cemeteries, residential and greenbelt

4.4 Conclusion

It can be concluded that the reuse of untreated greywater for the irrigation purpose has several issues on the soil and plants. The main concern lies in the accumulation of heavy metals ions in the plants which are then transmitted to the human. Similar concerns are associated with the pathogens which might have the potential to survive in the hard environment and then transmitted by the food chain into the human. It has to be mentioned that the greywater has less microbial loads in comparison to the black water. However, the health risk associated with those pathogens is not their virulence factor since also their ability to transmit the antimicrobial resistance gene into the pathogens and thus increase their pathogenicity.

Acknowledgements The authors wish to thank the Ministry of Higher Education (MOHE) for supporting this research under FRGS vot 1574 and also the Research Management Centre (RMC) UTHM for providing grant IGSP U682 for this research.

References

Abu-Zreig M, Rudra RP, Dickinson WT (2003) Effect of application of surfactants on hydraulic properties of soils. Biosys Eng 84:363–372
Al-Gheethi AA, Mohamed RR, Efaq AN, Hashim MA (2016a) Reduction of microbial risk associated with greywater by disinfection processes for irrigation. J Water Health 14(3):379–398
Al-Gheethi AA, Mohamed RM, Efaq AN, Norli I, Halid AA, Amir HK, Ab Kadir MO (2016b) Bioaugmentation process of secondary effluents for reduction of pathogens, heavy metals and antibiotics. J Water Health 14(5):780–795
Al-Jayyousi OR (2003) Greywater reuse: towards sustainable water management. Desalination 156(1–3):181–192

ANQIP (2011) Building reuse systems and recycling of gray water (SPRAC). National Association for Quality of Building Installations; ETA 0905 (Version 1)

Anwar AH (2011) Effect of laundry greywater irrigation. J Environ Res Devel 5(4)

Armon R, Dosoretz CG, Azov Y, Shelef G (1994) Residual contamination of crops irrigated with effluent of different qualities: a field study. Water Sci Technol 30(9):239–248

Belanger SE, Bowling JW, Lee DM, LeBlanc EM, Kerr KM, McAvoy DC, Davidson DH (2002) Integration of aquatic fate and ecological responses to linear alkyl benzene sulfonate (LAS) in model stream ecosystems. Ecotoxicol Environ Saf 52(2):150–171

Blumenthal UJ, Peasey A (2002) Critical review of epidemiological evidence of the health effects of wastewater and excreta use in agriculture. Unpublished document prepared for World Health Organization, Geneva. www.who.int/water_sanitation_health/wastewater/whocriticalrev.pdf

Bruins MR, Kapil S, Oehme FW (2000) Microbial resistance to metals in the environment. Ecotoxicol Environ Saf. 45(3):198–207

BS-8525-1 (2010). British standards part 1: greywater systems: code and practices. Issued by the British Standards Institution, London, UK, ISBN 978 0 580 63 475 8. BSI, p 54

BS-8525-2 (2011). Greywater systems part 2: domestic greywater treatment equipment, requirements and test methods. Issued by the British Standards Institution, London, UK, ISBN 978 0 580 63 476 5. BSI, p-32

Chorus EI, Bartram J (1999) Toxic cyanobacteria in water: a guide to their public health consequences, monitoring and management. ISBN 0-419-23930-8

Crites RW, Middlebrooks EJ, Robert K (2014) Natural wastewater treatment systems, 2nd ed, CRC Press, Italy

Dahdoh MSA, El-Hassanin AS (1994) Combined effects of organic source, irrigation water salinity and moisture level on the growth and mineral composition of barley grown on calcareous soil. The Desert Institute Bulletin, Egypt

Datnoff EL, Snyder HG, Komdorfer HG (2001) Silicon in Agriculture. Elsevier, Netherlands

Dixon AM, Butler D, Fewkes A (1999) Guidelines for greywater re-use: health issues. Water Environ J 13(5):322–326

Duqqa M (2002). Treated sewage water use in irrigated agriculture. Theoretical design of farming systems in Siel Al zarqa and the Middle Jordan Valley in Jordan.Ph.D. thesis. University of Wageningen, Wageningen/The Netherlands

Egyptian Regulation (1994) Egyptian Environmental Association Affair (EEAA), Law 48, No. 61–63, Permissible values for wastes in River Nile (1982) and Law 4, Law of the Environmental Protection

Finley S, Barrington S, Lyew D (2009) Reuse of domestic greywater for the irrigation of food crops. Water Air Soil Poll 199(1–4):235–245

Giller KE, Witter E, Mcgrath SP (1998) Toxicity of heavy metals to microorganisms and microbial processes in agricultural soils: a review. Soil Biol Biochem 30(10):1389–1414

Gross A, Wiel-Shafran A, Bondarenko N, Ronen Z (2008) Reliability of small scale greywater treatment systems and the impact of its effluent on soil properties. Int J Environ Stud 65(1):41–50

Jackson S, Rodda N, Salukazana L (2006) Microbiological assessment of food crops irrigated with domestic greywater. Water SA 32(5):700–704

Jais NM, Mohamed RM, Al-Gheethi AA, Hashim MA (2017) The dual roles of phycoremediation of wet market wastewater for nutrients and heavy metals removal and microalgae biomass production. Clean Technol Environ Policy 19(1):37–52

Jakobi G, Lohr A (1987) Detergents and textile washing. VCH Publisher, Weinheim

Jong J, Lee J, Kim J, Hyun K, Hwang T, Park J, Choung Y (2010) The study of pathogenic microbial communities in greywater using membrane bioreactor. Desalination 250:568–572

Karagunduz A, Pennell KD, Young MH (2001) Influence of a nonionic surfactant on the water retention properties of unsaturated soils. Soil Sci Soc Am J 65(5):1392–1399

Katukiza AY, Ronteltap M, Steen P, Foppen JWA, Lens PNL (2014) Quantification of microbial risks to human health caused by waterborne viruses and bacteria in an urban slum. J Appl Microbiol 116(2):447–463

Kuhnt G (1993) Behaviour and fate of surfactants in soil. Environ Toxicol Chem 12:1813–1820
Lange KR (1994) Detergents and cleaners, a handbook for formulators. SchoderDruck GmbH & Co.KG, New York
Lazarova V, Asano T (2005) Challenges of sustainable irrigation with recycled water. In: Lazarova V, Bahri A (ed) Water reuse for irrigation, agriculture, landscapes and turf grass, CRC Press, London New York, pp 1–30
Logan EM, Pulford ID, Cook GT, Mackenzie AB (1997) Complexation of Cu^{2+} and Pb^{2+} by peat and humic acid. Eur J Soil Sci 48(4):685–696
Mohamed RM, Kassim AHM, Anda M, Dallas SA (2013) Monitoring of environmental effects from household greywater reuse for garden irrigation. Environ Monit Assess 185 (10):8473–8488
Morel A, Diener S (2006) Greywater management in low and middle-income countries, review of different treatment systems for households or neighbourhoods. Swiss Federal Institute of Aquatic Science and Technology (EAWAG), Dübendorf, Switzerland
Najafi P, Shams J, Shams A (2015) The effects of irrigation methods on some of soil and plant microbial indices using treated municipal wastewater. Int J Recycl. Org Waste Agric 4(1):63–65
Nolde E (1999) Grey water reuse systems for toilet flushing in multi-storey buildings—over ten years' experience in Berlin. Urb Water 1(4):275–284
Ottoson J, Stenström TA (2003) Faecal contamination of greywater and associated microbial risks. Water Res 37(3):645–655
Peasey A (1999) A review of policy and standards for wastewater reuse in agriculture: a Latin American perspective. London, Water and Environmental Health at London and Loughborough Resource Centre, London School of Hygiene and Tropical Medicine, and Water, Engineering and Development Centre (WEDC), Loughborough University
QMRA (2015) Quantitative microbial risk assessment Wiki. http://qmrawiki.canr.msu.edu/index.php/Quantitative_Microbial_Risk_Assessment_(QMRA)_Wiki. Accessed 1 June 2017
NSW (2000) Greywater reuse in sewered single domestic premises. Published by the Government of NSW, Australia p 19
Robertson L, Nocker A (2010) Giardia. Retrieved from http://waterbornepathogens.susana.org/menuprotozoa/giardia on 12 Oct 2016
Rodda N, Salukazana L, Jackson SA, Smith MT (2011) Use of domestic greywater for small-scale irrigation of food crops: effects on plants and soil. Phys Chem Earth Parts A/B/C 36(14):1051–1062
Rusin PA, Rose JB, Haas CN, Gerba CP (1997) Risk assessment of opportunistic bacterial pathogens in drinking water. In reviews of environmental contamination and toxicology. Springer, New York
Sewell DL (1995) Laboratory-associated infections and biosafety. Clin Microbiol Rev 8: 389–405
Sivongxay A (2005) Hydraulic properties of Toowoomba soils for laundry water reuse. Thesis BEng Environmental, University of Southern Queensland
Smith JE, Gillham RW (1999) Effects of solute concentration-dependent surface tension on unsaturated flow: laboratory sand column experiments. Water Res Res 35(4): 973–982
Travis MJ, Alit W, Noam W, Adar E, Gross A (2010) Greywater reuse for irrigation: effect on soil properties. Sci Total Environ 408:2501–2508
U.S. EPA (2004) Guidelines for Water Reuse. Report EPA/625/R-04/108, United States Environmental Protection Agency, USEPA, Washington, DC, USA
Van der Hoek L, Sure K, Ihorst G, Stang A, Pyrc K, Jebbink MF, Überla K (2005) Croup is associated with the novel coronavirus NL63. PLoS Med 2(8):e240
WHO (1989) Health guidelines for the use of wastewater in agriculture and aquaculture. Technical Report Series 778, World Health Organization, Geneva, Switzerland
WHO (2005) Safe healthcare waste management-policy paper by the World Health Organisation. Waste Manag 25:568–569
WHO (2006) Guidelines for the safe use of wastewater, excreta and greywater. World Health Organization, Geneva, Switzerland
Yaron B, Dror I, Berkowitz B (2012). Soil-subsurface change: chemical pollutant impacts. Springer Science & Business Media

Chapter 5
Xenobiotic Organic Compounds in Greywater and Environmental Health Impacts

Efaq Ali Noman, Adel Ali Saeed Al-Gheethi, Balkis A. Talip, Radin Maya Saphira Radin Mohamed, H. Nagao and Amir Hashim Mohd Kassim

Abstract One of the most common organic compounds which represents real challenges in the environmental pollution treatment is the xenobiotic organic compounds (XOCs). They are complex organic compounds which have high persistence in the environment extend for several years due to their chemical structure. Meanwhile, its hazards risk lies in tier active poisons which directly affect aquatic life within a short exposure time. XOCs in the greywater are generated from utilisation of detergents and personal body care products and they have the potential to persist in nature for a long time and thus have long-term effects to the environment including toxicity and bioaccumulation in the organism's cells. There are many literatures discussing about the types of XOCs of greywater. For instance, some types of XOCs in greywater are toxic for aquaculture. This chapter will discuss the occurrence of XOCs in the greywater, chemical structure and bioassay for the toxicity of these compounds.

Keywords XOCs · Toxicity · Health risk · Regulations · Greywater

E. A. Noman
Faculty of Applied Sciences and Technology (FAST), Universiti Tun Hussein Onn Malaysia (UTHM), Pagoh, Johor, Malaysia

E. A. Noman
Department of Applied Microbiology, Faculty Applied Sciences, Taiz University, Taiz, Yemen

A. A. S. Al-Gheethi (✉) · R. M. S. Radin Mohamed · A. H. Mohd Kassim
Micro-Pollutant Research Centre (MPRC), Department of Water and Environmental Engineering, Faculty of Civil and Environmental Engineering, Universiti Tun Hussein Onn Malaysia (UTHM), 86400 Parit Raja, Batu Pahat, Johor, Malaysia
e-mail: adel@uthm.edu.my

B. A. Talip (✉)
Faculty of Applied Sciences and Technology (FAST), Universiti Tun Hussein Onn Malaysia (UTHM), 84000 KM11, Jalan Panchor, Pagoh Muar, Johor, Malaysia
e-mail: balkis@uthm.edu.my

H. Nagao
School of Biological Sciences, Universiti Sains Malaysia (USM), 11800 Penang, Malaysia

© Springer International Publishing AG, part of Springer Nature 2019   89
R. M. S. Radin Mohamed et al. (eds.), *Management of Greywater in Developing Countries*,
Water Science and Technology Library 87,
https://doi.org/10.1007/978-3-319-90269-2_5

5.1 Introduction

The concept of XOCs has been raised since the 1990s alongside the development of more advanced analytical technologies such as gas chromatography (GC), mass spectrometry (MS) and liquid chromatography (LC-MS) which have high potential in detecting nanograms per litre (Togunde et al. 2012). In the soil, the accumulation of XOCs is generated from the utilisation of pesticides and fertilisers. In contrast, accumulation of XOCs in the greywater commonly results from personal care products, shampoos, hair conditioners, oils and foodstuffs, moisturising oils and the food additive. Very few studies have been conducted on the XOCs in greywater due to the unavailability of chemical analytical quality as well as the difficulties in the identification of concomitant exposures. To date, many researches focusing on the XOCs mainly conducted in developed countries including Korea, China, Sweden, UK, Spain, Taiwan, USA, Greece, Denmark, Germany, Japan, Belgium and Switzerland. Meanwhile, no work has been reported on the XOCs in the developing countries. Limited studies have been reported by Etchepare and Van der Hoek (2015), Leal (2010) and Eriksson et al. (2002) on the occurrence of XOCs in sewage effluents and surface water in Netherlands, Sweden and Denmark. No studies report on the presence of these compounds in Malaysia including their toxicity on the biodiversity would represent a real contribution to reveal their occurrence following their removal technology. This chapter aims at highlighting the occurrence and health risk associated with the XOCs in the greywater.

5.2 Xenobiotic Organic Compounds in the Greywater

The XOCs are aromatic compounds, which are divided into three groups polyaromatic hydrocarbons (PAHs), benzene/toluene/ethyl benzene/xylene (BTEX) and the synthetic-substituted aromatics typified by the chlorophenols (Harvey and Thurston 2001). PAHs are among these compounds which have carcinogenic effect and accumulated with high concentration in the industrial sites. XOCs include pesticides, polyfluoroalkyl substances (PFASs), pharmaceuticals and personal care products (PPCPs), endocrine disrupting chemicals (EDCs), active pharmaceutical ingredients (APIs) and phosphorus-containing flame retardants (PFRs). The aromatic structure of XOCs includes two or more benzene rings. Figure 5.1 illustrated the chemical structure for XOCs most investigated and reported in the greywater. The source of organic micro-pollutants (OMPs) or xenobiotic organic compounds (XOCs) in greywater includes personal care products, shampoos, hair conditioners, oils and foodstuffs, moisturising oils and the food additive. Eriksson et al. (2003) revealed that the greywater in Denmark contained more than 200 types of OMPs which included plasticisers, surfactants, antioxidants, fragrances and dyes. Ying (2006) reported that the greywater from baths contains high concentrations of surfactants.

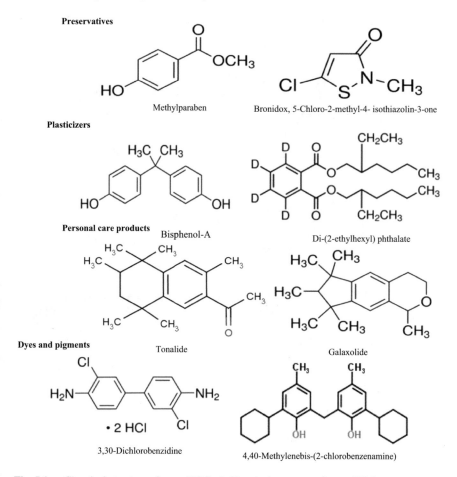

Fig. 5.1 **a** Chemical structure of some XOCs, **b** Chemical structure of some XOCs

The list of XOCs available in greywater is presented in Table 5.1. The sources of hydrocarbons and alcohols/ethers in greywater include skin cleaners, oils and foodstuffs, fragrance, food flavouring agents and shampoos.

Phenols, one of the XOCs that is present in greywater resulted from disinfectants such as UV stabiliser and antioxidants for hydrocarbon-based products. Other XOCs in the greywater include preservatives, biocide/surfactants, pharmaceutical residues, amphoteric, anionic, cationic and non-ionic detergents as well as dyes and pigments. The XOCs in the greywater might be similar to that in the original products or secondary products generated due to partial degradation after mixing with greywater (Deblonde et al. 2011; Luo et al. 2014; Etchepare and Van der Hoek 2015; Grčić et al. 2015).

Pharmaceutical products

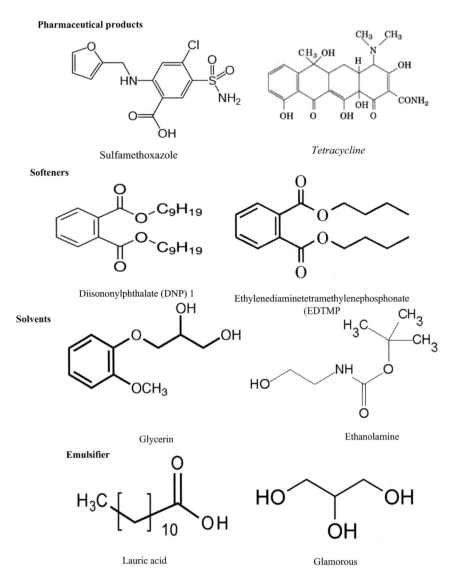

Sulfamethoxazole

Tetracycline

Softeners

Diisononylphthalate (DNP) 1

Ethylenediaminetetramethylenephosphonate (EDTMP

Solvents

Glycerin

Ethanolamine

Emulsifier

Lauric acid

Glamorous

Fig. 5.1 (continued)

Dyes in greywater have gained the attention of researchers in the few last years. This is because dyes have high tectorial values and a discharge of less than 1 ppm of dye into the water might cause changes in the physical characteristics of water. The main sources of dyes in greywater are food additives and hair colourants such as p-phenylenediamine, toluene- 2,5-diamine and 2,5-diaminoanisole (Grčić et al. 2015). Triclosan (TCS) is widely used in shampoo, soap, deodorant, disinfectants, detergent,

Table 5.1 Xenobiotic organic compounds (XOCs) in greywater

Source	Xenobiotic organic compounds (XOCs)
Skin care products, hair conditioners, shampoos, food additives, moisturising oils	Hydrocarbons
Oils and foodstuffs, skin cleaners, fragrance, food flavouring agents	Alcohols/ethers
Disinfectants; flavouring substance in foodstuffs, UV stabiliser and antioxidant for hydrocarbon-based products	Phenols
Foodstuffs and essential oils, Fragrance	Aldehydes and ketones
Soap, emulsifier for facial creams and lotions, shaving cream formulations, mould cheese	Carboxylic acids
fragrances, household dust, PVC floor wipes, food additive, emollient	Carboxylic acid esters
Flame retardant	Phosphoric acid ester
Cough syrup, dishwashing agents, body care, baby care, toiletries, artificial and natural jasmine oil, coffee, tobacco	Nitrogenous organics
Emulsifier	Deceth-3, Oleth-30, Lauric acid, Glycol distearate Ceteareth-25, Glamorous
Fragrances	Tonalide, Galaxolide, HCA, Hexyl cinnamic, aldehyde, flavours AHTN, HHCB, Styrene, Benzene-1,3-diol, p-Cresol
UV filters	BP, 4-methylbenzylidene-camphor (4MBC), octocrylene, octrocrylene, 2-ethylhexyl-4-methoxycinnamate (EHMC), avobenzone, 2-phenyl-5-benzimidazolesulfonic acid (PBSA), 2EHS, benzophenone-3 (BP3), 4-tert-Butyl-4′-methoxy-dibenzoylmethane (avobenzone) and 2-ethylhexyl salicylate (2EHS)
Softeners	Bis-(2-ethylhexyl)phthalate (DEPH), Diisononylphthalate (DNP) 1, Ethylenediaminetetramethylenephosphonate (EDTMP), Dibutylphthalate (DBP) 2, Diethylphthalate (DEP) 3, Nitrilotriacetic acid (NTA)
Solvents	Heptane, 1,2,4-Trichlorobenzene, Propylene glycol Diethanolamine, Ethanolamine, Glycerin, Isopropanol, Phenol, Xylene
Plasticizer	Bisphenol-A, Butylbenzyl phthalate, Di-(2-ethylhexyl) phthalate, Dibutyl phthalate, Diethyl phthalate, Di-isobutyl phthalate, Dimethyl phthalate

(continued)

Table 5.1 (continued)

Source	Xenobiotic organic compounds (XOCs)
Preservatives	Methylparaben, Ethylparaben, Propylparaben, Butylparaben, Bronopol, Bronidox, 5-Chloro-2-methyl-4- isothiazolin-3-one, Imidazolidinyl urea Triclosan, Quaternium
Biocide/surfactants	Triclosan, BaCl, Nonylphenol
Endocrine disrupting chemicals (EDCs)	Bisphenol-A (BPA), nonylphenol (NP) and triclosan (TCS)
Pharmaceutical products	Acetaminophen, Salicylic acid, Hormones, Antibiotics, Lipid regulators, Nonsteroidal anti-inflammatory drugs, Beta-blockers, Antidepressants, Anticonvulsants, Antineoplastics, Diagnostic Contrast Media
Food additives/stimulants	Caffeine
Personal care products	Benzophenone, Galaxolide, Tonalide, Triclosan
Surfactants (Amphoteric detergents)	4-Nonylphenol, 4-Octylphenol, Cetearyl alcohol Trideceth-2 carboxamide MEA, Cocamidopropyl betaine, Alkylamide betaines, Alkylamidopropyl betaines, Alkyl betaines, Amidopropyl betaines Amphoglycinates, Lauriminodipropionates, Lauroamphodiacetates
Surfactants (Anionic detergents)	-Methylestersulphonate, a-Olefinsulphonate Alkyl benzene sulphonates, Sulphonates Alkane sulphonates, Alkyl ether sulphates Alkyl sulphates, Alkyl sulphosuccinates Isotridecanolethoxylates, Panthenol
Surfactants (Cationic detergents)	Benzalkonium chloride, N-Hexadecyltrimethyl ammonium chloride, DHTDMAC, DSDMAC DTDMAC, Alkyltrimethylammonium, Chloride
Surfactants (Non-ionic detergents)	Alkylphenolethoxylates (APEO), Nonylphenol (NPE), Alcohol ethoxylates (AEO), Alkyl amide ethoxylates, Alkyl amine ethoxylates, Fatty alcohols (EO/PO) polymers, Fatty alcohol ethoxylates(AEO) Coconut diethanolamide, Ethylene glycol
Dyes and pigments	3,30-Dichlorobenzidine, 4,40-Methylenebis-(2-chlorobenzenamine), o-Aminoazotoluene Benzidine, o-Anisidine, CI77891 (TiO_2), CI77491 (iron oxides), Mica

Source Adams and Kuzhikannil (2000), Eriksson et al. (2002), Straub (2002), Simonich et al. (2002), Eriksson et al. (2003), Seo et al. (2005), Palmquist and Hanaeus (2005), Kupper et al. (2006), Andersen et al. (2007), U.S. EPA (2009), Pal et al. (2010), Deblonde et al. (2011), Luo et al. (2014), Etchepare and Van der Hoek (2015), Grcic et al. (2015)

mouthwash and toothpaste as a broad-spectrum antimicrobial (Dann and Hontela 2011). Bisphenol-A (BPA) is commonly used in the synthesis of plastics, which is generated from baby bottles, food containers and reusable water bottles (Vandenberg et al. 2009). Nonylphenol (NP) is used in the lubricating oil additives, manufacture of antioxidants and production of nonylphenol ethoxylates surfactants, which are used as emulsifiers, detergents, solubilizers, dispersing and antistatic agents, wetting and demulsifiers (Soares et al. 2008).

Surfactants (surface active agent) represent the major XOCs in greywater. It is used in the production of detergents and hygiene products which are utilised extensively for showers and laundry. It has been reported that the worldwide production of surfactants was 6.7 million metric tonnes, whereby 67.16% were anionic surfactants, 25.37% were non-ionic and 7.46% were cationic surfactants (Brackmann and Hager 2004). Heberer estimated the production of galaxolide and tonalide between 1000 and 5000 tonnes/year in Europe. The surfactants include the compounds generated from amphoteric, cationic, anionic and non-ionic detergents. Among these classes, there are anionic and cationic surfactants such as methylestersulphonate, olefinsulphonate, alkyl benzene sulphonates, alkyl ether sulphates, isotridecanolethoxylates, benzalkonium chloride, n-hexadecyltrimethy and ammonium chloride. The utilisation of these detergents depends on their potential to provide cleaning action, disinfection and cost (Jakobi and Löhr 1987; Lange 1994; Belanger et al. 2002). Types of detergents used contribute to the physical, chemical and microbiological characteristics of greywater. For instance, the usage of disinfection agents might reduce the density of microorganisms available while the utilisation of ammonium chloride might lead to increase TN concentrations. Other XOCs groups in greywater include fragrances, solvents, flavours and preservatives (Eriksson et al. 2002). Palmquist and Hanæus (2005) have detected 46 XOCs in greywater from ordinary Swedish households which include octylphenol and nonylphenol ethoxylates, phthalates, brominated flame retardants, PAH, organotin compounds, triclosan and monocyclic aromatics.

Etchepare and van der Hoek (2015) have detected 89 XOCs including surfactants, fragrances, flavours, plasticisers, preservatives, solvents, organotin compounds, UV filter, PAH and miscellaneous compounds in greywater. However, these compounds did not reflect entire compounds in greywater which are listed in Table 5.1. Only 41 compounds (out of 900) were classified as dangerous chemical substances by European Water Framework Directive (WFD) (EU 2000). Currently, pharmaceutical wastes are classified as hazardous and regulated under Subtitle C of the Resource Conservation and Recovery Act (RCRA). The classification of pharmaceutical wastes as hazardous substances is due to their ability to exhibit hazardous characteristics such as ignitability, corrosively, reactivity and/or toxicity. The identification of hazards of pharmaceutical wastes is quite difficult and thus should be subjected to stringent management and disposal requirements. However, there are 31 hazardous chemicals classified by USEPA which are used in the production of pharmaceutical products. The chemical determination of XOCs in water and wastewater depends on the use of liquid chromatography–tandem mass spectrometry (LC/MS/MS) and gas chromatography–tandem mass spectrometry (GC-MS/MS) which have high efficiency in determining low concentrations of residues in water and wastewater (Wang and

Wang 2016). Cimetiere et al. (2013) used ultra performance liquid chromatography coupled with tandem mass spectrometry (UPLC-MS/MS) and solid-phase extraction (SPE) to determine concentration of 5 ng and less per litre of the XOCs.

5.3 Environment and Human Health Impacts of XOCs

The high diversity and concentrations of XOCs in the discharged laundry, bathing and dishwashing greywater into natural water bodies have adverse effects on ecosystems and biodiversity (Beltrán-Heredia and Sánchez-Martín 2009; Braga and Varesche 2014). These compounds are known as persistent organic pollutants due to their chemical structure and continuous introduction into the environment from different sources including greywater and black water (Wang and Wang 2016). They are available in the surface water with concentrations ranging from µg/L to less than ng/L (Clara et al. 2005). The disposal of greywater with a high content of XOCs might lead to the lowering of the surface tension of water and thus create hard environmental conditions for aquatic life. These compounds are called persistent organic pollutants (POPs) which have long periods persistence in the environment.

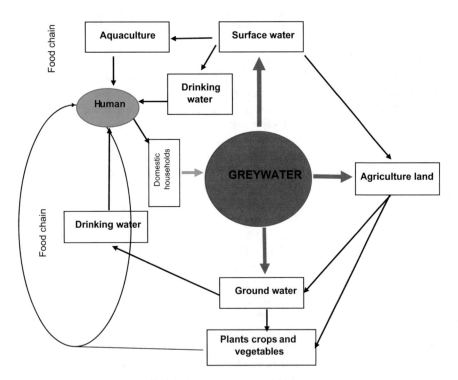

Fig. 5.2 Transmission route of XOCs from the greywater to human

The bioaccumulation of XOCs in the environment and reach to human via food chain, the chronic exposures for these compounds in human bodies may cause unknown health effects (Kim and Zoh 2016). The transmission route of XOCs from the greywater to human is presented in Fig. 5.2, while Table 5.2 illustrates the concentrations for some of XOCs reported in various water matrixes by previous studies.

The health risk of XOCs becomes more frequent especially in the countries which depend on surface water as a drinking water source. In some countries, surface water is subjected to a treatment process in order to increase water quality and subsequently be used as drinking water. However, the treatment process depends on the basic techniques which achieve the main requirements such as reduction of salinity, total suspended solids (TSS), chemical oxygen demand (COD) and biological oxygen demand (BOD). These techniques are not effective for the removal of XOCs. In other countries which depend on groundwater as the main resource of drinking water, the contaminated water might occur if wastewater is discharged into the land. The movement of XOCs into the groundwater might be possible although its occurrence is rare compared to surface water. There are a number of reports published in the European Union and in the USA which indicate the presence of XOCs in drinking water originating from surface water and groundwater (Loos et al. 2010; Morasch et al. 2010; Schriks et al. 2010; Fram and Belitz 2011).

XOCs are heterogeneous compounds available in different concentrations in greywater and often result from detergents, personal care products such as cosmetics and skincare as well as hair colourants (Grčić et al. 2015). These compounds have the ability to persist in the environment for different periods based on their chemical structure. Malathion, dibutyl phthalate, hexyl cinnamic aldehyde and triclosan pose a potential threat towards aquatic and terrestrial environments (Eriksson et al. 2002). Some of these compounds are available in low concentrations or persist for a short time. However, there are health risks associated with the occupational exposure and direct contact with these compounds (Zhang et al. 2008; Handa et al. 2012). Therefore, the presence of these compounds even in low concentrations might be hazardous because they can accumulate to reach higher concentrations and possess a high resistance towards biodegradation (Daughton and Ternes 1999).

5.4 Assay of XOCs Toxicity

The special risk groups for these compounds consist of old or young people and pregnant women, while the level of risk depends on direct or indirect exposure (Dorne et al. 2005). The assessment of hazard risks of XOCs has increased in the European Union and in the USA which aims to maintain the quality of surface water in terms of good ecological and chemical status (European Commission 2000). The best procedure used for the identification of hazard risk of these compounds is still under development. There are a few protocols which have been used for the identification process but the toxicity of these compounds towards water managers and specialists should be considered.

Table 5.2 Range of some XOCs concentrations in different water systems and their toxicological values (Kim and Zoh (2016)

XOCs compounds	Concentrations (ng L^{-1}) in water samples		Toxicological values (Toxic/effect)	References
	Surface water	Sewage influent		
Acetaminophen	4.1–170	1.9–54,000	NR	Kim et al. (2007), Choi et al. (2008), Grujić et al. (2009)
Atrazine	1.8–3600	52–59	NR	Klečka et al. (2009), Morasch et al. (2010), Yoon et al. (2010)
Bisphenol-A	10–6000	NR[*]	NR	Benotti et al. (2008), Klečka et al. (2009)
Caffeine	0–6,798	28–113,000	NR	Choi et al. (2008), Conkle et al. (2008), Kosma et al. (2010), Matamoros et al. (2012)
Carbamazepine	4.5–595	5–1680	24 μg/d, 36 weeks	Bendz et al. (2005), Collier (2007), Kim et al. (2007), Choi et al. (2008), Kim and Tanaka (2009), Schriks et al. (2010)
Diclofenac	8–147	8.8–3600	1 μg/L, 24 months Effect on the pregnant women, hemodynamic changes, carcinogenetic	Collier (2007), Gómez et al. (2007), Kim et al. (2007), Zhao et al. (2009); Schriks et al. (2010)
Ibuprofen	5–414	10–17,933	50 μg/L Effect on the pregnant women, hemodynamic changes, diabetic patients, hypoglycemia, carcinogenetic	Collier (2007), Kim and Tanaka (2009), Kim et al. (2009), Lin and Tsai (2009), Schriks et al. (2010)
Metoprolol	24.8–90	NR	440 μg/L, Rats, thyroid tumours	Huerta-Fontela et al. (2011), Yu et al. (2011)
Naproxen	20–483	12–3650	NR	Bendz et al. (2005), Nakada et al. (2006), Kim et al. (2007)

(continued)

Table 5.2 (continued)

XOCs compounds	Concentrations (ng L^{-1}) in water samples		Toxicological values (Toxic/effect)	References
	Surface water	Sewage influent		
Nonylphenol	23.2–2,800	500–1100	NR	Li et al. (2004), Nakada et al. (2006), Wu et al. (2007)
Sulfamethoxazole	1.7–300	3.8–1335	0.03 μg/L in water	Brown et al. (2006), Kim et al. (2007), Collier (2007), Lin and Tsai (2009), Schriks et al. (2010)
Triclosan	0.45–2230	NR	NR	Young et al. (2008)
Mefenamic acid	89–326	NR	NR	Kim et al. (2009)
Atenolol	160–690	NR	NR	
Fluconazole	16–111	NR	NR	
Sulfadiazine	NR	0.12–0.53	NR	Gao et al. (2012)
Trimethoprim	NR	120–230	NR	Gulkowska et al. (2008)

*Non-reported

Baun et al. (2006) used the RICH procedure (Ranking and Identification of Chemical Hazard's tool) for evaluating the hazards of XOCs in stormwater. The method was described by Eriksson et al. (2005) and developed by Müller-Herold et al. (2005) to evaluate the effects of XOCs on the aquatic environment. Baun et al. (2006) have developed this method by using a filtration process to achieve an integrated system for identification and assessment of chemical hazards for 11 types of XOCs in stormwater. In this method, the filtration process consists of four filters to separate the XOCs into groups based on their colours (white, grey and black). In the same study, Baun et al. (2006) had claimed that the white compounds resulting from the filtration system are priority pollutants which have high persistence but low bioaccumulation and thus no more hazard assessment is needed. Examples of these compounds are di (2-ethylhexyl) phthalate (DEHP), tributyltin (TBT) and benzo(a)anthracene (BaA). Tert-butyl methyl ether (MTBE) is more soluble in water and was identified as a black compound which means that MTBE requires more hazard assessments to evaluate the chemical risk but it was not classified as a priority pollutant. Linear alkylbenzene sulphonate (LAS) was identified as a white and biodegradable compound under aerobic conditions and has a low potential for bioaccumulation, but this compound can persist longer under anaerobic conditions. Therefore, it was identified as a priority pollutant and requires more evaluation to determine its toxicity. The hazard risks of XOCs are evaluated based on three criteria which are persistence–bioaccumulation—toxicity (PBT). There are no further studies conducted on the hazard identification of XOCs except for the studies conducted by Eriksson et al. (2005) and Baun et al. (2006) who has described the critical methods for this purpose.

The assessment and identification of hazard risks of XOCs depend on the final fate of these compounds which rely on whether greywater is discharged into the natural water system (liquid phase) or reused for irrigation purposes and then discharged into the soil (solid phase). This is the idea behind the RICH procedure (Ranking and Identification of Chemical Hazard's tool). These scenarios were defined by Eriksson et al. (2005) and depend on the environmental risk assessment described in the European Technical Guidance Document (TGD). Effects assessment can be divided into two concepts including predicted effect concentrations (PECs) and predicted no effect concentrations (PNECs). The XOCs are predicted to have effects in the water if they are not removed from greywater before being discharged into the water. They are only diluted 100 times in the received water, while they have no effect if they are available in concentrations less than the toxicity for algae, crustaceans and fish where it is diluted 1000 times.

Eriksson et al. (2007) developed a procedure called Chemical Hazard Identification and Assessment Tool (CHIAT) in order to assess the environmental risk of XOCs in stormwater and wastewater. This tool is used to evaluate the hazards of XOCs which include acute and long-term toxicity, carcinogenicity, bioaccumulation, reproduction hazards, mutagenicity and endocrine disrupting effects on the aquatic organisms in natural water. In this method, *Pseudokirchneriella subcapitata* was used as a biosensor. CHIAT procedure described by Eriksson et al. (2007) included the source characterization, criteria and hazard identification and assessment as well as the expert conclusion. The method is used to identify potential pollutants in the sediment/soil phase and the aqueous phase as well as hazardous pollutants. It has also the potential to determine the persistence/bioaccumulation and toxicity as a short- or long-term chronic effect.

The bio-toxicity of XOCs on aquatic organisms is assessed based on algal toxicity such as *P. subcapitata* which is used as indicator organism due to their high sensitivity than other aquatic organisms such as Vibrio fisheries and invertebrates (*Brachinus calyciflorus* and *Daphnia magna*) (Nyholm and Källqvist 1989; Radix et al. 2000). The algal growth inhibition test method is described by ISO standard (ISO 1989). In this method, the algae grow with samples contaminated with XOCs and the toxicity is estimated based on the biomass yield and effective concentration (EC10 and EC50) values of extract from the culture medium (this method is described in detail by Eriksson et al. 2006). In general, the toxicity of XOCs is more dangerous for children and pregnant women. Chronic exposure to these compounds might lead to serious health effects (Kim and Zoh 2016).

The phytotoxicity of greywater containing XOCs on the wheat, lettuce and soybean root-associated bacteria has been confirmed Van Kerkhof et al. (2000). Bubenheim and Wignarajah (1997) had revealed that the discharge of the greywater with anionic surfactant contents (Igepon TC-42) leads to changes in the rhizosphere communities. The reusing of these waste for the irrigation of lettuce inhibited the growth by 40%. Eriksson et al. (2006) examined the phytotoxicity of XOCs in the bathroom, kitchen and laundry greywater against *P. subcapitata* and Willow tree. The results recorded a toxicity of bathroom greywater against *P. subcapitata* (EC10 = 36 − 375 mL/L). In contrast, both kitchen and laundry greywater exhibited a toxic effect against *P. subcapitata* and Willow tree (EC10 = 55 − 198 mL/L). A study by Lupica (2013) has shown that the discharge of greywater containing dyes into the sea effect negatively on the shape and size of the fish's red blood cells. The author also revealed the reproductive rat organs reduced by 44%, as well as cholesterol by 91% and total protein concentration by 70%. Giesy and Snyder stated that the dispose of EDCs into the natural water system leading to loss of secondary sexual characteristics of the fish, feminization of males, lessened egg fertility, intersex, and increased embryo mortality and deformities as well as defeminisation of females.

Biologically active compounds like pharmaceuticals are another type of XOCs in greywater which resulted from the utilisation of medicinal skin creams which contain antibiotics or from kitchen greywater due to the washing of chicken meat, where antibiotics are used as a growth promoter. The first report on the contamination of wastewater with pharmaceuticals and personal care products (PPCPs) was published in the 1960s in the United States and Europe by Stumm-Zollinger and Fair (1965). The concerns about their potential contamination of the environment and the risks for humans were raised in 1999 (Daughton and Ternes 1999). Among different types of PPCPs in wastewater, the antibiotics classes have the most important due to high consumption rates in medical treatment and extractions through the urine due to the nature of their pharmacokinetics in the human body (Toloti and Mehrdadi 2001; Deegan et al. 2011).

The presence of bacteria-resistant against antibiotics among the bacteria isolated from chicken meat has been reported. The environmental health risk associated with these compounds lies in the development of antimicrobial resistance. The presence of antibiotics in natural water which receives discharged wastewater has been reported in the literature (Al-Gheethi et al. 2016). These antibiotics are available in low concentrations. However, it has the potential to increase antimicrobial resistance among indigenous bacteria which might play an important role in the transmission of the antimicrobial resistance gene to pathogenic bacteria. Therefore, the presence of bacteria-resistant against antibiotics in waste indicates the presence of antibiotics. However, unlike most XOCs, which have long period persistence in the environment, most antibiotics are sensitive towards environmental conditions and they are more degradable than XOCs except for some compounds such as ciprofloxacin and cephalexin which have resistance towards biodegradation in the environment (Al-Gheethi and Norli 2014). The determination of antimicrobial resistance among bacterial populations in an environment receiving wastewater contaminated with antibiotics residues has to be conducted using different methods. This is because of the differences in how bacterial cells interact with antibiotics and the potential of antibiotics gene transmission to other species living in the same environment (Al-Gheethi et al. 2013). The increase of antimicrobial resistance among pathogenic bacteria causes the medical treatment of infection caused by these pathogens more difficult. Even if these bacteria which carry the antibiotics resistance gene are human pathogens, they are still dangerous due to their ability to transfer their antibiotic resistance genes to other bacteria that are pathogenic for humans (Al-Gheethi et al. 2015, 2016).

Campylobacter spp., *E. coli* and *Salmonella* spp. isolated from farm animals have exhibited a resistance against antibiotics such as fluoroquinoline (Aarestrup et al. 1998; Aarestrup 2000). The microbial resistance among pathogens against antibiotics has been considered a global problem. The overuse of antibiotics is associated with the distribution of resistance among bacteria in the environment (Lukaova and Sustakova 2003). In contrast, the presence of antibiotics in the natural ecosystem even in very low concentrations might cause bacterial cells to resist these antibiotics more than if present in high concentrations (Reinthaler et al. 2003). A high percentage of bacteria might develop a mechanism to resist antibiotics if there is a low concentration of antibiotics, while a very low percentage of bacteria might exhibit persistence for a high concentration of antibiotics. Transfer of antibiotics resistance plasmid genes between microbial populations in wastewater is possible and would occur in two directions: either antibiotics resistance gene present in non-pathogenic bacteria could be spread over to the pathogenic strains or pathogenic bacteria could acquire resistance genes from bacterial flora (Rahube and Yost 2010). Soda et al. (2008) reported that the transfer of plasmids containing multiple resistance factors encoding resistance to ampicillin, kanamycin and tetracycline might take place from E. *coli* to other non-resistant *E. coli.*

5.5   Regulations of XOCs in the Greywater

The regulations for XOCs have been adopted in some of developed countries such as the USA, Canada, Australia and Denmark but none of the developing countries have legislations for XOCs in the water, wastewater and environment. The European Parliament through Directive 2008/105/EC has regulated the environmental quality standards for a small number of XOCs including nonylphenol, diclofenac, 17α-ethinylestradiol, diiron, bisphenol-A and 17β-estradiol (EU, 2008), while nonylphenol ethoxylates and nonylphenol are recognised by the Canadian government. Atrazine is among the XOCs which has international regulations (NJMRC 2011; WHO 2011; USEPA 2009; Health Canada 2012). The standard concentration of XOCs which are required to be available in the water is different based on the country. The international standard concentration of atrazine is required to be less than $100 \, \mu g \, L^{-1}$. In spite, the country-based standards for Australia, Canada, USA and Europe are $<20 \, \mu g \, L^{-1}$, $<5 \, \mu g \, L^{-1}$, $<3 \, \mu g \, L^{-1}$ and $<0.6 \, \mu g \, L^{-1}$, respectively. These differences might be related to the absence of accurate information on the toxicity of atrazine.

To date, there are no regulations for pharmaceutical and personal care products (PPCPs) wastes. USEPA is proposing to regulate hazardous pharmaceutical waste under the Universal Waste Rule. Nonetheless, the occurrence of these compounds in the greywater is not subjected to the consistent emission guidelines and standards. Currently, pharmaceutical wastes are classified as hazardous and regulated under

Subtitle C of the Resource Conservation and Recovery Act (RCRA). This classification is due to their ability to exhibit hazardous characteristics such as ignitability, corrosively, reactivity and/or toxicity. The identification of hazards of pharmaceutical wastes is quite difficult and thus should be subjected to stringent management and disposal requirements. However, there are 31 hazardous chemicals classified by USEPA which are used in the production of pharmaceutical products. USEPA is proposing to regulate hazardous pharmaceutical waste under the Universal Waste Rule due to unavailability of regulation for pharmaceutical waste.

In Malaysia, the EQA 1974 regulated the standards for disposal of domestic and industrial effluents into the water systems in terms of COD, BOD, TSS and heavy metals. There is little number of researchers focusing on the studies of nutrient removal from the wastewater but the country has not adopted legislations for these parameters yet. Neither studies nor regulations on the XOCs in the greywater have been adopted by the government.

5.6 Conclusion

The presence of XOCs in water is also contributed by non-biodegradable chemicals available in the water itself. The toxicity of the contributed compounds has been reported in the literature which indicated that the XOCs should be considered prior reusing or disposing the greywater into the environment. The regulations for the XOCs in many of the countries are still not available and require more attention to adopt a specific regulation in order to protect human health. In developing countries, more researches are necessary to list the most common XOCs in the greywater and then to select the appropriate treatment method.

Acknowledgements The authors wish to thank the Ministry of Higher Education (MOHE) for supporting this research under FRGS vot 1574 and also the Research Management Centre (RMC) UTHM for providing grant IGSP U682 for this research.

References

Aarestrup FM (2000) Occurrence, selection and spread of resistance to antimicrobial agents used for growth promotion for food animals in Denmark. APMIS 108(Suppl 101):5–48

Aarestrup FM, Bager F, Jensen NE, Madsen M, Meyling A, Wegener HC (1998) Surveillance of antimicrobial resistance in bacteria isolated from food animals to antimicrobial growth promoters and related therapeutic agents in Denmark. Apmis 106(1–6):606–622

Adams CD, Kuzhikannil JJ (2000) Effects of UV/H_2O_2 preoxidation on the aerobic biodegradability of quaternary amine surfactants. Water Res 34(2):668–672

Al-Gheethi AA, Lalung J, Efaq AN, Bala JD, Norli I (2015) Removal of heavy metals and β-lactam antibiotics from sewage treated effluent by bacteria. Clean Technol Environ Policy 17(8):2101–2123

Al-Gheethi AA, Mohamed RMS, Efaq AN, Norli I, Amir Hashim, Ab Kadir MO (2016) Bioaugmentation process of sewage effluents for the reduction of pathogens, heavy metals and antibiotics. J Water Health 14(5):780–795

Al-Gheethi AA, Norli I (2014) Biodegradation of pharmaceutical residues in sewage treated effluents by *Bacillus subtilis* 1556WTNC. J Environ Process 1(4):459–489

Al-Gheethi AA, Norli I, Lalung J, Azieda T, Efaq N, Ab Kadir MO (2013) Susceptibility for antibiotics among faecal indicators and pathogenic bacteria in sewage treated effluents. Water Pract Technol 8(1): 1–6

Andersen HR, Lundsbye M, Wedel HV, Eriksson E, Ledin A (2007) Estrogenic personal care products in a greywater reuse system. Water Sci Technol 56(12):45–49

Baun A, Eriksson E, Ledin A, Mikkelsen PS (2006) A methodology for ranking and hazard identification of xenobiotic organic compounds in urban storm water. Sci Total Environ 370(1):29–38

Belanger SE, Bowling JW, Lee DM, LeBlanc EM, Kerr K, McAvoy DC, Davidson DH (2002) Integration of aquatic fate and ecological responses to linear alkyl benzene sulfonate (LAS) in model stream ecosystems. Ecotoxicol Environ Saf 52(2):150–171

Beltrán-Heredia J, Sánchez-Martín J (2009) Removal of sodium lauryl sulphate by coagulation/flocculation with Moringaoleifera seed extract. J Hazard Mater 164(2):713–719

Bendz D, Paxeus NA, Ginn TR, Loge FJ (2005) Occurrence and fate of pharmaceutically active compounds in the environment, a case study: Höje River in Sweden. J Hazard Mater 122:195–204

Benotti MJ, Trenholm RA, Vanderford BJ, Holady JC, Stanford BD, Snyder SA (2008) Pharmaceuticals and endocrine disrupting compounds in US drinking water. Environ Sci Technol 43(3):597–603

Brackmann B, Hager CD (2004) The statistical world of raw materials, fatty alcohols and surfactants. In CD proceedings 6th world surfactant congress CESIO, Berlin, Germany

Braga JK, Varesche MBA (2014) Commercial laundry water characterization. Am J Anal Chem 5(1):8

Brown KD, Kulis J, Thomson B, Chapman TH, Mawhinney DB (2006) Occurrence of antibiotics in hospital, residential, and dairy effluent, municipal wastewater, and the Rio Grande in New Mexico. Sci Total Environ 366:772–783

Bubenheim DL, Wignarajah K (1997) Recycling of inorganic nutrients for hydroponic crop production following incineration of inedible biomass. Adv Space Res 20(10):2029–2035

Choi K, Kim Y, Park J, Park CK, Kim M, Kim HS, Kim P (2008) Seasonal variations of several pharmaceutical residues in surface water and sewage treatment plants of Han River. Korea Sci Total Environ 405(1):120–128

Cimetiere N, Soutrel I, Lemasle M, Laplanche A, Crocq A (2013) Standard addition method for the determination of pharmaceutical residues in drinking water by SPE–LC–MS/MS. Environ Technol 34(22):3031–3041

Clara M, Strenn B, Gans O, Martinez E, Kreuzinger N, Kroiss H (2005) Removal of selected pharmaceuticals, fragrances an endocrine disrupting compounds in a membrane bioreactor and conventional wastewater treatment plants. Water Res 39:4797–4807

Collier AC (2007) Pharmaceutical contaminants in potable water: potential concerns for pregnant women and children. Eco Health 4:164–171

Conkle JL, White JR, Metcalfe CD (2008) Reduction of pharmaceutically active compounds by a lagoon wetland wastewater treatment system in Southeast Louisiana. Chemosphere 73:1741–1748

Dann AB, Hontela A (2011) Triclosan: environmental exposure, toxicity and mechanisms of action. J Appl Toxicol 31(4):285–311

Daughton CG, Ternes TA (1999) Pharmaceuticals and personal care products in the environment: agents of subtle change? Environ Health Perspect 107(6):907–938

Deblonde T, Cossu-Leguille C, Hartemann P (2011) Emerging pollutants in wastewater: a review of the literature. Int J Hygiene Environ Health 214(6):442–448

Deegan AM, Shaik B, Nolan K, Urell K, Oelgemöller M, Tobin J, Morrissey A (2011) Treatment options for wastewater effluents from pharmaceutical companies. Int J Environ Sci Technol 8(3):649–666

Dorne JLCM, Walton K, Renwick AG (2005) Human variability in xenobiotic metabolism and pathway-related uncertainty factors for chemical risk assessment: a review. Food Chem Toxicol 43:203–216

Eriksson E, Auffarth K, Eilersen AM, Henze M, Ledin A (2003) Household chemicals and personal care products as sources for xenobiotic organic compounds in grey wastewater. Water SA 29(2):135–146

Eriksson E, Auffarth K, Henze M, Ledin A (2002) Characteristics of grey wastewater. Urb Water 4:85–104

Eriksson E, Baun A, Henze M, Ledin A (2006) Phytotoxicity of grey wastewater evaluated by toxicity tests. Urb Water J 3(1):13–20

Eriksson E, Baun A, Mikkelsen PS, Ledin A (2005) Chemical hazard identification and assessment tool for evaluation of stormwater priority pollutants. Water Sci Technol 51:47–55

Eriksson E, Baun A, Mikkelsen PS, Ledin A (2007) Risk assessment of xenobiotics in stormwater discharged to Harrestrup Å Denmark. Desalination 215(1):187–197

Etchepare R, van der Hoek JP (2015) Health risk assessment of organic micropollutants in greywater for potable reuse. Water Res 72:186–198

EU (European Union) (2000) Directive 2000/60/EC of the European Parliament and the Council of 23 October 2000 establishing a framework for community action in the field of water policy. J Eur Comm 22 Dec 2000, L327/1eL327/73

European Commission (2000) Proposal for a Directive of the European Parliament and of the Council on Environmental Quality Standards in the Field of Water Policy and Amending Directive 2000/60/EC, 2006. Accessed Nov 2016. Available from: http://eur-lex.europa

Fram MS, Belitz K (2011) Occurrence and concentrations of pharmaceutical compounds in groundwater used for public drinking-water supply in California. Sci Total Environ 409:3409–3417

Gao L, Shi Y, Li W, Niu H, Liu J, Cai Y (2012) Occurrence of antibiotics in eight sewage treatment plants in Beijing, China. Chemosphere 86:665–671

Gómez MJ, Lacorte S, Fernández-Alba A, Agüera A (2007) Pilot survey monitoring pharmaceuticals and related compounds in a sewage treatment plant located on the Mediterranean coast. Chemosphere 66:993–1002

Grčić I, Vrsaljko D, Katančić Z, Papić S (2015) Purification of household greywater loaded with hair colorants by solar photocatalysis using TiO_2-coated textile fibers coupled flocculation with chitosan. J Water Process Eng 5:15–27

Grujić S, Vasiljević T, Laušević M (2009) Determination of multiple pharmaceutical classes in surface and ground waters by liquid chromatography–ion trap–tandem mass spectrometry. J Chromatogr A 1216:4989–5000

Gulkowskaa A, Leunga HW, Soa MK, Taniyasub S, Yamashitab N, Yeunga L, Richardsona BG, Lei AP, Giesya JP, Lama KS (2008) Removal of antibiotics from wastewater by sewage treatment facilities in Hong Kong and Shenzhen, China. Water Res 42:395–403

Handa S, Mahajan R, De D (2012) Contact dermatitis to hair dye: an update. Indian J Dermatol Venereol Leprol 78(5):583–590

Harvey PJ, Thurston CF (2001) The biochemistry of ligninolytic fungi. In British Mycological Society Symposium Series, vol 23, pp 27–51)

Health Canada (HC) (2012) Guidelines for Canadian drinking water quality-summary table. Water, air and climate change bureau, healthy environments and consumer safety branch, Ottawa, Ontario

Huerta-Fontela M, Galceran MT, Ventura F (2011) Occurrence and removal of pharmaceuticals and hormones through drinking water treatment. Water Res 45:1432–1442

Jakobi G, Löhr A (1987) Detergents and textile washing: principles and practice. VCH Publishers

Kim I, Tanaka H (2009) Photodegradation characteristics of PPCPs in water with UV treatment. Environ Int 35:793–802

Kim MK, Zoh KD (2016) Occurrence and their removal of micropollutants in water environment. Environ Eng Res (in press)

Kim SD, Cho J, Kim IS, Vanderford BJ, Snyder SA (2007) Occurrence and removal of pharmaceuticals and endocrine disruptors in South Korean surface, drinking, and waste waters. Water Res 41:1013–1021

Kim JW, Jang HS, Kim JG, Ishibashi H, Hirano M, Nasu K, Arizono K (2009) Occurrence of pharmaceutical and personal care products (PPCPs) in surface water from Mankyung River South Korea. J Health Sci 55(2):249–258

Klečka GM, Staples CA, Clark KE, Van der Hoeven N, Thomas DE, Hentges SG (2009) Exposure analysis of bisphenol A in surface water systems in North America and Europe. Environ Sci Technol 43:6145–6150

DA KosmaCI Lambropoulou, Albanis TA (2010) Occurrence and removal of PPCPs in municipal and hospital wastewaters in Greece. J Hazard Mater 179:804–817

Kupper T, Plagellat C, Braendli RC, de Alencastro LF, Grandjean D, Tarradellas J (2006) Fate and removal of polycyclic musks, UV liters and biocides during wastewater treatment. Water Res 40(14):2603–2612

Lange KR (1994) Detergents and cleaners, a handbook for formulators. SchoderDruck GmbH & Co.KG, New York, p 1994

Leal HL, Temmink H, Zeeman G, Buisman CJ (2010) Comparison of three systems for biological greywater treatment. Water 2(2):155–169

Li D, Kim M, Shim WJ, Yim UH, Oh JR, Kwon YJ (2004) Seasonal flux of nonylphenol in Han River, Korea. Chemosphere 56:1–6

Lin AYC, Tsai YT (2009) Occurrence of pharmaceuticals in Taiwan's surface waters: Impact of waste streams from hospitals and pharmaceutical production facilities. Sci Total Environ 407:3793–3802

Loos R, Locoro G, Comero S, Contini S, Schwesig D, Werres F, Balsaa P, Gans O, Weiss S, Blaha L (2010) Pan-European survey on the occurrence of selected polar organic persistent pollutants in ground water. Water Res 44:4115–4126

Lukaova J, Sustakova A (2003) Enterococci and antibiotic resistance. J Acta Vet Brno 72:315–323

Luo Y, Guoa W, Ngo HH, Nghiemb L, Hai FI, Zhang J, Liang S, Wang XC (2014) A review on the occurrence of micropollutants in the aquatic environment and their fate and removal during wastewater treatment. Sci Total Environ 473–474:619–641

Lupica S (2013) Effects of textile dyes in wastewater, eHow contributor. Available online: www.ehow.com/info_8379849_effectstextile-dyes-wastewater.html (accessed 29.06.17)

Matamoros V, Arias CA, Nguyen LX, Salvadó V, Brix H (2012) Occurrence and behavior of emerging contaminants in surface water and a restored wetland. Chemosphere 88:1083–1089

Morasch B, Bonvin F, Reiser H, Grandjean D, de Alencastro LF, Perazzolo C, Chèvre N, Kohn T (2010) Occurrence and fate of micropollutants in the Vidy Bay of Lake Geneva, Switzerland. Part II: Micropollutant removal between wastewater and raw drinking water. Environ Toxicol Chem 29:1658–1668

Müller-Herold U, Morosini M, Schucht O (2005) Choosing chemicals for precautionary regulation: a filter series approach. Environ SciTechnol 39:683–691

Nakada N, Tanishima T, Shinohara H, Kiri K, Takada H (2006) Pharmaceutical chemicals and endocrine disrupters in municipal wastewater in Tokyo and their removal during activated sludge treatment. Water Res 40(17):3297–3303

NJMRC(2011) National health and medical research council. Australian drinking water guidelines

Nyholm N, Källqvist T (1989) Methods for growth inhibition toxicity tests with freshwater algae. Environ Toxicol Chem 8(8):689–703

Pal A, Gin KY, Lin AY, Reinhard M (2010) Impacts of emerging organic contaminants on freshwater resources: review of recent occurrences, sources, fate and effects. Sci Total Environ 408:6062–6069

Palmquist H, Hanæus J (2005) Hazardous substances in separately collected grey-and blackwater from ordinary Swedish households. Sci Total Environ 348(1):151–163

Radix P, Léonard M, Papantoniou C, Roman G, Saouter E, Gallotti-Schmitt S, Thiébaud H, Vasseur P (2000) Comparison of four chronic toxicity tests using algae, bacteria, and invertebrates assessed with sixteen chemicals. Ecotoxicol Environ Saf 47(2):186–194

Rahube TO, Yost CK (2010) Antibiotic resistance plasmids in wastewater treatment plants and their possible dissemination into the environment. Rev. Afr J Biotechnol 9:9183–9190

Reinthaler FF, Posch J, Feierl G, Wust G, Haas D, Ruckenbauer G, Mascher F, Marth E (2003) Antibiotic resistance of E. coli in sewage and sludge. Water Res 37:1685–1690

Schriks M, Heringa MB, Van der Kooi MM, de Voogt P, van Wezel AP (2010) Toxicological relevance of emerging contaminants for drinking water quality. Water Res 44:461–476

Seo DC, Cho JS, Lee HJ, Heo JS (2005) Phosphorus retention capacity of filter media for estimating the longevity of constructed wetland. Water Res 39(11):2445–2457

Simonich SL, Federle TW, Eckhoff WS, Rottiers A, Webb S, Sabaliunas D, De Wolf W (2002) Removal of fragrance materials during US and European wastewater treatment. Environ Sci Technol 36(13):2839–2847

Soares A, Guieysse B, Jefferson B, Cartmell E, Lester JN (2008) Nonylphenol in the environment: a critical review on occurrence, fate, toxicity and treatment in wastewaters. Environ Int 34(7):1033–1049

Soda S, Otsuki H, Inoue D, Tsutsui H, Sei K, Ike M (2008) Transfer of antibiotic multi-resistant plasmid RP4 from E. coli to activated sludge bacteria. J Biosci Bioeng 106:292–296

Straub JO (2002) Concentrations of the UV lterethylhexylmethoxycinnamate in the aquatic compartment: a comparison of modelled concentrations for Swiss surface waters with empirical monitoring data. Toxicol Lett 131(1–2):29–37

Stumm-Zollinger E, Fair GM (1965) Biodegradation of steroid hormones. Res J Water Poll C 37(11):1506–1510

Togunde OP, Oakes KD, Servos MR, Pawliszyn J (2012) Determination of pharmaceutical residues in fish bile by solid-phase microextraction couple with liquid chromatography-tandem mass spectrometry (LC/MS/MS). Environ Sci Technol 46(10):5302–5309

Toloti A, Mehrdadi N (2001) Wastewater treatment from antibiotics plant (UASB Reactor). Int J Environ Res 5(1):241–246

USEPA (2009) Exposure Factors Handbook: 2009 Update (EPA/600/ R-09/052A Office of Research and Development. National Center for Environmental Assessment, U.S. Environmental Protection Agency, Washington, DC

Van Kerkhof P, Govers R, dos Santos CMA, Strous GJ (2000) Endocytosis and degradation of the growth hormone receptor are proteasome-dependent. J Biol Chem 275(3):1575–1580

Vandenberg LN, Maffini MV, Sonnenschein C, Rubin B, Soto AM (2009) Bisphenol-A and the great divide: a review of controversies in the field of endocrine disruption. Endocr Rev 30(1):75–95

Wang J, Wang S (2016) Removal of pharmaceuticals and personal care products (PPCPs) from wastewater: a review. J Environ Manage 182:620–640

WHO (2011) World Health Organization. Guidelines for drinking-water quality, 4th ed. 2011

Wu Z, Zhang Z, Chen S, He F, Fu G, Liang W (2007) Nonylphenol and octylphenol in urban eutrophic lakes of the subtropical China. Fresen Environ Bull 16:227–234

Ying GG (2006) Fate, behavior and effects of surfactants and their degradation products in the environment. Environ Int 32(3):417–431

Yoon Y, Ryu J, Oh J, Choi BG, Snyder SA (2010) Occurrence of endocrine disrupting compounds, pharmaceuticals, and personal care products in the Han River (Seoul, South Korea). Sci Total Environ 408:636–643

Young TA, Heidler J, Matos-Pérez CR, SapkotaA Toler T, Gibson KE, Halden RU (2008) Ab initio and in situ comparison of caffeine, triclosan, and triclocarban as indicators of sewage-derived microbes in surface waters. Environ Sci Technol 42(9):3335–3340

Yu Y, Huang Q, Wang Z, Zhang K, Tang C, Cui J, Feng J, Peng X (2011) Occurrence and behavior of pharmaceuticals, steroid hormones, and endocrine-disrupting personal care products in wastewater and the recipient river water of the Pearl River Delta, South China. J Environ Monit 13:871–878

Zhang Y, De Sanjose S, Bracci PM, Morton LM, Wang R, Brennan P (2008) Personal use of hair dye and the risk of certain subtypes of non-Hodgkin lym-phoma. Am J Epidemiol 167(11):1321–1331

Zhao JL, Ying GG, Wang L, Yang JF, Yang XB, Yang LH, Li X (2009) Determination of phenolic endocrine disrupting chemicals and acidic pharmaceuticals in surface water of the Pearl Rivers in South China by gas chromatography–negative chemical ionization–mass spectrometry. Sci Total Environ 407(2):962–974

Chapter 6
A Potential Reuse of Greywater in Developed and Developing Countries

Radin Maya Saphira Radin Mohamed, Adel Ali Saeed Al-Gheethi,
Amir Hashim Mohd Kassim, Anda Martin, Stewart Dallas
and Mohd Hairul Bin Khamidun

Abstract The interest in greywater reuse as an alternative water supply is increasing in most part of the World. In Perth, Western Australia (WA), an industrial association to promote greywater reuse named Grey Water Industry Group (GWIG) has been established. Malaysia is a country seemingly endowed with abundant water resources with an annual average rainfall of more than 2000 mm. Despite its high rainfall and water resources compared to other regions in the world, Malaysia still suffers water problems (both excesses and deficits). The present work describes the suitability of greywater reuse in water supply strategy and wastewater management in Malaysia in comparison to that applied in Australia. Greywater should not be seen as a waste product, but as a valuable resource in wastewater management. Based on the comparison study between Australia and Malaysia, it appeared that the adoption of greywater treatment in Malaysia is more feasible and meaningful than the reuse approach, which creates problems in some instances when the greywater system is inappropriately designed for the type of environment. However, proper legislation, awareness and environmental considerations in terms of geochemistry characteristics, selection of the treatment method and the need for a paradigm shift are essential keys to ensuring optimum utilization of greywater as a future water resource in Malaysia.

Keywords Greywater management · Greywater · Malaysia · Perth
Water resources

R. M. S. Radin Mohamed (✉) · A. A. S. Al-Gheethi · A. H. Mohd Kassim · M. H. B. Khamidun
Micro-Pollutant Research Centre (MPRC), Department of Water and Environmental Engineering,
Faculty of Civil and Environmental Engineering, Universiti Tun Hussein Onn Malaysia (UTHM),
86400 Parit Raja, Batu Pahat, Johor, Malaysia
e-mail: maya@uthm.edu.my

A. A. S. Al-Gheethi
e-mail: adel@uthm.edu.my

A. Martin · S. Dallas
Environmental Engineering, School of Engineering, Murdoch University, Perth, Australia

© Springer International Publishing AG, part of Springer Nature 2019
R. M. S. Radin Mohamed et al. (eds.), *Management of Greywater in Developing Countries*,
Water Science and Technology Library 87,
https://doi.org/10.1007/978-3-319-90269-2_6

6.1 Introduction

Wastewater management is an increasingly serious issue demanding attention in both developing and developed nations worldwide. It is one of the problems that developing countries, such as Malaysia, have experienced due to its rapid development and urbanization. It has been demonstrated that six million tonnes of wastewater are generated annually by its 26 million inhabitants. Juahir et al. (2011) stated that 60% of the main rivers are used mainly for domestic, agricultural and industrial activities. Therefore, the main sources of pollution concerning rivers are sewage disposal, and discharges from small and medium-sized industries that are yet to be equipped with appropriate effluent treatment facilities. Juahir et al. (2011) also reported that the river basins are polluted with 42% of suspended solids (SS) ensuing from poorly planned and abandoned land clearing activities, biological oxygen demand (BOD) with 30% from industrial release, and 28% with ammoniacal nitrogen from animal husbandry activities and domestic sewage disposal. The wastewater is treated to varying levels and discharged into the rivers from which most of Malaysia's freshwater supply originates. Therefore, a more effective and sustainable wastewater management is required.

In the late 1990s, water reuse received great attention in Australia, due to the major national drought, and new water policies and resource protection legislation which were adopted to conserve more water (Ryan 2014). Most of the cities have been experiencing pressure from water restrictions, as well as unpredictable climate and declining rainfall. The stress on water supply in Perth, Western Australia (WA) will be further exacerbated by the predicted population increase from 1.5 million people to as many as 4.2 million by 2056 (Weller 2009). Despite the low rainfall, Australia is one of the world's largest water consumers. In 2010, the Australian Government issued the Water for the Future Framework a 10-year, AUS$12.9 billion investment in strategic programmes aimed at improving water management schemes and a renewed commitment to deliver a range of water policy reforms nationwide.

Greywater is wastewater drained from showers, bathtubs, washing machines and kitchen sinks excluding toilet waste, i.e. blackwater (Simon and Elisa 2013; Santos et al. 2012; Bino et al. 2010; Allen 2010; Mohamed et al. 2016). Greywater represents a resource, since it is generated in every household daily independent of the weather (Mohamed et al. 2013a). In fact, local authorities in WA have placed importance on water recycling with a projected target of 30% in 2030 which aimed to reduce the pressure on the finite water resources (Water Corporation of WA 2009). In comparison, greywater treatment and reuse have been gaining popularity in Australia due to the need for resource substitution during periods of low rainfall. However, does this practice fit the Malaysian condition? Australia and Malaysia are located in different geographical and climate zones, and face different issues in water resources. This paper explores the applicability of greywater and its necessary treatment for the situation in Malaysia. The reuse of greywater is not a key interest in Malaysia due to its abundant rainfall. Whether or not greywater treatment is a suitable approach to assist Malaysia's water management problems is discussed and compared.

Table 6.1 Summary of the weather conditions for Perth, Western Australia and Kuala Lumpur, capital city of Malaysia

	Perth[1]	Kuala Lumpur[2]
Climate type	Mediterranean	Hot and humid
Average annual rainfall (mm)	600–900	2500
Evaporation rates (mm/day)	1650–1180	2.0–6.0
Humidity (%)	51	74

[1]Bureau of Meteorology (BOM), Western Australia, Ali et al. (2012)
[2]Malaysian Meteorological Department (2015), Lim et al. (2013)

6.2 Weather of Perth, Australia and Kuala Lumpur, Malaysia

Perth and Malaysia's weather conditions are summarized in Table 6.1. Perth's weather is characterized by cool wet winters and hot dry summers in a temperate zone with a Mediterranean climate. Perth Metro has recorded rainfall ranging from 600 to 900 mm (BOM 2013). In comparison, Kuala Lumpur (Capital city of Malaysia) has a hot and humid weather in the tropics all year round. The temperature averages at 28 °C and the average humidity is 74% with an average rainfall between 2500 and 3000 mm (Azizul et al. 2011; UNEP 2015). Malaysia's climate is dominated by the effect of two monsoons or 'rainy seasons', which affect different parts of Malaysia to varying degrees. From November to February, the east coast Peninsular of Malaysia is affected by the northeast monsoon, which brings heavy rainfall, strong winds and huge waves along the entire coast. From April to September, the west coast of Peninsular of Malaysia is affected by the southwest monsoon, which is weaker compared to the northeast monsoon and from March to October, which are the transition period between the monsoons which is characterized by light wind.

6.3 Residential Water Consumption in Perth

In Perth, the water consumption issue has been heightened due to the declining rainfall by 12% since 1991. According to Water Corporation (2010), the domestic water consumption in Perth is approximately 52.8% inside the homes, 43.4% outside the homes and 3.77% is lost through private plumbing leaks. It is estimated that the average person uses around 112.33 L of scheme water per day to irrigate household lawns and gardens. Large residential lot sizes, sandy soils and hot windy conditions over summer are major contributors to the high outdoor consumption (Burton et al. 2009). It has been indicated that Perth was the highest residential water user in 2011–2012 despite the water scarcity crisis (Lehane 2014). However, there has been a reduction in the trend of residential water use in the city owing to the government approach to implement a scheme that will reduce water consumption across the state. One of the government strategies to embark on water use reduction is the release of its 50-year plan 'Water Forever' (Water Corporation 2009). It highlighted that reducing water consumption per person in conjunction with increasing water recycling and developing new sources were pivotal in achieving this aim. Together with the community, the Water Corporation sets a goal of reducing the consumption rate per person by a further 15% by 2030. In most of the developed world, water conservation measures are being mandated for new residential and commercial buildings such as the installation of dual flush toilets and filtration systems as well as refurbishments (Micou et al. 2012). For instance, the UK Code for Sustainable Housing includes an excellent calculator for house water use rating (BREEAM 2010).

6.4   Residential Water Consumption in Malaysia

In Malaysia, the problem of improper utilization of water resource in some states has limited the availability of water supply in other states. Consequently, water shortage has resulted in water rationing in most of the big cities (Ithnin 2007). Water consumption in Penang city, in the north of Malaysia, has increased by 11.8%, from 262 litres/capita/day (L/pc/d) in 2001 to 293 L/c/d for 2014 (PBA, 2015). In Selangor, one of the industrial states in Malaysia, a person in Selangor consumes an average of 226 L/pc/d—far greater than many developed countries, as shown in Fig. 6.1. A Malaysian needs about 17.2 L of water daily. The study found that the average person uses 5.2 L per day for toilet flushing and 4.8 L for bath or shower, accounting for approximately 58% of overall water use. The potential for high water consumption may be climate induced. The country's yearlong hot and humid weather results in a higher frequency of shower and change of clothes (hence laundry), as there is higher tendency to sweat more intensely.

6.5   Water Scarcity

6.5.1   Quantity Issues in Perth

In recent years according to the statistics, a decrease in rainfall has led to the forecast of 20% lower rainfall in Perth by 2030 compared to the 1990 levels (Water Corporation 2009). Perth is situated in the temperate southwest of Australia, with an estimated resident population of 2, 021, 200, which was recorded on 30 Jun 2014, an increase of 2.5% from the year of 2013 (Australian Bureau of Statistics 2015). In the Perth region, these measures include limiting garden irrigation with mains water to two nominated days per week, and three times per week when irrigating with water from a private bore.

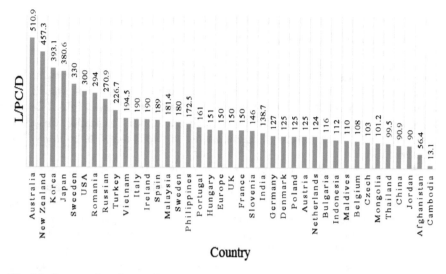

Fig. 6.1 Consumption of water litre per capita per day (L/pc/d) in Selangor, Malaysia compared to a number of countries in the world

6.5.2   Quality Issues in Malaysia

The state of the river has been aesthetically reduced due to the deterioration of the surface water as a result of pollution. Domestic sewage discharge has been reported to be the largest contributor of organic pollutant load (Al-Gheethi et al. 2016). In Johor Bahru, the average concentration of chemical oxygen demands (COD) of greywater was 1400 mg/L (Ujang and Henze 2006). In Sarawak, the rivers are seriously polluted with coliform bacteria derived from faecal matter due to the discharge of partially treated or untreated wastewater from the city into the river (Larsen and Lynghus 2004).

On 19 February 2014, Malaysia experienced a water crisis in Kuala Lumpur and the state of Selangor. A water shortage was reported in seven reservoir dams due to a substantial drop in water level caused by the hot and dry seasons in Peninsular Malaysia. The water crisis took a turn for the worse, with 2.2 million consumers in Selangor, Kuala Lumpur and Putrajaya experiencing water rationing from 10 March 2014. One of the reasons stated was the closure of two water treatment plants at 11th Mile, Cheras and Bukit Tampoi, due to the ammoniacal nitrogen content in the raw river water supply. This indicates that even though Malaysia experiences abundance of rainfall annually, problem of water pollution limited the availability of potable water. Hence, it becomes difficult for the water regulators to guarantee constant water supply. These results serve as an alarming warning for the authorities to implement a more effective wastewater management system for a sustainable future.

6.6 Greywater Treatment and Its Reuse Practice from a Global Perspective

Today, a wide range of greywater treatment technologies has been investigated to restore and maintain the physical, chemical and biological integrity of polluted greywater. Various forms of treatment systems have been developed to treat different types of greywater types ranging from low, medium and high. The primary treatment is mainly used to treat low and medium greywater which aimed to reduce BOD, COD and TSS.

Many researchers have sought to develop filtration-based treatment during the primary treatment as reported by researchers with different sources of greywater, including residential quarters (Nnaji et al. 2013), village houses (Mohamed et al. 2014a; Wurochekke et al. 2014), residential college hostel (Parjane and Sane 2011) and house kitchen wastewater (Mohamed et al. 2013b). The primary treatment can provide high reduction of BOD, COD and TSS (37–98, 74–90.8 and 40–95%, respectively) (Sahar et al. 2012; Parjane and Sane 2011). Sahar et al. (2012) revealed that pine bark filter achieves 98% of BOD reduction due to its high absorption capacity, as well as easy transportation of greywater. Al-Jayyousi (2003) found that the greywater treatment system consisted of a sand filtration unit has achieved 40 and 74% reduction of SS and BOD.

In order to purify high-strength greywater and for reuse purposes, secondary treatment is required. Several biological methods, such as anaerobic sludge blanket (UASB) (Lucia et al. 2010), membrane bioreactors (MBR) (Merz et al. 2007) and many more, have been applied to treat greywater. However, the treatment requires high-energy use, it is capital intensive and has high maintenance cost. Gunes et al. (2012) performed a free water surface flow-constructed wetland (FWS-CW) with a three-compartment septic system combined in series for high-strength greywater treatment. Moreover, if the system works independently, poor performance of the septic system in terms of TSS and nutrient removal has been observed. As a hybrid system, however, both performed very well, attaining up to 86, 91, 91, 57 and 45% reduction for TSS, BOD, COD, TN and TP, respectively. Dallas et al. (2004) showed that the reed bed system with ecological sanitation design presented a low-cost treatment on-site at Santa Elena-Monteverde, Costa Rica, Central America. This system has demonstrated the ability of low-cost reed beds to treat greywater from several houses to a level that meets the Costa Rican guidelines for wastewater reuse. Such constructed wetlands have been considered to be eco-friendly and financially acceptable for greywater treatment.

6.6.1   Greywater Treatment and Its Reuse Practice in Perth, Western Australia

In Western Australia, greywater treatment and its reuse systems range from simple direct diversion systems (either gravity-fed or pumped) that redirect untreated greywater from wastewater pipes to the garden, to coarse filtration and disinfection treatment systems and more sophisticated technologies, such as microfiltration. Greywater can be reused without treatment, such as bucketing bath water to the garden. However, direct disposal of untreated greywater may pose an unnecessary health risk. Greywater treatment systems (GTS) are therefore recommended where higher effluent quality is needed before reuse in irrigation (DOH 2010).

In Australia, the installation costs of these units typically range from $2000 for simple direct diversion systems up to $15,000 for higher end systems with storage capacity. To a certain degree, the variation in the cost of these systems correlates with their effectiveness and reliability in providing irrigation.

The examples of the greywater treatment and reuse application in Western Australia are presented in Table 6.2. Four houses were selected on the basis of their characteristics which might influence greywater quality. These characteristics included house type, number of occupants, presence or absence of children and pets, and landscape characteristics. Houses A and B were located at the Bridgewater Lifestyle Village (BWLV), while House C and D were located in White Gum and Hamilton Hill, respectively.

Greywater from House A and B are collected from both the laundry and bathroom, and directed to a dual sponge filter to remove hair, lint and other suspended solids. Thereafter, the greywater is pumped to the dripper system to irrigate the lawn or garden.

In House C and D, the greywater diverted from the normal waste stream is passed through a sedimentation tank for the treatment, disinfected by using ozone and then dispersed through an interconnecting substrata dripper system. This enables larger particles such as hair, lint, soap flakes and sand to settle at the bottom of the tank, thus preventing blockage of the pipe and/or soil as the greywater is dispersed through the infiltration field.

The evaluation of the greywater system was monitored for performance and reliability to meet the regulatory standards with three trial sites under the Premier's Water Foundation (PWF) research grant funded by the Department of Water, Western Australia in response to the State Water Strategy released in February 2003. The research team from Murdoch University of Western Australia conducted an assessment of the case studies, which led to the development of a new regulatory framework known as the decentralized wastewater treatment and recycling systems or DeWaTARS (Anda et al. 2010). The framework includes six main criteria which needed to be assessed when initiating a decentralized wastewater recycling project. The Technical Elements Model (TEM) was developed to determine the technical requirements and appropriate technologies for decentralized wastewater recycling for the public open space of urban villages. In order to make the technology selection process easier,

Table 6.2 Four case studies on greywater treatment for and its reuse for irrigation in Western Australia

House details		Greywater treatment		Greywater reuse for irrigation		
House/Suburb	Occupancy	Greywater System and Date installed	Greywater technology	Total block size (m^2)	Size of greywater irrigation area (m^2)	Vegetation
House A Mandurah	2 adults	AWWS Grey flow 00 Installed: July 2008	GDD	280	25	Native vegetation
House B Mandurah	2 adults + 2 pets	Land and Water Technology, Land and Water Greywater Reuse System Installed: July 2008	GDD	280	25	Native vegetation
House C White Gum Valley	2 adults + 2 children	GRS WaterSave Tank and Dripper System Installed: June 2007	GDD with sedimentation tank	596	52.5	Fruit trees and ornamental garden beds; roses
House D (Community) Pinakarri	6 adults + 5 teenage children	GRS WaterClear and Dripper System Installed: July 2008	GTS with sedimentation tank and ozone generation	2000	133	Lawn and fruit trees

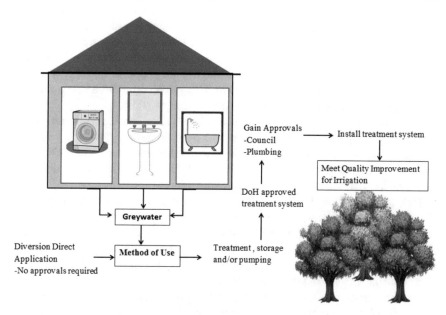

Fig. 6.2 The sequence of the household greywater reuse system in Western Australia

a decentralized wastewater treatment technology database was developed with over 150 sourced wastewater treatment products. The LaDeRS-H20 rating tool was developed for determining the water use performance efficiency of land developments.

The selection of the most appropriate greywater system depends on the source water, desired end use application and other characteristics. A greywater treatment system must be formally approved by the Department of Health before it can be installed in Western Australia. The flow of sequence of the household's greywater reuse system in WA can be simplified, as shown in Fig. 6.2. The implementation of greywater in WA requires five key processes, of which four are discussed briefly including local and state government approval, installation in each home, rebate acquisition, application and maintenance.

6.6.2   Greywater Treatment and Its Reuse Practice in Malaysia

Greywater treatment and/or combined with reuse at the household scale is still in the early stages of introduction in Malaysia. However, research on the treatment method of greywater has intensified recently. The knowledge can be incorporated to boost the setup of greywater treatment and its reuse in Malaysia. In Malaysia, most of the published articles derive from the pilot project on greywater ecological treatment in Kuching City, Sarawak, which was established in 2003. Mah et al. (2009)

explored the modelling process using a mathematical representation to evaluate a proposed greywater system. By using network simulation, Mah's modelling has shown potential savings of up to 40% through greywater reclamation for reuse as secondary sources of water for non-consumptive purposes. In Perth, a study has shown that scheme water savings from 9 to 37% can be achieved across a range of house types fitted with greywater reuse systems (Evans 2009).

Krishnan et al. (2008) explored the effectiveness of an aerobic sequencing batch reactor in treating nutrient-deficit and nutrient-spiked dark greywater for agricultural reuse. Hence, a treatment of nutrient-added dark greywater at a COD: N: P ratio 100:3.5:0.75 and 100:5:1 for 36 h HRT complies with the Malaysian discharge standards for agricultural activities in rural and urban areas.

Teck et al. (2009) evaluated the efficiency of biofilters and a constructed wetland with two species of terrestrial ornamental plants in greywater treatment. The study revealed that biofilters contributed the most in terms of the total removal of BOD_5, COD, FC and NH_4–N. However, wetland and biofilters were equally efficient in TSS removal. Wetland species, such as *F. macrocarpa,* are recommended for inclusion in urban housing areas to reduce river pollution.

Wurochekke et al. (2014) studied the capability of a constructed wetland with *Lepironia articulata* plant species for removing household greywater in a village house. Greywater samples were collected from the effluent of a single house, at influent, using a pre-treatment model (particle material), mini wetland model (*L. articulata*) and control model (without plants) at two sampling periods. The mini wetland model showed high removal performance of 81.42% BOD, 84.57% COD, 39.83% AN, 54.70% SS and 45.01% turbidity. In general, the water quality parameters showed the effectiveness of the peat filter and mini wetland model in removing impurities, especially suspended solids and COD level in the greywater. However, Mohamed et al. (2013b) explored the use of gravel and peat as filter media to treat kitchen greywater. In the following year, Mohamed et al. (2014a, b) studied a similar method to treat household greywater, and, specifically, bathroom greywater. The study showed the potential of peat for reducing 72% SS, and 87% $NH_4^+ - N$ of effluent greywater. In a similar study by Chan and Mohamed (2013), greywater treatment with peat was used for the post-filtration compressibility test on greywater parameter reduction.

6.7 Greywater Treatment as a Means of Pollution Control

Greywater is a major contributor to river surface water pollution; hence, effective measures to ensure pollution control from its source are very significant. This is in agreement with Hughes et al. (2006), in which onsite or source treatment is a paramount approach for water quality control, before it mixes into different wastewater streams. The treatment of greywater as a decentralized system, which occurs before being combined with other wastewater, is often less costly and more effective and decreases the pressure on the local wastewater treatment plant (WTP) in its daily operations. The Director-General of the Department of Environment

(DOE) Malaysia noted that wastewater treatment plants in Malaysia are outdated in terms of technology and identified them as the major reason for the bulk pollution of rivers (The Star 2006). In this context, domestic greywater treatment has the potential to address the issue of contaminants entering water bodies with nitrates, microbes, heavy metals (Mohammed et al. 2013) and technological stagnancy. Furthermore, such an approach allows greywater treatment at the household level and thus reduces the pollution load on the river.

6.8 Potential of Greywater Reuse in Malaysia

In Malaysia, the first sustainable approach to greywater reuse was implemented in Kuching City. This project was a joint collaboration between the Sarawak Government and the Danish International Development Assistance (DANIDA) using an Ecological Sanitation (Ecosan) Wastewater Treatment System at Hui Sing Garden in Kuching City. Kuching has no sewerage system, and, as in most Malaysian houses, the greywater is released untreated into the stormwater drains. Therefore, the water quality of drains, streams and rivers in Kuching is heavily polluted. However, it is possible to collect and reuse greywater as it can be treated to a less health-hazardous standard (Mah et al. 2008). A row of nine terrace houses was selected for carrying out the pilot project (Bjerregaard 2004). The terrace houses were linked to the ECOSAN system, which is constructed in a large recreational park adjacent to the backyards of the houses. From the pilot project, it was found that the trap is 88% efficient in removing the oil and grease, and effective in removing the bacteriological parameters and organic matter (Bjerregaard 2004; Teck et al. 2009).

The reuse of greywater in Malaysia is increasing the possibility of reducing environmental pollution, especially in rivers. However, it may create problems when the reuse system is inappropriately designed for the type of environment. Therefore, proper planning and strategic management are essential to make the system effective and reliable. For sustainable reuse of greywater, households must recognize the need to limit their use of products with high level of sodium or phosphorus or both combined, as water conservation may not be advantageous in this instance. It is recommended that for clothes washing, products low in Na should be selected (Madungwa and Sakuringwa 2007) because greywater quality is site specific except for certain recommended household products (Mohamed et al. 2013a). If Malaysians changed their attitude and use environmental-friendly detergents, it would also assist in protecting against river pollution and will have far-reaching improvement on the river water quality. The development of a greener environment for the future would also provide business for the detergent industry in Malaysia. With these boundaries in mind, a suitable zero tension lysimeter (ZTL) was developed as a device that could be used to collect leachate from greywater irrigation in household gardens (Mohamed et al. 2012). Based on abovementioned, the studies are being monitored with lysimeters under treated effluent irrigation areas and control samples are being taken from sampling bores at the perimeters of the developments. Over time, it will be determined if the following measures are adequate to redress excessive nutrient leaching to the environment.

6.9   Juggling the Differences in Adopting the Good Practice

Adopting the 'greywater treatment and reuse' system from the experience of a developed country like Australia without taking into consideration the geographical conditions, funding availability, the development of technological know-how of environmental awareness among the general public is less likely to be successful for expanding the treatment of greywater in Malaysia. Perth and Malaysia have many differences in terms of weather and soil type. In Perth, the major consideration for reusing greywater relates to the nutrient vulnerability of the groundwater due to the major sandy soil type. However, in Malaysia, with a predominant clayey soil profile, consideration should be made for appropriate greywater applications, as inappropriate usage can potentially alter the soil composition. It also has a detrimental effect on plants which are sensitive to salt, and thus, the low infiltration rate of these soils can lead to ponding and wastewater runoff.

6.10   Institutional and Legal Issues

In Malaysia, the city of Kuching provides a good example of successful application of greywater reuse. However, to widen its practical scope over a larger scale around the country, proper and thoughtful guidelines for greywater reuse are needed. The major challenge to an integrated water management arises from the lack of inter-agency coordination. According to Moorthy and Jeyabalan (2012), a lack of inter-agency coordination often occurs between the state and federal governments over water supply and sewerage matters. In 2009, the Malaysian government set up a new ministry called the Ministry of Energy, Green Technology and Water, (MEGTW) among others, to restructure the national water framework as problems arise with the federal governments over the control of state water assets. In Malaysia, with the changing political climate since 2007, water matters have been heavily politicized between the federal and state governments in which some have seriously questioned the lack of ethics and social ethics.

In addition, Chan et al. (2002) claimed that the lack of attention in securing water resources for the long term is regretful. Such concern is needed as water demands are growing every year as a result of the booming world population. Therefore, this is no longer the question of resource availability, but a problem that is attributed to effective resource management. The greywater use guidelines from Perth, which are embedded in the Code of Practice for the Use of Greywater (DOH 2005, 2010), could be a starting point for Malaysian authorities to design and implement Malaysian greywater management strategies. The author's view on the code of practice was that it provides user-friendly assistance concerning the use of greywater reuse, which wisely includes a wide range of issues related to greywater and conservation guidelines pertaining to the quality of groundwater and water supplies. In Malaysia, there is no specific standard for household greywater. The Environmental Quality (Sewage and Industry Effluences) 1979 is used in respect of the discharge of effluent into any inland water other than the effluent discharged from prescribed premises.

6.11 Conclusions

Even though Malaysia is blessed with an abundant supply of water resources, the authorities, industries and society should not take for granted that there will always be sufficient supply to meet the demand. The water regulators should provide adequate measures to reduce the effect of water pollution which may result in water scarcity. However, in some of the reported cases of water rationing, the Malaysian authorities can learn from the Perth experience in greywater treatment and reuse. The key criteria which were identified include clear guidelines concerning the installation of treatment and/or reuse systems, appropriate rebates and awareness-raising activities to enhance residents' knowledge about the reuse of greywater. In Malaysia, as most of the river pollution comes from domestic wastewater, proper planning to reduce wastewater discharge from households is an immediate task in this region. Among several management plans, treatment is preferable compared to reuse when dealing with greywater, and it is one of the suggestions and alternative plans to reduce the pollution load to the river.

Acknowledgements The authors wish to thank the Ministry of Higher Education (MOHE) for supporting this research under FRGS vot 1574 and also the Research Management Centre (RMC) UTHM for providing grant IGSP U682 for this research.

References

Al-Gheethi AA, Mohamed R, Efaq AN, Norli I, Halid AA, Amir HK, Ab Kadir MO (2016) Bioaugmentation process of secondary effluents for reduction of pathogens, heavy metals and antibiotics. J Water Health 14(5):780–795

ABS (Australian Bureau of Statistics) (2015) Regional population growth. Australia, 2013–14, viewed 14 Jan 2016. http://www.abs.gov.au/ausstats/abs@.nsf/mf/3218.0/

Ali R, McFarlane D, Varma S, Dawes W, Emelyanova I, Hodgson G, Charles S (2012) Potential climate change impacts on groundwater resources of south-western Australia. J Hydrol 475:456–472

Allen L (2010) Overview of greywater reuse: the potential of greywater system to aid sustainable water management. Pacific Institute, USA. ISBN 1-893790-27-4

Al-Jayyousi OR (2003) Greywater reuse: towards sustainable water management. Desalination 156:181–192

Anda M, Mohamed RMSR, Mathew K, Dallas S, Ho G (2010) Decentralised wastewater treatment and recycling in Urban Villages. Water Practice Technol 5(3):1–11

Bino M, Al-Beiruti S, Ayesh M (2010) Greywater use in rural home gardens in Karak, Jordan. In: McIlwaine, Redwood (eds) Greywater use in the Middle East, IDRC

Bjerregaard D (2004) Urban ecological sanitation Kuching is paving the way. In: Natural Resources and Environment Board (NREB), Kuching, Malaysia

BOM (2013) Australia. Western Australia in 2013: Warmest year on record, viewed 11 July 2014. http://www.bom.gov.au/climate/current/annual/wa/archive/2013.summary.shtml

BREEAM Centre (2010) Code for sustainable homes-technical guide. Communities and Local Government Publications UK

Burton R, Currie R, Wong L (2009) Managing pressure to save water in Perth. Paper read at 6th International Water Sensitive Urban Design Conference and Hydropolis #3, 5–8 May, 2009, at Perth, Western Australia

Chan CM, Mohamed RMSR (2013) Post-filtration compressibility characteristics of peat used as greywater filter media. Middle-East J Sci Res 17(5):647–654

Chan NW, Ibrahim AL, Hajar AR (2002) The role of non-governmental organization in water resources management in Malaysia. Paper read at regional symposium on environment and natural resources, 10–11th April 2002

Dallas D, Scheffe B, Ho G (2004) Reedbeds for greywater treatment—case study in Santa Elena-Monteverde, Costa Rica, Central America. Ecol Eng 23(1):55–61

DOH (2005) Code of Practice for the Use of Greywater in Western Australia. Water Corporation, Department of Health, Department of Environment

DOH (2010) Code of practice for the use of greywater in Western Australia. Water Corporation, Department of Health, Department of Environment

Evans C (2009) Greywater reuse: an assessment of the scheme water savings that can be achieved at a household scale, Honours thesis, School of Environmental Science, Murdoch University, Perth, Western Australia

Gunes K, Tuncsiper B, Ayaz S, Drizo A (2012) The ability of free water surface constructed wetland system to treat strength domestic wastewater: a case study for the Mediterranean. Ecol Eng 44:278–284

Hughes R, Ho G, Matthew K (2006) Settling effluent quality standards, In: Ujang Z, Henze M (eds) Municipal wastewater management in developing countries: principles and engineering, IWA Publishing

Ithnin H (2007) Rate of water use in Malaysian homes. Paper read at Geography Conference, 21–22 Ogos, 2007

Juahir H, Zain SM, Yusoff MK, Hanidza TT, Armi AM, Toriman ME, Mokhtar M (2011) Spatial water quality assessment of Langat River Basin (Malaysia) using environmetric technique. Environ Monit Assess 173(1–4):625–641

Krishnan VD, Ahmad Jeru J (2008) Influence of COD:N: P ratio on dark greywater treatment using a sequencing batch reactor. J Chem Technol Biotechnol 83(5):756–762

Larsen IB, Lynghus H (2004) Framework plan for integrated wastewater management for the city of Kuching Sarawak. Danish International Development Assistance (DANIDA), Kuala Lumpur, Malaysia. ISBN 9834054696

Lehane S (2014) Australia's water security part 2: water use. Future Directions International Pty Ltd., Dalkeith, WA Australia

Lim WY, Aris AZ, Praveena SM (2013) Application of the chemometric approach to evaluate the spatial variation of water chemistry and the identification of the sources of pollution in Langat River, Malaysia. Ara J Geosci 6(12):4891–4901

Lucia HL, Hardy T, Grietje Z, Cees JNB (2010) Comparison of three systems for biological greywater treatment. Water 2(2):155–169

Madungwa E, Sakuringwa S (2007) Greywater reuse: A strategy for water demand management in Harare? Phys Chem Earth 32(15–18):1231–1236

Malaysia Meteorological Department (Met) (2015) viewed 13 Nov 2015. http://www.met.gov.my/index.php?lang=english

Mah DYS, HIn CJB, Putuhena JF, Said S (2009) A conceptual modeling of ecological greywater recycling system in Kuching city, Sarawak, Malaysia. Res Conserv Recycl 53(3):113–121

Mah DYS, Putuhena FJ, Salim S, Lai SH (2008) Modelling of greywater reuse for Kuching, Malaysia. J Aust Water Assoc. Water Magazine: Technical features, August, pp. 31–34

Merz SR, El Hamouri B, Kraume M (2007) Membrane bioreactor technology for the treatment of greywater from a sports and leisure club. Desalination 215(1–3):37–43

Md. Azizul B, Rawshan AB, Abdul Hamid J, Zaharaton ZA, Pereira JJ (2011) Water use and their consumption pattern in Malaysia. Paper presented in the Climate Change Conference. Ministry of Science, Technology and Innovation Malaysia

Micou AP, Mitchell G, McDonald A (2012) Sustainable homes: a methodology for assessing influence on regional water demand. Water Sci Technol: Water Suppl 12(2):140–147

Mohammed B, Abeer A, Theib O (2013) Assessing the efficiency of grey-water reuse at household level and its suitability for sustainable rural and human development. British J Appl Sci Technol 3(4):962–972

Mohamed RMSR, Al-Gheethi AA, Jackson AM and Amir HK (2016) Multi component filters for domestic greywater treatment in village houses. J Am Water Works Assoc (AWWA) (In Press)

Mohamed RMSR, Kassim AHM, Martin A, Stewart D (2012) Zero-tension lysimeter for use in greywater irrigation monitoring. Int J Integrated Eng 4(2):15–21 (UTHM Publisher)

Mohamed RMSR, Kassim AHM, Anda M, Dallas S (2013a) A monitoring of environmental effects from household greywater reuse for garden irrigation. Environ Monit Assess 185(10):8473–8488

Mohamed RMSR, Chan CM, Hasyimah G, Mohd AMY, Kassim AHM (2013b) Application of peat filter Media in Treating Kitchen Wastewater. Int J Zero Waste Gene 1:11–16

Mohamed RMSR, Anwaruddin AW, Chan CM, Kassim AHM (2014a) The use of natural filter media added with peat soil for household greywater treatment. GSTF Int J Eng Technol (JET) 2(4). https://doi.org/10.7603/s40707-013-0011-x

Mohamed RMSR, Chan CM, Senin H, Kassim AHM (2014b) Feasibility of the direct filtration over peat filter media for bathroom greywater treatment. J Mater Environ Sci 5(6):2021–2029

Moorthy R, Jeyabalan G (2012) Ethics and Sustainability: a review of water policy and management. Am J Appl Sci 9(1):24–31

Nnaji CC, Mama CN, Ekwueme A, Utsev T (2013) Feasibility of a filtration-adsorption greywater treatment system for developing countries. Hydrol Curr Res S 1:006. https://doi.org/10.4172/2157-7587.S1-006

PBA (Perbadanan Bekalan Air Pulau Pinang) (2015) Penang's Water Tariffs the Most "People Friendly" in Malaysia, media release 26th April 2015

Parjane SB, Sane MG (2011) Performance of grey water treatment plant by economical way for Indian rural development. Int J Chem Tech Res CODEN. USA 3(4):1808–1815

Ryan S (2014). Subsidiarity in principle: decentralization of water resources management. Utrecht Law Rev 10(2), viewed 2 Nov 2015. http://www.utrechtlawreview.org/index.php/ulr

Sahar SD, Mikael P, Björn V, Lars DH, Ingrid Ö, Håkan J (2012) Efficiency of bark activated charcoal, foam and sand filters in reducing pollutants from Greywater. Water Air Soil Pollut 223(7):3657–3671

Santos C, Taveira-Pinto F, Cheng CY, Leite D (2012) Development of an experimental system for greywater reuse. Desalination 285:301–305

Simon J, Elisa F (2013) Single household greywater treatment with a moving bed biofilm membrane reactor (MBBMR). J Membr Sci 446:277–285

Teck YLK, Apun Zainuddin SR (2009) Performance of a pilot-scale biofilters and constructed wetland with ornamental plants in greywater treatment. World Appl Sci J 6(11):1555–1562

The Star Malaysia (2006) The Sewage causing bulk of river pollution. Media release, 5 Dec 2006

Ujang Z, Henze M (2006) Municipal wastewater management in developing countries, IWA Publishing

UNEP (2015) United Nation Environment Programme: State of waste management in South East Asia. Storm water, viewed 10 Dec 2015. http://www.unep.or.jp/Ietc/Publications/spc/State_of_waste_Management/5.asp

Water Corporation (2009) Water forever: directions for our water future draft plan

Water Corporation (2010). Perth residential water use study 2008/2009. Water Corporation, viewed 12 Dec 2015. www.watercorporation.com.au/waterforever

Weller R (2009) Boomtown 2050: Scenarios for a rapidly growing city. Paper read at WSUD 09 Conference, 5–8 May 2009, at Perth

Wurochekke AA, Harun NA, Mohamed RMSR, Kassim AHM (2014) Constructed wetland of *Lepironia articulata* for household greywater treatment. APCBEE Procedia 10:103–109

Chapter 7
Treatment Technologies of Household Greywater

Efaq Ali Noman, Adel Ali Saeed Al-Gheethi, Siti Asmah Bakar, Radin Maya Saphira Radin Mohamed, Balkis A. Talip and Amir Hashim Mohd Kassim

Abstract The shortage of water resource in the developing countries induced the search for alternative sources. Greywater alongside storm and ablution water might represent the best source of water because these waters have less contaminant than sewage. However, the separation of this water from the source point is the first step in the proper management which facilitates the treatment process. The selection of treatment technologies for greywater depends on the economic status and standards limits required for disposal or reuse of greywater which differs from one country to the others. In many of the developing countries, the treatment of greywater aims at achieving the basic requirements which lie in the reduction of the main parameters of greywater. The utilisation of flocculation and coagulants process might be effective for this purpose. Many of the natural coagulants have been reported to reduce the main parameters of greywater. In the developed countries, advanced technologies are used for removing of xenobiotics organic compounds (XOCs) and to produce high quality of the treated greywater. This chapter focuses on the treatment technology used for the treatment of greywater and their efficiency in the reduction of XOCs.

Keywords Treatment technology · XOCs · Removal efficiency · AOPs

E. A. Noman
Faculty of Applied Sciences and Technology (FAST), Universiti Tun Hussein Onn Malaysia (UTHM), Pagoh, Johor, Malaysia

E. A. Noman
Department of Applied Microbiology, Faculty Applied Sciences, Taiz University, Taiz, Yemen

A. A. S. Al-Gheethi · S. A. Bakar · R. M. S. Radin Mohamed · A. H. Mohd Kassim
Micro-Pollutant Research Centre (MPRC), Department of Water and Environmental Engineering, Faculty of Civil and Environmental Engineering, Universiti Tun Hussein Onn Malaysia (UTHM), 86400 Parit Raja, Batu Pahat, Johor, Malaysia
e-mail: maya@uthm.edu.my

B. A. Talip (✉)
Faculty of Applied Sciences and Technology (FAST), Universiti Tun Hussein Onn Malaysia (UTHM), 84000 KM11, Jalan Panchor, Pagoh Muar, Johor, Malaysia
e-mail: balkis@uthm.edu.my

© Springer International Publishing AG, part of Springer Nature 2019   125
R. M. S. Radin Mohamed et al. (eds.), *Management of Greywater in Developing Countries*,
Water Science and Technology Library 87,
https://doi.org/10.1007/978-3-319-90269-2_7

7.1   Introduction

The high deficiency of water resource in the developing countries induced the governments to look for alternative resources. The greywater represents one of the best resources of portable water due to the less contamination in comparison to the sewage. The separation of greywater from blackwater is the critical step for the proper management which is aimed at facilitating the treatment process. There are many benefits for the segregation process of the greywater from black water which includes the quantity of the consumed energy by approximately 11.8–37.5% of that energy in comparison to the sewage treatment processes. This process is consistent in guaranteeing fulfilling water demand and meets the quality standards established. The wetlands, rainwater reuse, anaerobic digesters for greywater might represent an alternative strategy for the generation of energy on a small scale in homes and business. Besides, the segregation process of greywater might improve the treatment efficiency to remove inorganic and organic matter as well as infectious agents. The main aim for the treatment processes is to remove faeces and urine from sewage and greywater, since they might cause the growth of pathogenic microorganisms and contaminate the environment. In the absence of huge amounts of organic matter in the greywater, the treatment process becomes more efficient. However, the main challenges in the greywater lie in the xenobiotic organic compounds (XOCs), which are not completely removed by the traditional treatment process. The disinfection of greywater and ablution water might be performed using the natural processes such as solar disinfection which is more appropriate for Middle East countries. It is due to the sunlight intensity between 5.2 and 6.8 kWh/m^2/day, which have been reported as effective for the reduction of pathogenic bacteria to less than detection limits (Al-Gheethi et al. 2015a).

Separation of greywater from black water is a crucial step in the wastewater treatment. It is more effective by incorporating the separation step and disinfecting process prior final disposal. The utilisation of treated greywater for irrigation contributes to the reduction of ground and freshwater. In this chapter, the improvements of greywater quality in terms of chemical oxygen demand (COD), biological oxygen demand (BOD), total suspended solids (TSS), turbidity as well as removal of XOCs are discussed. Further discussion of disinfection technologies for reduction of microorganism are reviewed in Chap. 10.

7.2   Current Practice of Greywater Treatment System

Nowadays, the treatments of greywater are ranging from simple coarse filtration to the advanced treatment processes. The current technologies include the filtration and ultrafiltration (UF) membrane systems (physical processes); coagulation/flocculation and ion exchange resins (chemical processes) and rotating biological reactor (RBC), constructed wetlands (CW), sequencing batch reactor (SBR) and membrane bioreactor (MBR) (biological processes) (Barışçı and Turkay 2016). The advantages and disadvantages, as well as the efficiency of each technology, had been reviewed by Abu Ghunmi (2009). The most appropriate technology is selected based on the scale of operation, final usage of the greywater, cost and regional customs and practices as well as the strength and flow rate of the greywater (Pidou et al. 2008). In this chapter, the aerobic and anaerobic system has been used are greywater treatment for lowering energy consumption. However, additional disinfection process is required in order to inhibit the growth of infectious agents. On the other hand, anaerobic treatment is not highly recommended for the greywater due to the low reduction of COD (Li et al. 2009). Indeed, this might be due to the presence of high concentrations of XOCs, which lead to an increase of the COD concentrations and are more resistant to the anaerobic degradation. Therefore, the aerobic treatment might be more efficient to degrade XOCs by the oxidation of the phenol ring of these compounds and then reduce the COD. It has been claimed that the XOCs contribute by 30% of the COD in the greywater (van de Wijst and Groot-Marcus 1998). Leal (2010) reported that the method to estimate the surfactants contribution in the greywater COD is by measuring the theoretical COD value of the reference compounds used for chemical analysis. The same study had also indicated that the specific COD value of dodecyl benzene sulphonate is 2.4 g COD g^{-1}, cetyl trimethyl ammonium bromide 5.3 g COD g^{-1} and 2.6 g COD g^{-1} of triton x 100.

In the wastewater treatment plants, the primary treatment is conducted by the physical process and followed by the secondary treatment which is a function of biological processes. In the secondary process, the decomposing microorganism available in the wastewater is the key for the degradation of organic matters. These organisms get the energy and carbon source for the nutrient contents (total nitrogen and total phosphorus) of the wastewater. In contrast, the high COD: BOD ratio as well as the low concentrations of the nutrients which are associated with low decomposers organism affect negatively on the efficiency of the biodegradation processes and thus biological process in the greywater (Jefferson et al. 2001).

In fact, both physical and biological processes might take place in nature, the natural sedimentation of suspended solids lead to separate the greywater into two layers: in the upper layer with the presence of oxygen (aerobic conditions) the biological process takes place by aerobic and facultative microorganism and in the lower layer with anaerobic conditions the reduction in the organic content is due to the anaerobic degradation of biodegradable compounds. The anaerobic process led to the generation of foul smel due to the release of methane from degradation of organic compounds by the anaerobic microorganisms. Subsequently, it becomes a fertile media for attracting of insect and transmission of infectious agents to human. Therefore, the greywater should be treated immediately after discharge or at least before reaching the anaerobic conditions (Allen et al. 2010).

In the rural regions, the intention to build sewerage networks connected to a small greywater treatments system is coming from the civil society (Abdel-Shafy et al. 2014). This intention aims to reduce the main pollutants in the greywater to be acceptable for the irrigations of public gardens. However, these efforts need maintenance and operation unit. The alternative is to depend on the individual treatment unit.

This option is cost-effective and more flexible in operation and maintenance since no experience is required to operate the treatment system, which is relatively affordable and practical for the household (Halalsheh et al. 2008). The single-domestic system consists of coarse filtration to remove hair or others tiny particles via metal strainer and simple chemical disinfection such as chlorination. Nonetheless, it has to mention that these treatments are not enough to generate high quality of greywater or at least acceptable by the local strands regulations. Recently, the concept of xenobiotic organic compounds (XOCs) has complicated the dependence on the simple treatment processes, since these processes are insufficient for the removal of XOCs which are classified as toxic substances for the natural biodiversity.

7.3 Storage of Greywater

The storage system is one of the selected options which are used for domestic wastewater before and after the treatment process. Storage systems are used mainly for the accumulation of the wastewater before subjecting to the treatment process and for regulating the reuse in the irrigation. In many cases, the treated wastewater is stored during the rainy season for the disposal into the environment where the dilution rate is higher (U. S. EPA 1989). The storage system might improve the microbiological quality of the greywater in terms of pathogenic bacterial reduction due to the deficiency in the favourable conditions necessary for their growth such as nutrients and increase the competition between the microorganisms in the greywater (Al-Gheethi et al. 2013). However, the disadvantage of this system lies in the deterioration of the quality of the stored greywater in comparison to the raw greywater. The deterioration is accelerated after the storage period of 24 h. Tal et al. (2011) showed that the storage of greywater for more than 24 h leads to an increase of the total coliform (TC). However, Al-Gheethi et al. (2013) revealed that the storage of effluents for 3 weeks reduced the concentrations of TC, but the resistance of this organism against antibiotics has increased. These differences might be related to the conditions of the storage system and type of wastewater.

Schneider (2009) had reported that the concerns associated with the storage of the greywater and then discharge to the environment lie in the low level of dissolved oxygen (DO), which leads to the death of many aquatic organisms due to anaerobic conditions. The reduction in the level of DO might be due to the oxidation and degradation processes which consume the oxygen available in the greywater. It has been reported in the same study that the decomposition in the greywater might reach 90% compared to 40% in the blackwater. Therefore, one of the benefits of the storage system is the degradation of organic matter to simple substances and thus their more availability for the plants. Storage of the greywater might also facilitate the disinfection process due to the reduction in the total suspended solids during the storage. In spite, the degradation in the organic substances during the storage might improve the efficiency of the disinfection process, because these substrates react with disinfectant and thus greater initial dose is required to inactivation of pathogenic bacteria (Ronen et al. 2010).

7.4 Treatment Technologies of Greywater

The most common treatment technologies used for the treatment of greywater are presented in Table 7.1. The efficiency of these technologies in reducing the main parameters of greywater is reviewed in this section.

7.4.1 Filtration System

Filtration media is a constructed material made up of gravel or other appropriate media. The efficiency of these systems in removing pollutants from greywater depend on the greywater parameters such as pH, temperature as well as characteristics of the raw material used in the design of the filter such as specific surface area, particle size and distribution, porosity, and surface chemical composition. The simple filtration systems consist of charcoal, gravel, peat, sand, limestone, clamshell, ceramics, wetland bed and volcanic ash (Mohamed et al. 2014; Wurochekke et al. 2014). These systems are used as a primary treatment process for the greywater to remove SS and turbidity; they also contribute to the reduction of COD, BOD, heavy metals and bacteria.

The clamshell is one of the raw materials which are used for the design of filtration system (in a form of calcium carbonate ($CaCO_3$) which swaps out calcium atoms in favour of heavy metals locking them into a solid form. The clamshells exhibit efficiency in the removal of arsenic metal ions from water, so it has the potential application as a filter medium in individual households (Köhler et al. 2007). Steel slag is a by-product material generated from the steel industry. The slag has been reported as efficient tools to remove TP from different wastewater (Barca et al. 2014). Steel slag is mildly alkaline materials with the pH range from 8 to 10. However, the leachate from the filter system consists of steel slag which has pH 11 due to the corrosive of aluminium made by iron (Fe) and calcium oxide (CaO) as a result of the direct contact with the slag (Motz and Geiseler 2001).

Limestone is an alkaline agent available in different crystal forms of calcium carbonate ($CaCO_3$) and it consists of mineral calcite and aragonite. Limestone is highly effective for removing Fe and Mn ions from the water than sandstone filter (Ghaly et al. 2007). Ceramic is an inorganic and non-metallic material. It is used in most of the filtration methods, due to the efficiency in the disinfection process to remove bacteria, protozoa and microbial cysts but not viruses which have less size than the pore size of ceramic, so can pass through to the other clean side of the filter (Plappally et al. 2011). The ceramic filter is made from the cheap materials which include clay soil and rice bran, and it has been used/adopted in the rural

Table 7.1 Most common treatment technologies investigated for the treatment of greywater

Treatment process	Country	Efficiency (%)????	References
A mini constructed wetland system	Malaysia	81.42 for BOD, 84.57 for COD, 39.83 for NH_4, 54.70 of TSS, 45.01 of turbidity	Wurochekke et al. (2014)
Granular-activated carbon (GAC), Biofilm up-flow-expanded bed (UEB), Reactor and a slow down-flow-packed sand filter	Malaysia	COD (70), TSS (72)	Al-Mughalles et al. (2012)
Coagulation aluminium and ferric salts	UK	COD (63.71), turbidity (90.85), TN (12.77), TP (94.7)	Pidou et al. (2008)
Ion exchange resin processes		COD (65.61), turbidity (82.67), TN (15), TP (45.87)	
Vertical flow-constructed wetland (RVFCW)	Israel	TSS (98), COD (81)	Gross et al. (2007)
Slanted soil system	Japan	COD (85), SS (78), TN (78), TP (86)	Itayama et al. (2006)
Lab-scale membrane bioreactor (MBR)	Morocco	COD (85), BOD_5 (94), TKN (93), NH_4 (72), TN (19), Surfactants (97),	Merz et al. (2007)
Slanted soil system	Japan	TSS (60–94), COD (90), LAS (93–96)	Ushijima et al. (2013)
Multicomponent filter consisted of steel slag, clamshell, limestone and sand media	Malaysia	COD (74), BOD (87.7), Turbidity (96), TSS (96.2)	Mohamed et al. (2016)

Chemical oxygen demand (COD), biological oxygen demand (BOD_5), total suspended solids (TSS), total nitrogen (TN), ammonium (NH_4), total phosphorous (TP), alkylbenzene sulfonate (LAS)

area of developing countries such as Bangladesh (Shafiquzzaman et al. 2011). The applicability of the ceramic filter belongs to their pores size which is between 1 and 5 μm (Shafiquzzaman et al. 2011). In the greywater, the ceramic filtration is used as a primary treatment followed by the wetlands and the UASB-hybrid reactor (Halalsheh et al. 2008).

The absorption process is a mechanism used in the filtration system to remove organic matter such as TP and TN as well as bacterial cells (Stevik et al. 1999). The porosity, which is defined as a number of voids available in the filter surface, plays an important role in the determination of water retention, hydraulic residence time (HRT) and hydraulic conductivity in the filter. These parameters detect the adsorption and biofilm activities and thus the transformation of organic matter, nitrogen and phosphorus as well as the retention of solids and bacteria (Dalahmeh et al. 2014). The porous ceramic filters are made in micron or sub-micron pore sizes to enhance bacterial and viruses' removal. However, the filtration system should be subjected to the frequent cleaning to remove the suspended solids which might reduce the removal efficiency. Therefore, the ceramic filtration system is most common for improving the drinking water with fewer particle molecules. Different filtrations systems are available in the markets; these filters are made up of clays, glass, diatomaceous earth and other fine particles. For instance, the ceramic filters made of fired clay, lime, limestone and calcium sulphate and used for water filtration in Pakistan, have the potential to reduce turbidity by 90% and bacteria by 60% from the drinking water (Jaafar and Michałowski 1990). Nevertheless, this efficiency might be not enough for the greywater since the density of microorganism in the greywater might be 100 times than that in the drinking water.

Ushijima et al. (2013) examined the efficiency of slanted soil system for the treatment of greywater and removal of COD and TSS as well as XOCs such as linear alkylbenzene sulfonate (LAS) and biodegradable dissolved organic carbon (BDOC). Two types of the soil were used including Kanuma soil consisting of alumina and hydrated silica and crushed baked mud brick, while the treatment system consisted of four soil chambers. The maximum removal of TSS is ranging from 60 to 94%, 90% for COD and BDOC. Meanwhile, the removal rate of LAS is 93–96%. These findings reflected the high efficiency of the slanted soil system to improve the quality of treated greywater and then to be used for irrigation. One disadvantage observed in this system was the clogging in the fine soil after 3–5 weeks operation which consisted of coarse soil and fine soil chamber hence extended their effectiveness to 8 weeks. The slanted soil system with several chambers provides a low-cost and simple operating option for the treatment of the greywater. Several studies had indicated that the greywater generated from this system did not meet the standard limits for the irrigation, but it might contribute in avoiding eutrophication in the natural water system which receives the greywater (Itayama et al. 2006).

Dalahmeh et al. (2014) developed a filter system consisting of activated charcoal, pine bark (bark), and sand to treat of greywater and removal of organic matter and nitrogen. The bacterial diversity at different times and depths in the filters was also investigated. The study revealed that the charcoal (97%) and bark (98%) filters exhibited higher removal of BOD_5 than the sand (75%). The bacterial community on the charcoal and sand filters was more diverse and dynamic than that recorded in the bark filters, while the bark filter showed the highest richness. These differences might be due to the lignocellulosic composition of the bark and organic content, which provide the nutrients required for bacterial growth. In contrast, the composition of the greywater defined the bacterial community in the charcoal and sand filter. Therefore, the low bacterial biomass in these layers allows a diversified bacterial community to develop. The most dominant bacteria detected in this study were *Rhizobium* spp., *Pseudomonas* spp., and *Acinetobacter* spp.

Mohamed et al. (2016) investigated a multicomponent filter for the treatment of greywater in the village houses. The filtration system consisted of steel slag, clamshell, limestone and sand media. The study revealed that this system exhibited high efficiency for the treatment of household greywater with 74.0% of removal for COD, 87.70% for BOD5, 98% for turbidity and 96.2% for TSS. The filtrated greywater met the Malaysian standard limits required by EQA 1974, Regulations 2009, Standards A for disposal into upstream of the drain. The using natural materials such as the sand bed, fine particles, charcoal bed, course size bricks bedded, the bed of coconut shell cover and wooden sawdust bed have been used in developing a natural filter system for individual users (Lalander et al. 2013; Mohamed et al. 2014). Mohamed et al. (2014) found that the filtration system consisted of charcoal, peat and gravel exhibited efficiency for the greywater treatment.

It appears that the filtration systems have effectiveness in the greywater treatment due to the low concentrations of solids materials in these wastes in comparison to the blackwater. However, the challenges are the clogging of the filter, therefore the raw materials used in the development of filter system should be replaced frequently.

7.4.2 Constructed Wetland

Constructed wetland is one of the alternatives for the conventional system which has many issues such as high energy required for the operation process and cost of maintenance, especially in the developing countries. Wetland system has less cost due to the natural flow of the water from the beginning to the end of the system. It has also less construction costs and maintenance as well as easy operation without the need for well-trained technicians (Saroj and Mukund 2011). In the wetlands, the sand and gravel media are the main materials used, besides that the specific aquatic plants such as duckweed or hyacinths are provided to reduce the pollutant content such as organic matter. Wurochekke et al. (2014) designed a mini constructed wetland for the treatment of greywater used *Lepironiaarticulata*. It was performed in three containers made from high-density polyethene (HDPE) plastic and connected via polyvinyl chloride (PVC) pipes. The first container consisted of three layers included gravity (diameter <25 mm), charcoal and fine sand (0.2 mm of diameter). The capacity of this container was 20 L, and was used for the separation of food particles and suspended solids. The filtered grey water generated from this stage was passed to the second container containing fine sand and gravel layer as well as a tube sedge with Lepironia articulates as biological treatment. The third container which contained sand and gravel was used as a control model for the treatment (without *Lepironiaarticulata*). The results revealed that the reduction percentage of the greywater parameters were 81.42% for BOD, 84.57% for COD, 39.83% for ammonia nitrogen, 54.70% of SS and 45.01% of the turbidity.

Fowdar et al. (2017) developed a low-energy and maintenance technology called living wall system for the treatment of greywater. This technology depends basically on employing ornamental plants which include *Lonicera japonica*, *Canna lilies* and ornamental grape vine and grown in a sand filter on a side of a building in Melbourne, Australia. The study revealed an effective removal of nitrogen (>80%), phosphorus (-13%–99%), >90% for BOD and >80% for TSS. This study provided a good alternative strategy for the greywater treatment. However, more research on their effects on the building should be conducted. The treatment system performed in this study was biofiltration systems, which have the similar function of constructed wetland with a smaller footprint and might increase the sustainability and liveability of urban cities (Bratieres et al. 2008; Payne et al. 2015; Pérez-Urrestarazu et al. 2015). The processes which take place in the living wall include sedimentation and straining as a physical process, precipitation and adsorption as a chemical process, as well as microbial and plant assimilation as a biological process (Fowdar et al. 2017).

7.4.3 Chemical and Electrochemical Treatment Systems

The electrochemical treatment systems include the usage of electrooxidation, electrocoagulation (EC) and electro-flotation. In the traditional coagulation treatments, the aluminium sulphate and ferric chloride have been used as coagulants which dissolve in the water to forms Al^{3+}, Fe^{2+} and Fe^{3+}. These cations form metal hydroxides which lead to the destabilisation of the suspended solids and this process takes place as a function of adsorption and charge neutralisation. However, the process leads to generate a toxic sludge which needs further treatment to desorb the metal ions before the final disposal into the environment. In contrast, in the electrocoagulation, the electrodes used is start to dissolve the water to generate the similar metal ions Al^{3+}, Fe^{2+} and Fe^{3+}, which acts in a similar mechanism of chemical coagulants but without chemical additives (Barışçı and Turkay 2016). The efficiency of this technique has achieved more than 70% of COD removal according to the studies conducted by Tchamango et al. (2010) and Yavuz et al. (2011).

The using of natural zeolites is one of the treatment technologies for the greywater, the mechanism of pollutants removal depend on the ion exchange and adsorption capability. The economic cost of this technologies belongs to the possible regeneration of natural zeolites for several times using NaCl or KCl solutions (Widiastuti et al. 2008). The chemical composition of zeolites is a three-dimensional framework of aluminosilicate tetrahedral, which gives the negative charges of zeolites surface. The zeolites contain also K^+, Na^+, Mg^{2+} and Ca^{2+} which balance the negative charges. On the other hand, these ions can be easily exchanged with the certain cations in wastewater such as NH^{4+}, as well as radioactive cations such as Sr^{2+} and C^{s+} (ion exchange mechanism). It has been reported that the natural zeolites exhibited 80–100% in the removal of 137Cs, 95% of COD, 100% of BOD, 10 to 100% of heavy metals, 30 to 98% of NH_4^+ and 90% of PO_4^{3-} (Garcia et al. 1992; Nunez 1998; Kallo and Ming 2001; Englert and Rubio 2005).

7.4.4 Hybrid Treatment Systems

The hybrid system is designed with the combination of two treatment processes (physical/chemical processes, physical/biological processes/chemical/biological processes) or more. Abdel-Shafy et al. (2014) investigated the efficiency of different hybrid treatment processes including a sedimentation process at different times, an aeration system and the addition of effective microorganism (EM) (a mixture of lactic acid bacteria, photosynthetic bacteria, yeast, fermenting fungi and actinomycetes) to generate an unrestricted greywater for reuse in Egypt. The optimisation for each stage was performed on the lab scale, while the whole treatment process at the optimal operating parameters was implemented in the pilot plant. The findings recorded removal rates by 70.8% for TSS, 63.1% for COD, 70.6% for BOD5 and 63.5% for oil and grease, after settling for 3 h and aeration for 90 min. However, the addition of EM (1.5 mg/L) enhanced the removal rates to 98.1% of TSS, 91.1 of COD, 96.1 of BOD5 and 96.2% of oil and grease after 4.5 h settling time and 90 min of the aeration process. The utilisation of selected microorganisms for the treatment of wastewater is called bioaugmentation. Recently some of the companies like Hydra International Ltd in the UK have products called Bio Start (Bacteria/Enzyme start-up treatment For STPs). These products contain a range of carefully selected and acclimatised natural microorganisms (such as *Bacillus* sp. and *Pseudomonas* sp.) to produce high levels of lipase, amylase, cellulose and protease enzymes (Hydra International Ltd, 2010).

It can be concluded from the studies abovementioned that the treatment technologies focused mainly on the reduction of COD, BOD TSS. It has to be mention that these technologies are insufficient to remove the XOCs from the greywater, which represents real hazards in the greywater.

7.5 Removal of XOCs from Greywater

The techniques used in the treatment of wastewater which include filtration systems with aerobic/anaerobic biological treatment unit, flocculation and ultrafiltration are the most effective in comparison to other systems in removing pollutants (Mohamed et al. 2016). However, these systems have higher capital, operations and maintenance costs. Some trends of designing a filtration system of greywater from natural and low-cost materials have been investigated during the past few years. These systems have exhibited potentials to pollutant removal as well as low-cost and simple operation (Mohamed et al. 2014). Nonetheless, these systems are classified as primary techniques for removing the main parameters of greywater such as BOD, COD and TSS, but they are not advanced techniques for degradation of XOCs and nutrients (Aracagok and Cihangir 2013). Schäfer et al. (2006) studied the efficiency of ultra-filtration (UF) of greywater in removing bisphenol A (BPA). The results found that the maximum removal of BPA by UF was 30–45%. The phytoremediation process by using microalgae is an effective technology for the removal of nutrients (Jais et al. 2017). However, it has no potential to degrade the XOCs compounds due to the absence of hydrolysis enzymes such as peroxidase and laccase enzymes. Hence, an alternative design for an effective degradation of XOCs is required for producing high quality of the treated greywater. The natural attenuation of the XOCs has been reported by Baun et al. (2003), who mentioned that the natural attenuation of XOCs took place under strongly anaerobic conditions in the leachate. However, the levels of degradation for chloride and non-volatile organic carbon (NVOC) were low. These findings suggested that XOCs are not completely degradable in nature. Therefore, the XOCs should be eliminated from the wastewater before the final disposal.

A new generation of sustainable treatment technology for greywater with low maintenance requirements and high treatment efficiency is needed to minimise the health risk associated with the greywater in terms of XOCs (Chong et al. 2015). Fungi might be the alternative organisms because they are the saprophytic organism, which produces peroxidase and laccase enzymes specified in the degradation of XOCs. Fungi break down the organic compounds and then utilise it as a carbon and energy source. Most of the studies conducted on the removal of XOCs focused on the removal of PPCPs from the wastewater with very few studies performed on the greywater. Therefore, this section scrutinises the most recent methods which have been investigated for the removal of XOCs from wastewater and their efficiency, advantages and disadvantages.

7.5.1 Physical Removal of XOCs

The conventional treatment processes such as sand filtration, coagulation/sedimentation and oxidation process by chlorination and ozonation which are performed for the treatment of wastewater might contribute to the removal of XOCs. The studies confirmed that the efficiency of these technologies might reach to more than 80% for some XOCs such as acetaminophen, diclofenac and sulfamethazine, but are insufficient for others such as metoprolol, carbamazepine, caffeine and bisphenol-A (Nam et al. 2014). In the sewage treatment plants, the removal of XOCs is due to the adsorption of the generated sludge. It has been reported that the efficiency of conventional treatment process of sewage removed naproxen, ibuprofen, diclofenac, ketoprofen, sulfamethoxazole, bezafibrate, and trimethoprim ranged from zero to 100% dependent on the characteristics of the activated sludge (Carballa et al. 2004; Lindqvist et al. 2005; Kasprzyk-Hordern et al. 2009). However, it has also to be mention that the treatment process of greywater does not lead to producing sludge. Therefore, the removal efficiency might be undetectable. Indeed, there is no sufficient information on the efficiency of these technologies for removing XOCs from greywater. However, in a view for their removal mechanism, it can be noted that the effectiveness of the treatment system for removing XOCs depends on the chemical structure and oxidabilities as well as the treatment conditions.

The most common adsorbent is activated carbon (AC) in form of powdered activated carbon (PAC) and granular-activated carbon (GAC). Both forms of activated carbon exhibited an efficiency nearly of 90% and more of diclofenac, carbamazepine, caffeine, trimethoprim, bisphenol A, nonylphenol and triclosan (Grover et al. 2011; Hernández-Leal et al. 2011; Yang et al. 2011; Kovalova et al. 2013). The mechanism of adsorption process to remove XOCs relies on the physical interaction of electrostatic (van der Waals interactions) which takes place in very short times. The optimal pH for the adsorption process is located within 6–7, the temperature might induce or reduce the adsorption process. The explanation for the effect of temperature on the adsorption process is inconsistent among different studies. It is due to the concept of adsorption process as endothermic or exothermic reactions. Habib-ur-Ruhman et al. (2006) had claimed that the adsorption process an endothermic process. Therefore, the increase of temperature induces the removal efficiency. In contrast, Selatnia et al. (2004) had reported that the adsorption increased with the decrease of temperature because the process is a physical adsorption reaction which is defined as exothermic.

Many factors which affect the adsorption process are the nature of XOCs as hydrophilic or hydrophobic compounds and their affinity to the adsorbent in a competition with dissolved organic matter (DOM) (Nam et al. 2014). The hydrophilic XOCs dissolved in the water make their functional groups more able to form electrostatic bonds with the adsorbent. Several studies have also indicated that the pretreatment of adsorbent by chemical or physical process might enhance the adsorption process of XOCs (Al-Gheethi et al. 2016, 2017). Nonetheless, there are two disadvantages for applying CA in the adsorption process to remove XOCs which include the high cost of CA and the deterioration in the wastewater (Wang and Wang 2016). Regarding the high cost of CA, the utilisation of low-cost biomasses prepared from the natural materials and microorganisms such as bacteria, fungal and microalgae cells might provide the alternative absorbents (Al-Gheethi et al. 2015b). In the past few years, there was an interest to use carbon nanotubes (CNT) which have the large surface area with high adsorption capacity for XOCs such as pharmaceuticals and personal care products (PPCPs). The efficiency of CNT in removing carbendazim, triclosan, carbamazepine, ibuprofen, ketoprofen, acetaminophen, sulfamethoxazole, carbamazepine, caffeine, 4-acetylamino-antipyrine and their properties have been widely reported (Ji et al. 2009; Liu et al. 2014; Wang et al. 2016).

The utilisation of natural zeolite in the removal of XOCs has also been reported. The efficiency of these materials belongs to the net negative structural charge available on the large specific surface area. Among the XOCS removed by natural zeolite are the cationic surfactants hexadecyltrimethylammonium (HDTMA), tetramethylammonium, octadecyl-trimethylammonium, ethyl hexadecyl-dimethylammonium, 4-methylpyridinium and cetylpyridinium (Schulze-Makuch et al. 2002; Schulze-Makuch et al. 2004).

The main challenges in the adsorption techniques for removal of XOCs are the less affinity and weak competition for these compounds to the adsorbent in comparison with other contaminants in the wastewater such as heavy metals. Al-Gheethi et al. (2016) investigated the removal of cephalexin compounds from the effluents by consortium bacterial biomass and revealed that the removal percentage was more than 90% at the low concentration of cephalexin. However, the efficiency was reduced to less than 20% in the presence of heavy metal ions. Many studies have been performed on the removal of XOCs by the adsorption processes conducted on few numbers of XOCs. Hence, more studies should be conducted on different types of XOCs to evaluate the adsorption techniques for removing of XOCs from the wastewater, since the adsorption of XOCs depends mainly on their physiochemical properties. For instance, the XOCs with high molecular weight have less affinity to be adsorbed compared with the micro-molecular pollutants such as heavy metals.

7.5.2 Chemical Removal of XOCs

The removal of XOCs takes place via chemical and oxidation processes, therefore, the aerobic treatment of greywater appeared to be more efficient for removing these compounds in comparison to the anaerobic. It has been stated that the aerobic treatment of greywater contaminated with XOCs has achieved 90% of COD removal and 97% of anionic surfactants after 12 h and at 32 °C. In comparison, the anaerobic treatment conducted at the same conditions removed only 51% of COD and 24% of anionic surfactants. On the other hand, the combined anaerobic–aerobic system has not given an advantage in comparison with the aerobic treatment (Hernandez Leal et al. 2010a, b)

The coagulation and flocculation process is one of the methods investigated for removal of XOCs from the water. This process depends on using chemical coagulants such as aluminium sulphate ($Al_2(SO_4)_3$ and ferric chloride ($FeCl_3$) which are ionised in the water into ions with the positive charge and then interacting electrostatically with the negative charges of XOCs. The electrical aggregation used in the coagulation and flocculation process takes place in response to pH and dose. Several studies have revealed that the removal percentage of XOCs by the coagulation process has ranged from less than that reported for removing of sulfamethoxazole by $Al_2(SO_4)_3$ at 25 mg L^{-1} and pH to more than 90% for nonylphenol by $FeCl_3$ at 100, 200 mgL^{-1} and pH 4, 7 and 9 (Suárez et al. 2009; Asakura and Matsuto 2009). It has to be mention that the coagulation–flocculation processes are not removing the XOCs by a degradation process. It depends mainly on the chemical reaction between coagulants and XOCs to generate a sludge which should be considered for further treatment before the final disposal into the environment due to the presence of active XOCs.

The chemical oxidation processes are mainly used for the disinfection of water and wastewater and inactivation of pathogens in that environment. The chemical oxidation process acts by releasing free radicals, which are high oxidative agents and lead to destruct or damage the microorganism cell contents. Therefore, the effectiveness in the degradation of XOCs has been investigated. Chlorine molecules generated from the chlorination of wastewater recorded higher reactivity to aromatic compounds such as β-lactam antibiotics, and higher electro-affinity for XOCs functional groups (Kosjek and Heath 2008). However, the chlorination technique has less efficiency towards several types of XOCs during degradation. This technique is easily applied, readily available and cheaper than other oxidising agents. Nevertheless, the chlorination limits for the degradation of XOCs are the toxicity of free and combined chlorine residues for the aquatic organisms. Besides, the potential health hazards of nitrosodimethylamine (NDMA) which is generated secondary by-products for the chlorination and has been reported as a probable human carcinogen (Pehlivanoglu-Mantas et al. 2006). Chlorine dioxide (ClO_2) is used as an alternative to chlorine, because ClO2 does not form chlorogenic and chloramines compounds as toxic by-products. Despite this, the compound has higher oxidation effects than chlorine and it can act within a wide range of pH (Hofmann et al. 1999; Junli et al. 1997). However, the toxicity of ClO_2 on the biodiversity in the environment has been immensely reported in the literature. Chhetri et al. (2016) assessed the toxic effect of ClO_2, peracetic acid (PAA) and performic acid (PFA) as well as chlorite and hydrogen peroxide (H_2O_2) which are by-products for these disinfectants on *Pseudokirchneriella* sub.*capitate*. The study has revealed the toxicity of the investigated disinfectants to *P.* sub *capitate*. The EC50 value, which is defined as the concentration of a drug that gives a half-maximal response, ranging from 0.16 to 2.9 mg/L. More toxicity was recorded for chlorine dioxide followed by performic acid, peracetic acid and chlorite. The minimum effects were noted for hydrogen peroxide. Moreover, the study revealed that the stability of disinfectants reduces as the toxicity increases, which does mean that the ClO2 is more applicable for the disinfection process due to biodegradability in the environment.

The degradation of XOCs such as sulphamethoxazole in wastewater by ozonation has also been investigated. Yargeau and Leclair (2007) revealed that the ozonation process of wastewater removed 99.24% of XOCs within 5 min of the treatment. Lajeunesse et al. (2013) indicated that the ozonation achieved 92% of removal for paroxetine, fluoxetine, fluvoxamine, sertraline, desmethyl mirtazepine, mirtazapine, O-desmethylvenlafaxine, desmethylsertraline, carbamazepine and nortriptyline after 5 min of the treatment with 5 mg L^{-1} of ozone. The study also revealed that the removal efficiency for other compounds such as citalopram, venlafaxine and amitriptyline were less than 50%. However, the findings of the study has indicated that the removal efficiency depends mainly on the concentrations of ozone, the high concentrations (13 mg L^{-1}) achieved more than 94% of the removal for most investigated compounds. Sui et al. (2010) revealed that the ozonation recorded 95% degradation of diclofenac, carbamazepine, trimethoprim and indomethacin, with 5 mg/L of dosage. Regardless, the efficiency of ozonation in the degradation of XOCs, the concerns for application of ozone lies in the production of toxic by-products (Joss et al. 2008). Another limitation for ozonation is the biological regrowth in the wastewater due to increase assimilation of organic carbon (AOC) by ozonation (Vital et al. 2010).

Fenton oxidation is another technique which has been investigated for the degradation of XOCs in the wastewater. This technique uses iron salts and hydrogen peroxide under acidic conditions to generate hydroxyl radicals and then degrade XOCs. Some of XOCs which has been degraded in the wastewater by Fenton oxidation include ofloxacin (100%), bezafibrate (100%), diclofenac (20–100%), oxazepam (10%) and diazepam (100%) (Bautitz and Nogueira 2010; Bae et al. 2013; Mackulak et al. 2015). However, the environmental fate of by-products of Fenton oxidation is still unknown and need more studies.

The gamma irradiation is an ionising irradiation technology exhibited an efficiency technology for degradation of XOCs. The gamma irradiation depends on the hydroxylation, demethylation and cleavage of the chemical structure of XOCs. The application of this technology has been evaluated in the aqueous solution (Chu and Wang 2016). Nonetheless, the presence of intermediate products during the gamma irradiation treatment of XOCs has been detected by HPLC/MS. The toxicity of these compounds has not been investigated. Therefore, before considering gamma irradiation as the alternative technology for the degradation of XOCs, it has to assess the presence or absence the toxicity of the intermediate products.

The solar treatment for wastewater might have efficiency in the degradation of XOCs based on the photo-oxidation and photocatalysis processes. The use of photocatalytic treatment of XOCs such as dye and their derivatives, surfactants and fragrances compounds seem to be promising (Santiago-Morales et al. 2012). Grčić et al. (2015) reported the possibility of the solar photocatalysis and flocculation for removal of hair colourants from the household greywater. In this system, the greywater was placed in a reactor has the photocatalytic layer which was consisted of TiO_2-coated textile fibres and exposed to direct sunlight. The analysis of treated samples by using UV-Vis spectra recorded complete degradation of dye molecules within 4 h, with a significant reduction of the emulsifying and surfactant compounds. The flocculation process of the treated greywater revealed flocs formed during the photocatalytic and a complete sedimentation within 12 h. The analysis of photocatalytic layer surface by thermogravimetric analysis (TGA), Fourier transform infrared spectroscopy (FTIR), scanning electron microspore (SEM), Raman spectroscopy and contact angle measurements provided an evidence of Ti(III) species. The study indicated that the solar photocatalysis is an efficient pretreatment step for degradation of XOCs in the greywater. The quantities of the flocs were less and has no toxicity compared to that generated when coagulation by the chemical substances is used.

The mechanism by which solar radiation degrades the XOCs is explained as a result of the destruction of chemical bonds by UV waves, which is called photolysis (Wang and Wang 2016). However, some of XOCs are resistant to UV destruction. Therefore, one option introduced by the researchers for the degradation of XOCs in the wastewater is the combination of oxidative agents and UV which is called advanced oxidation process (AOPs). Several of AOPs are used in the degradation of XOCs such as UV/H_2O_2, O_3/UV, $O_3/OH-$, $UV-A/Fe-2/H_2O_2$, $Fe(II)/H_2O_2$

and (TiO$_2$/UV, ZnO/UV) exhibited high efficiency in the degradation of XOCs (Klavarioti et al. 2009; Postigo and Richardson 2014; Rozas et al. 2016). The previous studies which have been performed on the degradation of XOCs by oxidation and UV processes in different water systems are presented in Table 7.2.

Chong et al. (2015) evaluated TiO$_2$ photocatalytic technology as a standalone system for the degradation kinetics of Reactive Black 5 (RB5) dye in real and synthetic greywater. The study optimises the best operating parameters for removing of RB5 based on the concentration of TiO$_2$ and RB5 as well as the pH. The maximum removal was 97% in synthetic greywater at 1 ppm of RB5, 0.1 g L^{-1} of TiO$_2$ and pH 5 achieved after 150 min of the photocatalytic reaction in synthetic greywater effluent. The photocatalytic degradation kinetics of RB5 dye studies showed that the degradation process in the real greywater was 76% after 330 min of the reaction. TiO$_2$ photocatalytic technology acts by forming electron–hole pairs on TiO$_2$ surfaces via a series of redox reactions, which lead to activate the TiO$_2$ surfaces. Subsequently, the reactive hydroxyl radicals degrade XOCs without generating secondary products, since the XOCs are completely degraded rather than adsorbed or phase transferred (Chong et al. 2010). So far, the study concluded by Chong et al. (2015) stated that the presence of other contaminants in the real greywater effluent effects negatively on the affinity of RB5 to the surface binding sites of TiO$_2$ photocatalytic reaction. Among the pollutants in the greywater, *E. coli* has reduced the efficiency of TiO$_2$ photocatalytic reaction for degrading of RB5 dye. It can be indicated that the removal of the main pollutant should be the first step in order to achieve high degradation of XOCs such as dye in the greywater.

Rozas et al. (2016) employed UV and UV/H$_2$O$_2$ for the transformation of carbamazepine (CBZ) (antiepileptic), atrazine (ATZ), triclosan (TCS) and diclofenac (DCL) in river water. The experiments were conducted at laboratory scale. Different UV doses (100, 200, 300, 600 and 900 mJ cm^{-2}) and the river water matrix effects were investigated. The initial H$_2$O$_2$ concentrations used in the study was 10 mg L^{-1}. The bioassays and toxicity tests were carried out with Daphnia Magna. The study revealed that the optimal UV dosage was 900 mJ cm^{-2} which high transformation profile for DCL, TCS, ATZ and CBZ. The findings also indicated that at 300 mJ cm-2 of UV dosage, a higher toxicity was noted. This might be due to the formation of toxic intermediates during the treatment process. However, the high UV doses reduced the toxicity level of the tested compounds.

The AOPs have high potential to transform of XOCs to hydrophilic substances with similar or lower molecular weights and lower biological activity (Altmann et al. 2014). AOPs might destroy the carcinogenic compounds generated by using the chemical disinfects (chlorination and ozonation). It has been mentioned that the high dose of UV might destroy N-nitrosodimethylamine (NDMA) formed as an intermediate product for the ozonation (Schmidt and Brauch 2008). Therefore, the combination of AOPs and disinfectants might provide efficient and alternative methods for degradation of XOCs in the wastewater. However, there are some disadvantages including the high required energy and toxic by-products which need additional post-treatment processes to be removed before final disposal into the environment (Kim and Zoh 2016).

Table 7.2 Previous studies conducted on the degradation of XOCs by advanced oxidation processes (AOPs) in different water systems

XOCs compound	Medium	Treatment process	Degradation percentage (%)	References
Diclofenac	Water	UV dose of 40 mJ cm^{-2}	27	Canonica et al. (2008)
Carbamazepine and diclofenac	Natural water	UV dose of 230 mJ cm^{-2}	8 and 100	Kim and Tanaka (2009)
Carbamazepine		1700 mJ cm^{-2} of UV dosage and 10 mg L^{-1} of H$_2$O$_2$	100	Pereira et al. (2011)
Atrazine	Surface water	UV and UV/H$_2$O$_2$ (UV dose of 1500 mJ cm^{-2})	60	Sanches et al. (2010)
Triclosan	Grade water	UV Dose of 700 mJ cm^{-2}	90	Carlson et al. (2015)
Triclosan	Wastewater	UV dosage 550 w/m^2, H$_2$O$_2$ ¼ 50 mg/l	100	De la Cruz et al. (2012)
Atrazine (ATZ), Triclosan (TCS) and Diclofenac (DCL)	River water	UV and UV/H$_2$O$_2$ (UV dosage, 100 mJ cm^{-2} and 10 mg L^{-1}-of H$_2$O$_2$	50	Rozas et al. (2016)
Trace organic contaminants (TOrCs), total estrogenicity	Effluents	O$_3$/H$_2$O$_2$ < 0.50 ~ < 25 ng/L	>90	Gerrity et al. (2011)
Ketoprofen, diclofenac and antipyrine	Aqueous solution	UV (254 nm)	90	Kim et al. (2009)
Metoprolol		UV/O$_3$	84	Šojić et al. (2012)

7.6   Conclusion

Technologies are varied in their efficiency and effectiveness to produce high quality of treated greywater. However, the main concern in the greywater is the occurrence of XOCs due to the toxicity of these compounds and the difficulties in their removal by using traditional methods. Therefore, researchers have shifted to find an effective and low-cost technology to remove or degrade the XOCs into simple substance without toxic by-products or secondary effects on the environment which receives the treated greywater. The AOPs appear to be the effective method for the degradation of XOCs, but there are still some concerns about these techniques. Therefore, more attention is still required in order to find an appropriate process has the simple requirements, footprints and effective.

Acknowledgements The authors wish to thank the Ministry of Higher Education (MOHE) for supporting this research under FRGS vot 1574 and also the Research Management Centre (RMC) UTHM for providing grant IGSP U682 for this research.

References

Abdel-Shafy HI, Al-Sulaiman AM, Mansour MS (2014) Greywater treatment via hybrid integrated systems for unrestricted reuse in Egypt. J Water Process Eng 1:101–107

Abu Ghunmi L (2009) Characterization and treatment of grey water; options for (re)use. Ph.D. thesis, Wageningen University, 200 p

Al-Gheethi AA, Norli I, Lalung J, Azieda T, Ab. Kadir MO (2013) Reduction of faecal indicators and elimination of pathogens from sewage treated effluents by heat treatment. Caspian J Appl Sci Res 2(2):29–45

Al-Gheethi AA, Lalung J, Noman EA, Bala JD, Norli I (2015a) Removal of heavy metals and antibiotics from treated sewage effluent by bacteria. Clean Technol Environ Policy 17(8):2101–2123

Al-Gheethi AA, Norli I, Efaq AN, Bala JD, Al-Amery RM (2015b) Solar Disinfection and lime treatment processes for reduction of pathogenic bacteria in sewage treated effluents and biosolids before reuse for agriculture in Yemen. Water Reuse Des 5(3):419–429

Al-Gheethi AA, Efaq AN, Mohamed RM, Norli I, Kadir MO (2016) Potential of bacterial consortium for removal of cephalexin from aqueous solution. J Ass Arab Univ Basic Appl Sci. Online

Al-Gheethi AA, Efaq AN, Mohamed RM, Norli Ismail, Ab Kadir MO (2017) Removal of heavy metals from aqueous solution using Bacillus subtilis biomass pretreated by supercritical carbon dioxide. CLEAN—Soil, Air, Water (Accepted)

Allen L, Christian-Smith J, Palaniappan M (2010) Overview of greywater reuse: the potential of greywater systems to aid sustainable water management. Pac Inst 654

Al-Mughalles MH, Rahman RA, Suja FB, Mahmud M, Abdullah SMS (2012) Greywater treatment using GAC biofilm reactor and sand filter system. Aus J Basic Appl Sci 6(3):283–292

Altmann J, Ruhl AS, Zietzschmann F, Jekel M (2014) Direct comparison of ozonation and adsorption onto powdered activated carbon for micropollutant removal in advanced wastewater treatment. Water Res 55:185–193

Aracagok YD, Cihangir N (2013) Decolorization of reactive black 5 by yarrowialipolytica NBRC 1658. Am J Microbiol Res 1:16–20

Asakura H, Matsuto T (2009) Experimental study of behavior of endocrine-disrupting chemicals in leachate treatment process and evaluation of removal efficiency. Waste Manage 29:1852–1859

Bae S, Kim D, Lee W (2013) Degradation of diclofenac by pyrite catalyzed Fenton oxidation. Appl Catal B Environ 134:93e102

Barca C, Meyer D, Liira M, Drissen P, Comeau Y, Andrès Y, Chazarenc F (2014) Steel slag filters to upgrade phosphorus removal in small wastewater treatment plants: removal mechanisms and performance. Ecol Eng 68:214–222

Barışçı S, Turkay O (2016) Domestic greywater treatment by electrocoagulation using hybrid electrode combinations. J Water Process Eng 10:56–66

Baun A, Reitzel LA, Ledin A, Christensen TH, Bjerg PL (2003) Natural attenuation of xenobiotic organic compounds in a landfill leachate plume (Vejen, Denmark). J Cont Hydrol 65(3):269–291

Bautitz IR, Nogueira RFP (2010) Photodegradation of lincomycin and diazepam in sewage treatment plant effluent by photo-Fenton process. Catal Today 151(1):94–99

Bratieres K, Fletcher T, Deletic A, Zinger Y (2008) Nutrient and sediment removal by stormwater biofilters: a large-scale design optimisation study. Water Res 42(14):3930–3940

Canonica S, Meunier L, von Gunten U (2008) Phototransformation of selected pharmaceuticals during UV treatment of drinking water. Water Res 42:121–128

Carballa M, Omil F, Lema JM, Llompart MA, García-Jares C, Rodríguez I, Gomez M, Ternes T (2004) Behavior of pharmaceuticals, cosmetics and hormones in a sewage treatment plant. Water Res 38(12):2918–2926

Carlson JC, Stefan MI, Parnis JM, Metcalfe CD (2015) Direct UV photolysis of selected pharmaceuticals, personal care products and endocrine disruptors in aqueous solution. Water Res 84:350–361

Chhetri RK, Baun A, Andersen HR (2016) Algal toxicity of the alternative disinfectants performic acid (PFA), peracetic acid (PAA), chlorine dioxide (ClO_2) and their by-products hydrogen peroxide (H_2O_2) and chlorite (ClO_2). Int J hygiene Environ Health 220(3):570–574

Chong M, Jin B, Chow C, Saint C (2010) Recent developments in photocatalytic water treatment technology: a review. Water Res 44(10):2997–3027

Chong MN, Cho YJ, Poh PE, Jin B (2015) Evaluation of Titanium dioxide photocatalytic technology for the treatment of reactive Black 5 dye in synthetic and real greywater effluents. J Cleaner Production 89:196–202

Chu LB, Wang J (2016) Degradation of 3-chloro-4-hydroxybenzoic acid in biological treated effluent by gamma irradiation. Radiat Phys Chem 119:194–199

Dalahmeh S, Jönsson H, Hylander LD, Hui N, Yu D, Pell M (2014) Dynamics and functions of bacterial communities in bark, charcoal and sand filters treating greywater. Water Res 54:21–32

De la Cruz N, Gimenez J, Esplugas S, Grandjean D, De Alencastro LF, Pulgarin C (2012) Degradation of 32 emergent contaminants by UV and neutral photo-fenton in domestic wastewater effluent previously treated by activated sludge. Water Res 46(6):1947–1957

Englert AH, Rubio J (2005) Characterization and environmental application of a Chilean natural zeolite. Int J Min Process 75(1):21–29

Fowdar HS, Hatt BE, Breen P, Cook PL, Deletic A (2017) Designing living walls for greywater treatment. Water Res 110:218–232

Garcia JE, Gonzalez MM, Notario JS (1992) Removal of bacterial indicators of pollution and organic matter by phillipsite-rich tuff columns. Appl Clay Sci 7(4):323–333

Gerrity D, Gamage S, Holady JC, Mawhinney DB, Quiñones O, Trenholm RA, Snyder SA (2011) Pilot-scale evaluation of ozone and biological activated carbon for trace organic contaminant mitigation and disinfection. Water Res 45(5):2155–2165

Ghaly MY, Farah JY, Fathy AM (2007) Enhancement of decolorization rate and COD removal from dyes containing wastewater by the addition of hydrogen peroxide under solar photocatalytic oxidation. Desalination 217(1–3):74–84

Grčić I, Vrsaljko D, Katančić Z, Papić S (2015) Purification of household greywater loaded with hair colorants by solar photocatalysis using TiO 2-coated textile fibers coupled flocculation with chitosan. J Water Process Eng 5:15–27

Gross A, Shmueli O, Ronen Z, Raveh E (2007) Recycled vertical flow constructed wetland (RVFCW)—a novel method of recycling greywater for irrigation in small communities and households. Chemosphere 66(5):916–923

Grover D, Zhou J, Frickers P, Readman J (2011) Improved removal of estrogenic and pharmaceutical compounds in sewage effluent by full scale granular activated carbon: Impact on receiving river water. J Hazard Mater 185:1005–1011

Habib-ur-Rehman SM, Ahmad I, Shah S (2006) Sorption studies of nickel ions onto saw dust of Dalbergiasissoo. J Chin ChemSoc 53:1045–1052

Halalsheh M, Dalahmeh S, Sayed M, Suleiman W, Shareef M, Mansour M, Safi M (2008) Grey water characteristics and treatment options for rural areas in Jordan. Biores Technol 99(14):6635–6641

Hernandez Leal L, Temmink H, Zeeman G, Buisman C (2010a) Comparison of three systems for biological greywater treatment. Water 2(2):155–169

Hernandez Leal L, Vieno N, Temmink H, Zeeman G, Buisman C (2010b) Occurrence of xenobiotics in gray water and removal in three biological treatment systems. Environ Sci Technol. https://doi.org/10.1021/es101509e

Hernández-Leal L, Temmink H, Zeeman G, Buisman C (2011) Removal of micropollutants from aerobically treated grey water via ozone and activated carbon. Water Res 45:2887–2896

Hofmann R, Andrews RC, Ye Q (1999) Impact of Giardia inactivation requirements on ClO2 by-products. Environ Technol 20:147–158

Hydra International Ltd, (2010) Bio Start; Bacteria/Enzyme start up treatment for sewage treatment plants. http://www.hydra-bio.com/MSDS/Bio_Start.pdf. Date 11.7.2017

Itayama T, Kiji M, Suetsugu A, Tanaka N, Saito T, Iwami N, Inamori Y (2006) On site experiments of the slanted soil treatment systems for domestic gray water. Water Sci Technol 53(9):193–201

Jaafar F, Michałowski S (1990) Modified BET equation for sorption/desorption isotherms. Drying Technol 8(4):811–827

Jais NM, Mohamed R, Al-Gheethi A, Hashim MA (2017) The dual roles of phycoremediation of wet market wastewater for nutrients and heavy metals removal and microalgae biomass production. Clean Technol Environ Policy 19(1):37–52

Jefferson B, Burgess JE, Pichon A, Harkness J, Judd SJ (2001) Nutrient addition to enhance biological treatment of grey water. Water Res 35(11):2702–2710

Ji L, Chen W, Zheng S, Xu Z, Zhu D (2009) Adsorption of sulfonamide antibiotics to multiwalled carbon nanotubes. Langmuir 25(19):11608–11613

Joss A, Siegrist H, Ternes TA (2008) Are we about to upgrade wastewater treatment for removing organic micropollutants? Water Sci Technol 57:251–255

Junli H, Li W, Nanqi R, Fang M, Juli Huang J, Wang L, Ren N, Ma F (1997) Disinfection effect of chlorine dioxide on bacteria in water. Water Res 31:607–613

Kallo D, Ming W (2001) Applications of Natural Zeolites in Water and Wastewater Treatment, Mineralogical Society of America: Washington, USA. 2001, 654 p

Kasprzyk-Hordern B, Dinsdale RM, Guwy AJ (2009) The removal of pharmaceuticals, personal care products, endocrine disruptors and illicit drugs during wastewater treatment and its impact on the quality of receiving waters. Water Res 43(2):363–380

Kim I, Tanaka H (2009) Photodegradation characteristics of PPCPs in water with UV treatment. Environ Int 35:793–802

Kim JW, Jang HS, Kim JG, Ishibashi H, Hirano M, Nasu K, Arizono K (2009) Occurrence of pharmaceutical and personal care products (PPCPs) in surface water from Mankyung River, South Korea. J Health Sci 55(2):249–258

Kim MK, Zoh KD (2016) Occurrence and their removal of micropollutants in water environment. Environ, Eng Res in press

Klavarioti M, Mantzavinos D, Kassinos D (2009) Removal of residual pharmaceuticals from aqueous systems by advanced oxidation processes. Environ Int 35:402–417

Köhler SJ, Cubillas P, Rodríguez-Blanco JD, Bauer C, Prieto M (2007) Removal of cadmium from wastewaters by aragonite shells and the influence of other divalent cations. Environ Sci Technol 41(1):112–118

Kosjek T, Heath E (2008) Applications of mass spectrometry to identifying pharmaceutical transformation products in water treatment. TrAC-Trend. Anal Chem 27:807–820

Kovalova L, Siegrist H, von Gunten U, Eugster J, Hagenbuch M, Wittmer A (2013) Elimination of micropollutants during post-treatment of hospital wastewater with powdered activated carbon, ozone, and UV. Environ Sci Technol 47:7899–7908

Lajeunesse A, Blais M, Barbeau B, Sauvé S, Gagnon C (2013) Ozone oxidation of antidepressants in wastewater-treatment evaluation and characterization of new by-products by LC-QToFMS. Chem Cent J 7:1–15

Lalander C, Dalahmeh S, Jönsson H, Vinnerås B (2013) Hygienic quality of artificial greywater subjected to aerobic treatment: a comparison of three filter media at increasing organic loading rates. Environ Technol 34:2657–2662

Leal LH (2010) Removal of micropollutants from grey water: combining biological and physical/chemical processes. Ph.D Thesis, Wageningen University, Netherland. ISBN 978-90-8585-701-3

Li FY, Wichmann K, Otterpohl R (2009) Review of the technological approaches for grey water treatment and reuses. Sci Total Environ 407(11):3439–3449

Lindqvist N, Tuhkanen T, Kronberg L (2005) Occurrence of acidic pharmaceuticals in raw and treated sewages and in receiving waters. Water Res 39(11):2219–2228

Liu F, Zhao J, Wang S, Du P, Xing B (2014) Effects of solution chemistry on adsorption of selected pharmaceuticals and personal care products (PPCPs) by graphenes and carbon nanotubes. Environ Sci Technol 48(22):13197–13206

Mackulak T, Mosný M, Grabic R, Golovko O, Koba O, Birosova L (2015) Fenton-like reaction: a possible way to efficiently remove illicit drugs and pharmaceuticals from wastewater. Environ Toxicol Pharmacol 39(2):483–488

Merz C, Scheumann R, El Hamouri B, Kraume M (2007) Membrane bioreactor technology for the treatment of greywater from a sports and leisure club. Desalination 215(1–3):37–43

Mohamed R, Al-Gheethi A, Miau JA, Kassim AM (2016) Multi-component filters for domestic graywater treatment in village houses. J Am Water Works Ass 108(7):E405–E415

Mohamed R, Wurochekke A, Chan C, Kassim AH (2014) The use of natural filter media added with peat soil for household greywater treatment. Int J Eng Technol 2:33–38

Motz H, Geiseler J (2001) Products of steel slags an opportunity to save natural resources. Waste Manage 21(3):285–293

Nam SW, Jo BI, Yoon Y, Zoh KD (2014) Occurrence and removal of selected micropollutants in a water treatment plant. Chemosphere 95:156–165

Nunez YR (1998) Removal of phosphates from water using tailored zeolites. University of Puerto Rico, San Juans

Payne EG, Hatt BE, Deletic A, Dobbie MF, McCarthy DT, Chandrasena GI (2015) Adoption guidelines for stormwater biofiltration systems—summary report. Coop Res Centre Water Sensitive Cities, Melbourne, Australia

Pehlivanoglu-Mantas E, Elisabeth L, Hawley R, Deeb A, Sedlak DL (2006) Formation of nitrosodimethylamine (NDMA) during chlorine disinfection of wastewater effluents prior to use in irrigation systems. Water Res 40(2):341–347

Pereira RO, de Alda ML, Joglar J, Daniel LA, Bardeló D (2011) Identification of new ozonation disinfection byproducts of 17β-estradiol and estrone in water. Chemosphere 84:1535–1541

Pérez-Urrestarazu L, Fernández-Cañero R, Franco-Salas A, Egea G (2015) Vertical greening systems and sustainable cities. J Urban Technol 22(4):65–85

Pidou M, Avery L, Stephenson T, Jeffrey P, Parsons SA, Liu S, Jefferson B (2008) Chemical solutions for greywater recycling. Chemosphere 71(1):147–155

Plappally A, Chen H, Ayinde W, Alayande S, Usoro A, Friedman KC, Malatesta K (2011) A field study on the use of clay ceramic water filters and influences on the general health in Nigeria. J Health Behavior Public Health 1(1):1–14

Postigo C, Richardson SD (2014) Transformation of pharmaceuticals during oxidation/disinfection processes in drinking water treatment. J Hazard Mater 279:461–475

Ronen Z, Guerrero A, Gross A (2010) Grey water disinfection with the environmentally friendly Hydrogen Peroxide Plus. Chemosphere 78:61–65

Rozas O, Vidal C, Baeza C, Jardim WF, Rossner A, Mansilla HD (2016) Organic micropollutants (OMPs) in natural waters: oxidation by UV/H_2O_2 treatment and toxicity assessment. Water Res 98:109–118

Sanches S, Barreto Crespo MT, Pereira VJ (2010) Drinking water treatment of priority pesticides using low pressure UV photolysis and advanced oxidation processes. Water Res 44:1809–1818

Santiago-Morales J, Gomez M, Herrera S, Fernandez-Alba A, Garcia-Calvo E, Rosal E (2012) Oxidative and photochemical processes for the removal of galaxolide and tonalide from wastewater. Water Res 46:4435–4447

Saroj BP, Mukund GS (2011) Performance of grey water treatment plant by economical way for Indian rural development. Int J Chem Tech Res, CODEN. USA, 3(4):1808–1815

Schäfer AI, Nghiem LD, Oschmann N (2006) Bisphenol A retention in the direct ultrafiltration of greywater. J Membrane Sci 283(1):233–243

Schmidt CK, Brauch HJ (2008) N, N-Dimethylsulfamide as precursor for N-Nitrosodimethylamine (NDMA) formation upon ozonation and its fate during drinking water treatment. Environ Sci Technol 42:6340–6346

Schneider L (2009) Grey water reuse in Washington state. Rule development committee issue research report final, Washington State Department of Health-Wastewater Management Program, pp 1–16

Schulze-Makuch D, Bowman RS, Pillai S (2004) Removal of biological pathogens using surfactant-modified zeolite. USA

Schulze-Makuch D, Pillai SD, Guan H, Bowman R, Couroux E, Hielscher F, Kretzschmar T (2002) Surfactant-modified zeolite can protect drinking water wells from viruses and bacteria. EOS Trans Am Geophys Union 83(18):193–201

Selatnia A, Boukazoula A, Kechid N, Bakhti MZ, Chergui A, Kerchich Y (2004) Biosorption of lead (II) from aqueous solution by a bacterial dead Streptomyces rimosus biomass. Biochem Eng J 19(2):127–135

Shafiquzzaman M, Azam MS, Nakajima J, Bari QH (2011) Investigation of arsenic removal performance by a simple iron removal ceramic filter in rural households of Bangladesh. Desalination 265(1):60–66

Šojić D, Despotović V, Orčić D, Szabó E, Arany E, Armaković S, Sajben-Nagy E (2012) Degradation of thiamethoxam and metoprolol by UV, O_3 and UV/O_3 hybrid processes: kinetics, degradation intermediates and toxicity. J Hydrol 472:314–327

Stevik TK, Ausland G, Jenssen PD, Siegrist RL (1999) Removal of E. coli during intermittent filtration of wastewater effluent as affected by dosing rate and media type. Water Res 33(9):2088–2098

Suárez S, Lema JM, Omil F (2009) Pre-treatment of hospital wastewater by coagulation–flocculation and flotation. Bioresour Technol 100:2138–2146

Sui Q, Huang J, Deng S, Yu G, Fan Q (2010) Occurrence and removal of pharmaceuticals, caffeine and DEET in wastewater treatment plants of Beijing, China. Water Res 44:417–426

Tal T, Sathasivan A, Krishna KB (2011) Effect of different disinfectants on grey water quality during storage. J Water Sustain 1(1):127–137

Tchamango S, Nanseu-Njiki CP, Ngameni E, Hadjiev D, Darchen A (2010) Treatment of dairy effluents by electrocoagulation using aluminium electrodes. Sci Total Environ 408:947–952

U.S. EPA (1989) Hazardous waste treatment, storage, and disposal facilities (TSDF). Air Emission Models, EPA-450/3-87-026. U.S. Environmental Protection Agency, Research Triangle Park, NC

Ushijima K, Ito K, Ito R, Funamizu N (2013) Greywater treatment by slanted soil system. Ecol Eng 50:62–68

van de Wijst M, Groot-Marcus A (1998) Household wastewater, calculation of chemical oxygen demand (huishoudelijkafvalwater, berekening van de zuurstofvraag). Technical report, Wageningen University, Household and consumer studies

Vital M, Stucki D, Egli T, Hammes F (2010) Evaluating the growth potential of pathogenic bacteria in water. Appl Environ Microbiol 67(19):6477–6484

Wang J, Wang S (2016) Removal of pharmaceuticals and personal care products (PPCPs) from wastewater: a review. J Environ Manage 182:620–640

Wang Y, Ma J, Zhu J, Ye N, Zhang X, Huang H (2016) Multi-walled carbon nanotubes with selected properties for dynamic filtration of pharmaceuticals and personal care products. Water Res 92:104–112

Widiastuti N, Wu H, Ang M, Zhang DK (2008) The potential application of natural zeolite for greywater treatment. Desalination 218(1–3):271–280

Wurochekke AA, Harun NA, Mohamed R, Kassim AH (2014) Constructed Wetland of *Lepironia Articulata* for household greywater treatment. APCBEE Procedia 10:103–109

Yang X, Flowers RC, Weinberg HS, Singer PC (2011) Occurrence and removal of pharmaceuticals and personal care products (PPCPs) in an advanced wastewater reclamation plant. Water Res 45:5218–5228

Yargeau V, Leclair C (2007) Potential of ozonation for the degradation of antibiotics in wastewater. Water Sci Technol 55(12):321–326

Yavuz Y, Öcal E, Koparal AS, Ö˘gütveren UB (2011) Treatment of dairy industry wastewater by EC and EF processes using hybrid Fe–Al plate electrodes. J Chem Technol Biotechnol 86:964–969

Chapter 8
Phycoremediation: A Green Technology for Nutrient Removal from Greywater

A. A. Wurochekke, Radin Maya Saphira Radin Mohamed,
Adel Ali Saeed Al-Gheethi, Efaq Ali Noman and Amir Hashim Mohd Kassim

Abstract Phycoremediation as a green technology relies on microalgae which have high potential to grow in greywater. The presence of high levels of nutrients is necessary for microalgae growth to improve the efficiency of this process. However, the main consideration of the phycoremediation process of greywater lies in the wastewater composition, the selection of microalgae strains with high potential to compete with the indigenous organisms in the greywater and remove nutrients and elements from greywater as well as microalgae, which possess the ability to survive under stressful environmental conditions. Besides, this process can be applied to individual houses. The cost of the phycoremediation process, source of microalgae and energy required are the main points which need to be discussed further. The study indicated that the phycoremediation process is most effective for the treatment of greywater. However, many aspects have to be evaluated in order to achieve the high-quality-treated greywater. In this chapter, the effectiveness of phycoremediation and the mechanism of nutrient removal are discussed. Most microalgae species exhibited greater efficiency in removing nitrogen compared to phosphorous due to the nature of the anabolic pathway of microalgae cells and the ability of nitrogen compounds to diffuse through the cell membrane faster than phosphorous compounds.

A. A. Wurochekke · R. M. S. Radin Mohamed (✉) · A. A. S. Al-Gheethi (✉)
A. H. Mohd Kassim
Micro-Pollutant Research Centre (MPRC), Department of Water and Environmental Engineering, Faculty of Civil and Environmental Engineering, Universiti Tun Hussein Onn Malaysia (UTHM), 86400 Parit Raja, Batu Pahat, Johor, Malaysia
e-mail: maya@uthm.edu.my

A. A. S. Al-Gheethi
e-mail: adel@uthm.edu.my

E. A. Noman
Faculty of Applied Sciences and Technology (FAST), Universiti Tun Hussein Onn Malaysia (UTHM), Pagoh, Johor, Malaysia

E. A. Noman
Department of Applied Microbiology, Faculty Applied Sciences, Taiz University, Taiz, Yemen

© Springer International Publishing AG, part of Springer Nature 2019
R. M. S. Radin Mohamed et al. (eds.), *Management of Greywater in Developing Countries*,
Water Science and Technology Library 87,
https://doi.org/10.1007/978-3-319-90269-2_8

149

Keywords Algae · Treatment · Nutrient removal · Alternative technology

8.1   Introduction

Due to the growth of the human population, people have begun to live in larger and more densely populated groups. Subsequently, the removal of human waste has become a serious issue. The lack of treatment plants for carbon, ammonia and nitrogen, as well as pathogen removal, is common in recent times due to the scarcity of land for treatment facilities which are mostly centralised with a broad range of consented pollutants. In a highly populated area, it is difficult to install individual treatment systems and this has led to the installation of pumps and pipes to efficiently transport waste to centralised outlets. Currently, European countries have set regulations to maintain the natural quality of wastewater streams without affecting biodiversity.

Domestic and industrial waste sewage discharges contain dissolved and suspended organic and inorganic constituents, faecal and other potentially pathogenic bacteria. Presently, wastewater has to be treated prior to disposal due to a number of reasons. For example, nutrients in wastewater such as nitrogen, phosphorus and sulphur pose a threat to the environment and its ecological sustainability. Besides that, wastewater also emits foul odours due to anaerobic digestion and can also cause potential health risks when it comes into contact with potable water. Reclaimed wastewater is now an important constituent of water supply (5–10%), which can be reused for non-drinking purposes (Wurochekke et al. 2016).

The operational characteristics of a common wastewater treatment plant consist of a few processes such as preliminary, physical, chemical and biological processes. However, the required effluent standard for the discharge of wastewater depends on the flow rate of receiving water. Wastewater preliminary treatment consists of two stages. First, a screening process will be carried out to reduce and remove coarse, medium and fine solids using different screen sizes. Second, the removal of grit and sand will occur in a grit chamber which settles dense materials easily and quickly. The organic matter remains suspended in downstream units for treatment (Al-Gheethi et al. 2015).

The purpose of primary treatment is to remove most of the suspended solids which thus reduces and regulates BOD (typically by 30%) before proceeding with the secondary treatment. This is achieved using gravitational settlement tanks where the residual part of solids is the raw primary sludge. The materials which possess less density than water such as detergents will move up to the surface of the sedimentation tank. They will be collected and removed prior to the next treatment. This process could be improved by using phosphate-precipitating chemicals like Al and Fe ions.

The secondary treatment is purposely designed to remove dissolved organic matter and other fine suspended organic matter. This treatment technique is predominantly controlled and enhances natural decomposition mechanisms in bioreactors. In this situation, the controlled condition ensures shorter treatment time. The organic matter is removed by bacteria that adapt to the environment such as ammonia oxidizers. The secondary process needs a sufficient supply of oxygen so that nitrifying microorganisms are kept in contact with the aerobic process. Besides the process described previously, there is a wide variety of secondary treatment processes available. The main ones include algal stabilisation ponds, land disposal systems, anaerobic reactors, activated sludge systems and aerobic biofilters (Jais et al. 2017).

Lastly, the tertiary treatment is designed to achieve low solid values and to remove pathogenic organisms, heavy metals and nutrients such as nitrogen and phosphorus. This tertiary treatment is a treatment process for wastewater that is expected to achieve the highest degree of effluent quality through the use of sand filters or UV lamps to remove or eliminate pathogenic organisms. These processes are capital intensive especially the backwashing of the sand filter. It requires large amounts of power. However, the chemical precipitation process is widely used for nutrient removal because biological nutrient removal organisms used in absorbing nutrients may not necessarily attain the consistency of chemical treatment even though they are more sustainable (Atiku et al. 2016).

Apparently, wastewater treatment processes are designed to achieve improvement in effluent quality. These processes perform well and meet the required quality standards. However, tertiary treatment such as nutrient removal faces some limitations due to the high costs it incurs. This is especially true for the removal of phosphorus.

For phosphorus, a normal 1 mg/L PO_4 is required to meet 95% of the time to meet UWWTD or WFD standards which are challenging and capital intensive compared to traditional chemical processes. Achieving N and P effluent standards is the key to eutrophication control and prevention. In terms of eutrophication, P is considered as the most problematic element in natural waters that causes point source pollution as the average concentration is below 0.5 mg/L. The eutrophic condition of natural waters and microalgae concentration agree with the phosphorus level. An environment that receives more nitrogen than what plants require for growth is called a nitrogen-saturated environment. However, phosphorus in the environment is less soluble in water compared to nitrogen. Therefore, it is considered as a much more significant growth-limiting nutrient in the aquatic system.

Microalgae growth in water bodies can lead to unwanted effects. Though waters with abundant nutrients (N and P) known as eutrophic waters grow algae well, waters with a lower quantity of nutrients limit growth. The eutrophic environment is characterised by cool, dark and deep waters with depleted levels of oxygen, especially in temperate and tropical areas. On the other hand, algae decolourise the water which affects the aesthetic and recreational values of water. Algae also change the taste of water and produce another toxin that is harmful to higher forms of life in the food chain. Yet, they are considered a good bio-treatment technique in the control treatment process. This lack of oxygen in the water makes it difficult for aquatic animals to survive.

The effect of oxygen-depleted water forms a microbial breakdown from the sediment of dead algae and animal waste in environments with excess nutrients such as eutrophic waters. Human activity also affects lakes and rivers through the discharge of P from detergents to surface waters although some waters are already eutrophic in nature. Besides that, the upper layer of water in a stratified lake in autumn cools to a lower temperature less than the deep lower layer of a stratified stagnant lake where mixing occurs in the stratified layers. Thus, the oxygen-depleted water from the bottom rises to the surface replacing the cooler water. Consequently, aquatic organisms that need oxygen sometimes suffocate and die during this process.

Bio-treatment with microalgae is an appropriate method because of their photosynthetic capabilities, changing solar energy into biomasses yields and embracing phosphorus and nitrogen content which inflicts eutrophication (Abdel-Raouf et al. 2012).

This chapter focused on the phycoremediation process as a green technology for nutrient removal from greywater. The selection of the most potent microalgae strain and the process used to enhance removal efficiency are discussed.

8.2   Nutrient Removal Techniques from Wastewater

Biomass is another technique used in wastewater treatment. Microalgae and duckweed are used in lagoon wastewaters, while reed beds are commonly used for non-bacterial options. The presence of nutrients in wastewater makes it an ideal place for the optimal growth of microalgae. Artificial wetlands planted with specific species of reeds are mostly used in rural areas where there are no wastewater treatment facilities. This is a biological process that treats wastewater completely. They do not require chemicals, are inexpensive to run, produce reeds for compost after harvest and are very effective in treatment (See "Reed Beds for the Treatment of Domestic Wastewater, Grant and Griggs"). Wastewater lagoons are also used for the removal of N and P. Typically, wastewater lagoons are involved in the final stage in wastewater treatment system. Wastewater lagoons refer to large ponds or tanks planted with plants which possess properties for nutrient removal like duckweed where the removal efficiency is proportional to plant biomass growth. According to a previous study by Al-Nozaily et al. (2000), N and P absorption rates and the growth rate of duckweed at high ammonia levels were inhibited. However, the tank depth does not affect N removal and P removal is less than that of algae-based pounds (Arceivala 1981). It was noted that duckweed is considerably easier to harvest than algae.

Nitrification is a process whereby reduced forms of inorganic and organic nitrogen, particularly ammonium, are oxidised to nitrate. Ammonia-oxidizing bacteria first oxidise ammonia to nitrite, followed by the oxidation of nitrite to nitrate by the nitrite-oxidizing bacteria. The oxidising bacteria *Nitrosomonas* and *Nitrobacter* are the main organisms responsible for the reaction, however, the process produces acid and lowers pH. Therefore, alkalinity needs to be added. Hence, gaseous nitrogen is formed through denitrification by the biological conversion of nitrate.

Biological treatment amid other physical and chemical treatments has more advantages. Chlorine is added to the wastewater stream to oxidise ammonia–nitrogen, which later becomes gaseous nitrogen. Ammonia stripping occurs when pH increases up to 11.5 during the treatment process with plant growth when CO_2 is reduced. The solution with high pH predominantly has ammonia as dissolved gas while enough air–water contact strips ammonia gas from the solution. Hence, the phycoremediation technique is viable for wastewater treatment.

8.3 Phycoremediation of Wastewater

The phycoremediation technique is defined as a bio-treatment process for wastewater using microalgae species. This process mainly aims to remove nutrients such as nitrogen and phosphorus from wastewater through the assimilation process of microalgae cells. The nutrients absorbed through the cell membrane of microalgae cells are transformed in the cytoplasm to be used in anabolic pathways (Atiku et al. 2016). The phycoremediation process is quite different from phycoremediation in which specific plants are used for the bioremediation of pollutants from wastewater. The potential of several microalgae species in the phycoremediation process of different types of wastewater has been reported in the literature. The phycoremediation process is recommended for the individual users due to the low operation costs, the absence of the toxic by-products, high efficiency in nutrient removal, increase in dissolved oxygen as well as the production of the microalgae biomass with high nutrient values. Besides, this process might also contribute to the inactivation of pathogenic bacteria due to the antibacterial properties of some microalgae species (Al-Gheethi et al. 2017).

Phycoremediation has been shown to exhibit high efficiency in the removal of nutrients from different types of the wastewater such as domestic, brewery and dairy wastewater (Mohamed et al. 2017). Therefore, it has high potential to produce high-quality-treated wastewater compared to those generated during the primary and secondary treatment processes (Abou-Shanab et al. 2013). Microalgae are photoautotrophic as they have the ability to obtain energy and carbon sources. However, these organisms still need some nutrients in form of nitrogen and phosphorus as well as trace elements for the metabolic and anabolic pathways. Hence, during the inoculation of microalgae in the wastewater, nutrients can be absorbed through the assimilation process. Moreover, microalgae growth in wastewater depends on environmental conditions. Therefore, the best option to achieve high efficiency in the removal of nutrients is by using indigenous strains since these strains have been acclimatised to the surrounding conditions and are able to survive and compete with other indigenous bacteria in greywater. Several species of microalgae such as *Scenedesmus dimorphic*, *Botryococcus braunii*, *Chlorella vulgaris*, *Spirulina* sp., and *Phormidium* sp., have exhibited high efficiency in the phycoremediation process due to their potential to tolerate harsh environmental conditions. Another effective option is by subjecting the microalgae to the starvation process in which the microalgae are harvested from the culture medium and then dried at room temperature. The deficiency in nutrients would induce the microalgae cells to transform from a vegetative state to form cysts (dormant state). Thereafter, the microalgae inoculum used in the phycoremediation process of greywater added as a dry biomass. The starvation process makes the microalgae biomass more effective in the uptake of nutrients and accelerates the removal process (Mohamed et al. 2017).

8.4 Controlled Eutrophication of Microalgae for Phycoremediation

To control the release of wastewater with high nutrient content especially for P, a standard limit needs to be achieved to prevent effluent from of causing eutrophication in water bodies. The growth of microalgae indicates water pollution since they typically respond to many types of ions and toxins. Green algae play a dual role in the treatment of wastewater as they enable the effective utilisation of different constituents essential for growth which further leads to enhanced biomass production for green products (Rawat et al. 2011).

A controlled environment can be a viable place for microalgae isolated specifically to grow in wastewater in which they consume P and other pollutants for their survival which is termed as phycoremediation. They are used for the removal of pollutants prior to discharge into rivers and lakes. These microalgae are not discharged together with the treated water as they are used for the continuous flow treatment of wastewater and are often harvested to sustain the optimal value of microalgae concentration. The collected biomass has the potential to produce valuable products such as biofuel, biodiesel and fertilisers due to the high content of carbon and nutrients.

Microalgae absorb a large amount of atmospheric CO_2 and release oxygen during its growth period. For additional microalgae growth, additional CO_2 is required though it affects the condition of bacterial growth. Waters that contain microalgae give out foul odour and may cause a change in how they taste. Nevertheless, freshwater microalgae have absorption abilities where they utilise nutrients and metals found in wastewaters for photosynthesis and thus purifies the water. However, this purified water is not suitable for drinking purposes until it undergoes natural purification after being discharged into rivers and lakes. Furthermore, the water can be drinkable if it undergoes another treatment process using activated carbon to further remove dissolved organic pollutants.

Furthermore, the role of microalgae is to accumulate and convert wastewater nutrients into biomass and lipids. The capability of microalgae to remove or degrade hazardous organic pollutants is well known (Abeliovich 1986). The usage of microalgae is summarised in Table 8.1. The examples given demonstrate that algae are indeed capable of contributing to the degradation of environmental pollutants, either by directly transforming the pollutant in question or by enhancing the degradation potential of the microbial community present. The biomass resulting from the treatment of wastewater can be easily converted into sustainable products. Depending on the species used for this purpose, the resulting biomass can be applied and used in many different ways. For instance, it can be used as an additive for animal feed as well as for the extraction of value-added products like carotenoids, bio-molecules or the production of biofuel. Microalgae biomass is, therefore, useful for both biofuel production as well as wastewater treatment. The successful implementation of this strategy would allow the use of freshwater to be minimised.

Table 8.1 Microalgae contributing to the degradation of environmental pollutants

Microalgae	Type of microalgae	Source of wastewater	Reference
Prototheca zopfii	Freshwater	Degraded petroleum hydrocarbons found in Louisiana crude and motor oils waste	Walker et al. (1975)
Chlamydomonas species	Freshwater	Meta cleavage in wastewater	Jacobson and Alexander (1981)
Chlorella pyrenoidosa	Fresh and brackish	Degradation of azo dyes wastewater	Jinqi and Houtian (1992)
Spirulina platensis	Freshwater and brackish water	Domestic wastewater treatment	Laliberte et al. (1997)
Chlorella sorokiniana	Freshwater	Wastewater treatment under aerobic dark heterotrophic condition	Ogbonna et al. (2000)
Scenedesmus sp.	Freshwater	Removal of ammonia from effluents containing high alkaline and ammonium levels	Jin et al. (2005)
Chlorella sp.	Fresh and marine water	Anaerobically digested dairy waste	Wang et al. (2010)
Ankistrodesmus and *quadricauda*	Freshwater	Olive-oil mill wastewaters and paper industry wastewaters	Tran et al. (2010)

8.5 Wastewater Treatment Potential of Microalgae

Microalgae have the potential to survive in a broad range of environmental conditions. Therefore, it represents a good source of biomass (Xin et al. 2010; Xu et al. 2013). The utilisation of wastewater as a production medium for microalgae biomass leads to a reduction in nitrogen and phosphate content (Mata et al. 2010; Xin et al. 2010). In addition, microalgae biomass can be used as feedstock for many industries (Spolaore et al. 2006; Harun et al. 2010).

It has been argued that biodiesel production with wastewater treatment is the most reasonable area to be commercialised in the short term. They provide a way to remove chemical and biological contaminants, heavy metals and pathogens from wastewater while producing biomass for biodiesel production. Wastewater rich in CO_2 provides a conductive growth medium for microalgae because CO_2 balances the wastewater by allowing higher microalgae production rates, reduced nutrient levels in the treated wastewater, decreased harvesting costs and increased lipid production (Brennan and Owende 2010).

Microalgae have the ability to capture sunlight and use that energy to store carbon. According to Alabi et al. (2009) the amount of carbon found in microalgae biomass is about 45–50%. Algae can have a doubling time of as little as 4 h to accumulate biomass. The microalgae growth will be limited if the culture is being supplied with CO_2 from the air because the amount of CO_2 found in air is small (0.033%). Therefore, extra CO_2 can be mixed with air and later added into the culture to facilitate microalgae growth (Mata et al. 2010). The efficiency of CO_2 can be improved by introducing deep level subaquatic through air stones, or through CO_2-rich industrial flue gas into the cultures.

In the photosynthesis process, O_2 is being generated and released into the air. However, if the O_2 gas concentration increases and gets trapped within the cell, it will cause damage to the chlorophyll reaction centres thus leading to a decrease in the production of biomass (Alabi et al. 2009). This problem may occur only in a closed system such as PBRs. In order to avoid such problems, a gas exchanger is needed to help release the O_2 gas. In open systems, however, this is not necessary because the O_2 gas can be automatically released into the atmosphere (Mata et al. 2010).

According to a previous study by Sawayama et al. (1995), *Botryococcus* sp. was used to remove nitrate and phosphate from sewage after primary treatment along with the production of hydrocarbon-rich biomass while Martinez et al. (2000) achieved a significant removal of phosphorus and nitrogen from urban wastewater using the microalgae *Scenedesmus obliquus*. They were able to eliminate 98% of phosphorus and completely remove ammonium (100%) in a stirred culture at 25 °C over 94 and 183 h retention time, respectively. To further strengthen the potential of microalgae in wastewater treatment, Hodaifa et al. (2008) recorded a 67.4% reduction in BOD_5 with *S. obliquus* cultured in diluted (25%) industrial wastewater from olive-oil extraction. The percentage of elimination reduced to 35.5% with undiluted wastewater because of low nitrogen content which inhibited microalgae growth during the exponential phase.

In addition, Yun et al. (1997) successfully grew *Chlorella vulgaris* in wastewater discharged from a steel plant to achieve an ammonia bioremediation rate of 0.022 g NH_3 per day. Munoz et al. (2009) found that the use of a biofilm attached to the reactor walls of flat plate and tubular photobioreactors improved BOD_5 removal rates by 19 and 40%, respectively, as compared with a control suspended bioreactor for industrial wastewater effluent. The retention of algal biomass showed remarkable potential in maintaining optimum microbial activity while remediating the effluent. To absorb heavy metal ions, Chojnacka et al. (2005) found that *Spirulina* sp. acted as a biosorbent where the biosorption properties of microalgae depended strongly on cultivation conditions with photoautotrophic species showing greater biosorption characteristics.

Mohamed et al. (2017) used *Botryococcus* sp. for the removal of BOD_5, COD, ammonia (NH_3), nitrate (NO_3^-) and orthophosphate (PO_4^{3-}) as well as K, Ca and Na ions from bathroom greywater by using a photo-reactor system at village houses. The phycoremediation process was conducted for 21 days. The study revealed that *Botryococcus* sp. reduced BOD_5, by 85.3 to 98%, COD by 71.22 to 85.47%, NH_3 by 86.21 to 99% and PO_4^{3-} by 39.12 to 99.3% after 21 days. The reduction of NO_3^- was recorded after 18 days of the treatment period at an efficiency of 98%. Moreover, *Botryococcus* sp. removed 97% of K after 3 days and 95% of Ca at the end of the phycoremediation process. Al-Gheethi et al. (2017) claimed that the kinetic coefficient of *Scenedesmus* sp. for the removal of NH_4^- was $k = 4.28$ mg NH_4^{-1} log_{10} cell mL^{-1} d^{-1} with 94% of the coefficient and $km = 52.01$ mg L^{-1} ($R^2 = 0.94$) while the removal of NH_4^- was $k = 1.09$ mg NH_4^- log_{10} cell mL^{-1} d^{-1} for PO_4^{3-} and $km = 85.56$ mg L^{-1}, with 92% of the efficiency. Both studies by Al-Gheethi et al. (2017) and Mohamed et al. (2017) indicated that microalgae species have high potential in the removal of the nutrients from wastewater. Most of the microalgae species exhibited more effectiveness in the removal of nitrogen compared to phosphorus. This is due to the role of nitrogen in anabolic activities and the building of amino acids. Therefore, nitrogen represents between 7 and 10% of the microalgae cells by dry weight (Richmond 2004). Phosphorus is also necessary for the microalgae genome (DNA and RNA), phospholipids, and ATP (Yao et al. 2015). The removal percentages for different wastewater parameters via the phycoremediation process using microalgae species are presented in Table 8.2.

Temperature condition has a strong correlation with the biochemical reactions which affect microalgae growth as maximal productivity can only be achieved when nutritional needs are met at correct optimal temperatures (Richmond 2004). Temperature optimal for the growth of microalgae are usually between 20 and 30 °C while temperatures lower than 16 °C decrease growth rate as many microalgae die at temperatures above 35 °C. There is a relationship between temperature and light intensity since lamps and sunlight emit heat (Richmond 2004).

The polyunsaturated fatty acids within the membranes and the fluidity of the membrane system increase at low temperature and this is essential for protecting the thylakoids and the photosynthetic machinery of microalgae cells from photoinhibition. Therefore, lipid classes and composition are affected by temperature instead of the total lipid content. However, in any algae media, the value of the pH that is suitable for cultivation ranges from 6 to 8 (Zeng et al. 2011). Although each media may have a different pH value since the values are not fixed during the cultivation process, it can be changed. Whitacre (2010) stated that the general pH value for most freshwater microalgae species ranges between 7 and 9 but the optimum pH is around 7.5. Microalgae species seem to be more tolerant of the broad range of pH values (Lam and Lee 2012). The growth of *Botryococcus* species has shown its tolerance to different pH conditions ranging from 6 to 8.5. The genus *Botryococcus* race A strains has been observed to yield biomass with high lipid content (15–35% by dry weight) and when the fatty acids were analysed, the oleic, linoleic, stearic and palmitic acids were the major fatty acids found with traces of pharmaceutically important alkyl substituted fatty acids such as 12-methyl hexadecanoic acid, 14-methyl tetradecanoic acid and 16-methyl heptadecanoic acid (Jin et al 2005). Thus, the algae *Botryococcus* appears to be a potential organism for lipid-rich biomass production under varying pH conditions. The failure to maintain the correct pH can lead to the slow growth of microalgae or eventual culture collapse because pH can affect the availability and solubility of CO_2 and minerals in the medium.

Therefore, one of the most important points in the phycoremediation process is the source of microalgae species. Microalgae species obtained from fresh water

Table 8.2 Removal of pollutants from different wastewater samples via the phycoremediation process using microalgae species

Microalgae species	Wastewater	Removal efficiency (%)	References
Botryococcus sp.	Household greywater	COD 88, BOD 82, TN 52, TOC 76	Gani et al. (2015)
	Men hostel greywater	COD 85.6, BOD 66.7, TSS 12.3	Gokulan et al. (2013)
	Aerated Swine wastewater	TP 93.3, TN 40.8	Liu et al. (2013)
	Greywater	BOD_5 85.3–98, COD 71.22–85.47, NH_3 86.21–99, $PO4^{3-}$ 39.12–99.3, NO_{3-} 98, K 97, Ca 95	Mohamed et al. (2017)
Chlorella sp.	Secondary wastewater from municipal	N 96, P 84	Jing (2009)
	Dairy wastewater	NO_3^- 49.09, NO_2^- 79.06, TP, 83.23, Fe 32.0	Khothari et al. (2012)
	Textile wastewater	BOD 81, NO_2 62, PO_4 87	Pathak et al. (2014)
	Wastewater	TN 83.2–88.1	Yao et al. (2015)
	Central municipal wastewater	COD 70, TN 61, TP 61	Min et al. (2011)
	Rubber latex concentrate processing wastewater	COD 93.4, TN 79.3	Bich et al. (1999)
	Drainage solution from greenhouse production	TP 99.7, TN 20.7	Hultberg et al. (2013)
	Leather processing chemical manufacturing facility	BOD 22, COD 38, NH_4 80, Ca 63, Mg 50, Na 14, K 18, Ni 89, TN 29–73	Rao et al. (2011)
Nostoc sp.	Dairy effluent	BOD 40.4, COD 40.3, PO_4 21, TSS 53	Kotteswari et al. (2012)
Pithophora sp.	Dairy wastewater	NH_3 99.01, NO_3^- 84.56, P = 97.98	Silambarasan et al. (2012)

(continued)

Table 8.2 (continued)

Microalgae species	Wastewater	Removal efficiency (%)	References
Scenedesmus sp.	Anaerobically digested palm oil mill effluent	TN 99.5, TP 98.8, COD 86, BOD 86.5	Kamarudin et al. (2013)
	Swine wastewater	TN 59, TP 24, IC 27	Abou-Shanab et al. (2013)
	Tannery wastewater	NO_3 44.3, PO_4 95, Cu 73.2–98, Zn 65–98, Pb 75–98	Ajayan et al. (2015)
	Animal wastewater	N 87, P 83.2	Kim et al. (2006)
	Wastewater	N>99, P>99	Li et al. (2010)
	Wet market wastewater	NH_{4-} 91, PO_4^{3-} 92.27	Al-Gheethi et al. (2017)
	Wet market wastewater	TOC 71.73, TN 73.01, PO_4^{3-} 87.60, Zn 79.65, Fe 59.33, Cu 100	Jais et al. (2015)
	Fermented swine wastewater	TP 83.2, TN 87, TC 12.9	Kim et al. (2007)
Spirulina platensis	Sago starchy wastewater	TN 99.9, TP 99.4, COD 98	Phang et al. (2000)

might be suitable to be used for the phycoremediation of wastewater. Laboratory observations indicated that most microalgae obtained from freshwater survive and grow in various type of wastewater. Based on the literature, it appears that most microalgae have high efficiency in reducing nitrogen and phosphorous in wastewater during the phycoremediation process. Therefore, phycoremediation as a treatment process could enhance the quality of wastewater which would then be reused for unrestricted irrigation purposes or discharged into surface waters.

8.6 Conclusions

Microalgae species possess high potential to be used in the phycoremediation process of greywater. However, the selection of microalgae strains should be considered based on its ability to survive, grow and remove nutrients from waste. Moreover, the use of indigenous strains might be the best option to achieve high effectiveness for the phycoremediation process. Moreover, the starvation process might enhance the efficiency of the phycoremediation process within a short time.

Acknowledgements The authors wish to thank the Ministry of Higher Education (MOHE) for supporting this research under FRGS vot 1574 and also the Research Management Centre (RMC) UTHM for providing grant IGSP U682 for this research.

References

Abdel-Raouf N, Al-Homaidan A, Ibraheem IBM (2012) Microalgae and wastewater treatment. Saudi J Biol Sci 19:257–275

Abeliovich A (1986) Algae in wastewater oxidation ponds. In: Richmond A (ed) Handbook of microbial mass culture. CRC Press, Boca Raton, pp 331–338

Abou-Shanab RA, Ji MK, Kim HC, Paeng KJ, Jeon BH (2013) Microalgal species growing on piggery wastewater as a valuable candidate for nutrient removal and biodiesel production. J Environ Manage 115:257–264

Ajayan KV, Selvaraju M, Unnikannan P, Sruthi P (2015) Phycoremediation of tannery wastewater using microalgae *Scenedesmus* species. Int J Phytoremed 17(10):907–916

Alabi AO, Bibeau E, Tampier M (2009) Microalgae technologies & processes for biofuels-bioenergy production in British Columbia: current technology, suitability & barriers to implementation. British Columbia Innovation Council, Vancouver

Al-Gheethi AA, Ismail N, Efaq AN, Bala JD, Al-Amery RM (2015) Solar disinfection and lime stabilization processes for reduction of pathogenic bacteria in sewage effluents and biosolids for agricultural purposes in Yemen. J Water Reuse Des 5(3):419–429

Al-Gheethi AA, Mohamed RM, Jais NM, Efaq AN, Abdullah AH, Wurochekke AA, Amir-Hashim MK (2017) Influence of pathogenic bacterial activity on growth of *Scenedesmus* sp. and removal of nutrients from public market wastewater. J Water Health. Online

Al-Nozaily F, Alaerts G, Veenstra S (2000) Performance of duckweed-covered sewage lagoons II. Nitrogen and phosphorus balance and plant productivity. Water Res 34(10):2734–2741

Arceivala SJ (1981) Wastewater treatment and disposal; engineering and ecology in pollution control, vol 15. M. Dekker, New York

Atiku H, Mohamed RMSR, Al-Gheethi AA, Wurochekke AA, Kassim AHM (2016) Harvesting of microalgae biomass from the phycoremediation process of greywater. Environ Sci Poll Res 23(24):24624–24641

Bich NN, Yaziz MI, Bakti NAK (1999) Combination of *Chlorella vulgaris* and *Eichhornia crassipes* for wastewater nitrogen removal. Water Res 33:2357–2362

Brennan L, Owende P (2010) Biofuels from microalgae—a review of technologies for production, processing, and extractions of biofuels and co-products. Renew Sustain Energy Rev 14(2):557–577

Chojnacka K, Chojnacki A, Gorecka H (2005) Biosorption of Cr^{3+}, Cd^{2+} and Cu^{2+} ions by blue-green algae *Spirulina* sp.: kinetics, equilibrium and the mechanism of the process. Chemosphere 59(1):75–84

Gani P, Sunar NM, Matias-Peralta HM, Abdul Latiff AA, Kamaludin NS, Parjo UK, Er CM (2015) Experimental study for phycoremediation of *Botryococcus* sp. on greywater. Appl Mech Mater 773:1312–1317

Gokulan R, Sathish N, Kumar RP (2013) Treatment of grey water using hydrocarbon producing *Botryococcus braunii*. Int J Chem Tech Res 5(3):1390–1392

Harun R, Singh M, Forde GM, Danquah MK (2010) Bioprocess engineering of microalgae to produce a variety of consumer products. Renew Sustain Energy Rev 14(3):1037–1047

Hodaifa G, Martinez ME, Sanchez S (2008) Use of industrial wastewater from olive oil extraction for biomass production of *Scenedesmus obliquus*. Biores Technol 99(5):1111–1117

Hultberg M, Carlsson AS, Gustafsson S (2013).Treatment of drainage solution from hydroponic greenhouse production with microalgae. Biores Technol 136:401–406

Jacobson SN, Alexander M (1981) Enhancement of the microbial dehalogenation of a model chlorinated compound. Appl Environ Microbiol 42:1062–1066

Jais NM, Mohamed RMSR, Apandi WA, Matias-Peralta H (2015) Removal of nutrients and selected heavy metals in wet market wastewater by using microalgae *Scenedesmus* sp. Appl Mech Mater 773–774:1210–1214

Jais NM, Mohamed RMSR, Al-Gheethi AA, Hashim MA (2017) The dual roles of phycoremediation of wet market wastewater for nutrients and heavy metals removal and microalgae biomass production. Clean Technol Environ Policy 19(1):37–52

Jin Y, Veiga MC, Kennes C (2005) Bioprocesses for the removal of nitrogen oxides from polluted air. J Chem Technol Biotechnol 80:483–494

Jing S (2009). Removal of nitrogen and phosphorus from municipal wastewater using microalgae immobilized on twin-layer system. Ph.D. thesis, Universitätzu Köln, Germany

Jinqi L, Houtian L (1992) Degradation of azo dyes by algae. Environ Pollut 75:273–278

Kamarudin KF, Yaakob Z, Rajkumar R, Takriff MS, Tasirin SM (2013) Bioremediation of palm oil mill effluents (POME) using *Scenedesmus dimorphus* and *Chlorella vulgaris*. Int J Adv Sci Lett 19:2914–2918

Kim MK, Park JW, Park CS, Kim SJ, Jeune KH, Chang MU, Acreman J (2007) Enhanced production of *Scenedesmus* spp. (green microalgae) using a new medium containing fermented swine wastewater. Biores Technol 98(11):2220–2228

Kothari R, Pathak VV, Kumar V, Singh DP (2012) Experimental study for growth potential of unicellular alga *Chlorella pyrenoidosa* on dairy waste water: an integrated approach for treatment and biofuel production. Biores Technol 116:466–470

Kotteswari M, Murugesan S, Ranjith Kumar R (2012) Phycoremediation of dairy effluent by using the microalgae *Nostoc* sp. Int J Environ Res Dev 2(1):35–43

Laliberte G, Olguin EJ, Noue JDL (1997) Mass cultivation and wastewater treatment using Spirulina. In: Vonshak A (ed) *Spirulina platensis*-physiology, cell biology and biotechnology. Taylor and Francis, London (UK), pp 59–73

Lam MK, Lee KT (2012) Microalgae biofuels: a critical review of issues, problems and the way forward. Biotechnol Adv 30(3):673–690

Li X, Hu H, Gan K, Sun Y (2010) Effects of different nitrogen and phosphorus concentrations on the growth, nutrient uptake and lipid accumulation of a freshwater microalga *Scenedesmus* sp. Biores Technol 101:5494–5500

Liu J, Ge Y, Cheng H, Wu L, Tian G (2013) Aerated swine lagoon wastewater: a promising alternative medium for *Botryococcus braunii* cultivation in open system. Biores Technol 139:190–194

Martinez ME, Sanchez S, Jimenez JM, Yousfi EF, Munoz L (2000) Nitrogen and phosphorus removal from urban wastewater by the microalga *Scenedesmus obliquus*. Biores Technol 73(3):263–272

Mata TM, Martins AA, Caetano NS (2010) Microalgae for biodiesel production and other applications: a review. Renew Sustain Energy Rev 14:217–232

Min M, Wang L, Li Y, Mohr MJ, Hu B, Zhou W, Chen P, Ruan R (2011) Cultivating *Chlorella* sp. in a pilot-scale photobioreactor using centrate wastewater for microalgae biomass production and wastewater nutrient removal. Appl Biochem Biotechnol 165:123–137

Mohamed RM, Al-Gheethi AA, Aznin SS, Hasila AH, Wurochekke AA, Kassim AH (2017) Removal of nutrients and organic pollutants from household greywater by phycoremediation for safe disposal. Int J Energy Environ Eng:1–14

Munoz R, Kollner C, Guieysse B (2009) Biofilm photobioreactors for the treatment of industrial wastewaters. J Hazard Mat 161(1):29–34

Ogbonna JC, Yoshizawa H, Tanaka H (2000) Treatment of high strength organic wastewater by a mixed culture of photosynthetic microorganisms. J Appl Phycol 12:277–284

Pathak VV, Singh DP, Kothari R, Chopra AK (2014) Phycoremediation of textile wastewater by unicellular microalga *Chlorella pyrenoidosa*. Cell Mol Biol 60(5):35–40

Phang SM, Miah MS, Yeoh BG, Hashim MA (2000) Spirulina cultivation in digested sago starch factory wastewater. J Appl Phycol 12:395–400

Rao HP, Kumar R, Raghavan BG, Subramanian VV, Sivasubramanian V (2011) Application of phycoremediation technology in the treatment of wastewater from a leather-processing chemical manufacturing facility. Water SA 37(1):07–14

Rawat I, Kumar RR, Mutanda T, Bux F (2011) Dual role of microalgae: phycoremediation of domestic wastewater and biomass production for sustainable biofuels production. Appl Energy 88:3411–3424

Richmond A (2004) Environmental effects on cell composition. In: Handbook of microalgae culture—biotechnology and applied phycology. Blackwell publishing, Oxford, UK, pp 83–93

Sawayama S, Inoue S, Dote Y, Yokoyama SY (1995) CO_2 fixation and oil production through microalga. Energy Convers Manage 36:729–731

Silambarasan T, Vikramathithan M, Dhandapani R, Mukesh DJ, Kalaichelvan PT (2012) Biological treatment of dairy effluent by microalgae. World J Sci Technol 2(7):132–134

Spolaore P, Joannis-Cassan C, Duran E, Isambert A (2006) Commercial applications of microalgae. J Biosci Bioeng 101:87–96

Tran NH, Bartlett JR, Kannangara GSK, Milev AS, Volk H, Wilson MA (2010) Catalytic upgrading of biorefinery oil from micro-algae. Fuel 89:265–274

Walker JD, Colwell RR, Petrakis L (1975) Degradation of petroleum by an alga, *Prototheca zopfii*. Appl Microbiol 30:79–81

Wang L, Li YC, Chen P, Min M, Chen YF, Zhu J, Ruan RR (2010) Anaerobic digested dairy manure as a nutrient supplement for cultivation of oil-rich green microalgae *Chlorella* sp. Biores Technol 101:2623–2628

Whitacre JM (2010) Degeneracy: a link between evolvability, robustness and complexity in biological systems. Theoretical Biol Med Model 7(1):6

Wurochekke AA, Mohamed RMS, Al-Gheethi AA, Atiku H, Amir HM, Matias-Peralta HM (2016) Household greywater treatment methods using natural materials and their hybrid system. J Water Health 14(6):914–928

Xin HS, Schaefer DM, Liu QP, Axe DE, Meng QX (2010) Effects of polyurethane coated urea supplement on in vitro ruminal fermentation, ammonia release dynamics and lactating performance of Holstein dairy cows fed a steam-flaked corn-based diet. Asian-Aust J Anim Sci 23:491–500

Xu Y, Purton S, Baganz F (2013) Chitosan flocculation to aid the harvesting of the microalga *Chlorella sorokiniana*. Biores Technol 129:296–301

Yao L, Shi J, Miao X (2015) Mixed wastewater coupled with CO_2 for microalgae culturing and nutrient removal. PLoS ONE 10(9):e0139117. https://doi.org/10.1371/journal.pone.0139117

Yun YS, Lee SB, Park JM, Lee CI, Yang JW (1997) Carbon dioxide fixation by algal cultivation using wastewater nutrients. J Chem Technol Biotechnol 69:451–456

Zeng J, Singh D, Chen S (2011) Thermal decomposition kinetics of wheat straw treated by *Phanerochaete chrysosporium*. Int Biodeterior biodegradation 65(3):410–414

Chapter 9
Bioremediation of Xenobiotic Organic Compounds in Greywater by Fungi Isolated from Peatland, a Future Direction

Efaq Ali Noman, Adel Ali Saeed Al-Gheethi, Balkis A. Talip,
Radin Maya Saphira Radin Mohamed, H. Nagao,
Amir Hashim Mohd Kassim and Junita Abdul Rahman

Abstract The conventional wastewater treatment processes aim to remove pathogens and priority pollutants in terms of chemical and physical characteristics such as chemical oxygen demand (COD), biological oxygen demand (BOD) and total suspended solids (TSS). Some of the technologies are used for reduction of nutrients such as the phycoremediation process which has high efficiency for the reduction of total nitrogen and phosphorus from the wastewater. Unfortunately, these techniques have no contribution to the removal of XOCs. The greywater with XOCs should be subjected to an advanced treatment process to remove xenobiotic organic compounds (XOCs) before the final disposal into the environment. The current treatment by the oxidation processes is insufficient and expensive as well as have many of toxic by-products. This gap offered the investigators greater opportunities to explore

E. A. Noman
Faculty of Applied Sciences and Technology (FAST), Universiti Tun Hussein Onn Malaysia (UTHM), Pagoh, Johor, Malaysia

E. A. Noman
Department of Applied Microbiology, Faculty Applied Sciences, Taiz University, Taiz, Yemen

A. A. S. Al-Gheethi (✉) · R. M. S. Radin Mohamed · A. H. Mohd Kassim · J. A. Rahman
Micro-Pollutant Research Centre (MPRC), Department of Water and Environmental Engineering, Faculty of Civil and Environmental Engineering, Universiti Tun Hussein Onn Malaysia (UTHM), 86400 Parit Raja, Batu Pahat, Johor, Malaysia
e-mail: adel@uthm.edu.my

B. A. Talip (✉)
Faculty of Applied Sciences and Technology (FAST), Universiti Tun Hussein Onn Malaysia (UTHM), 84000 KM11, Jalan Panchor, Pagoh Muar, Johor, Malaysia
e-mail: balkis@uthm.edu.my

H. Nagao
School of Biological Sciences, Universiti Sains Malaysia (USM), 11800 George Town, Penang, Malaysia

© Springer International Publishing AG, part of Springer Nature 2019
R. M. S. Radin Mohamed et al. (eds.), *Management of Greywater in Developing Countries*, Water Science and Technology Library 87,
https://doi.org/10.1007/978-3-319-90269-2_9

effective and eco-friendly alternative technologies for XOCs degradation in greywater. Moreover, many of the fungi from the peat soil especially that belong to white rot fungi have higher enzymatic activities and produce a lot of oxidative enzymes such as laccase, lignin and manganese peroxidases. These enzymes are the main factor in the bioremediation process of the pollutants in the contaminated environment such as wastewater. Among different types of the oxidative enzymes from the fungi, the peroxidase and laccase have high importance in the biodegradation of XOCs. The current chapter discusses the potential of fungi as an alternative green technology for the degradation of XOCs from the greywater.

Keywords XOCs · Fungi · Laccase · Peroxidase · Mechanism · Bio-carrier

9.1 Introduction

The microbial species especially those used in the environmental biotechnology and wastewater treatment have received more attention since many years ago due to their efficiency in achieving good results in comparison to the chemical and physical process. Fungi are among the different microorganisms which are used extensively in the environmental treatment such bioremediation and detoxification processes. Moreover, the fungal species inhabit everywhere due to their simple requirements for the growth and reproduction. Peatland represents one of the best media for fungal diversity due to the high contents of nutrients and organic matter as well as the salts required for fungal growth and pH value appropriate for their metabolic and anabolic activities. Fungi have no need for intermediate hosts in their life cycle as in the case of parasites and viruses. Bacterial cells also reproduce without the need for intermediate hosts but the fungi are more tolerant of the hard environmental conditions and they can survive more. In fact, fungi have the ambient environmental temperature for the growth. These organisms produce many of extracellular enzymes such as cellulase, chitinase, pectinase, lipase and protease as well as oxidative enzymes which include laccase lignin and manganese peroxidase. It has been reported that 68% of the fungi possess cellulase enzyme (Jahangeer et al. 2005). These enzymes have a higher importance in the industrial and environmental applications.

In the field of wastewater treatment, RMK-11 indicated that the developments in the field of city building will continue in four major cities such as Kuala Lumpur, Johor Bahru, Kuching and Kota Kinabalu to stimulate economic growth and improve habitation. By 2020, 99% of the population will have clean and treated water, 99% of rural houses will have electricity and 95% of populated areas will have access to broadband services at lower prices. Despite having several water resources, the rapidly growing population and urbanisation in Malaysia exhibit the increase of pollution levels of natural water resources since the year 2000. There are more than 17,633 water pollution point sources which are revealed in Malaysia. Hence, crucial steps have to be taken in order to improve the quality of water bodies and, therefore, can adhere to become a fully developed nation by the year 2020.

According to the Policy Directions for Core Area 2: Water Resources Sustainability in Target 9–11 (Strategy 15–18) the country aims to "adopt national criteria for water resources characterization and standards as well as to protect condition and state of water resources, catchment and bodies". To date, the degradation of xenobiotic organic compounds (XOCs) in the greywater has not been reported in the literature. Most of the studies focused on the degradation of XOCs in the soil, aqueous solution and industrial effluents such as textile wastewater. Greywater has different chemical, physical and biological compositions in comparison to the soil and textile wastewater in terms of nutrients, metal ions and XOCs generated from the personal care products and detergents. These substances might induce or inhibit the degradation of XOCs. Thus, more studies are required to investigate the potential of XOCs degradation in the greywater by fungi. In this chapter, the characteristics of fungi associated with peatland and their potential application in the degradation of XOCs are reviewed. The mechanism of XOCs degradation by fungal enzymes will also be discussed.

9.2 Characteristics and Identification of Fungi

Fungi are eukaryotic organisms belongs to the kingdom of Fungi. The kingdom has high diversity in the organism cell structures. The single cells with a simple structure are called the yeast. In contrast, several genus and species belong to the filamentous fungi which have complex structures; these are composed of filaments named hyphae. The hyphae is a prolong cells that may contain internal cross walls called septa. A mass of hyphae that is not a reproductive structure is called a mycelium. Hyphae represent the vegetative part of fungi which contains the nucleus, mitochondrion, ribosomes and all metabolic pathways of fungal cell. The hyphae are surrounded by the cell wall that protects fungus cell from changes in osmotic pressure and other environmental stresses. The cell wall is composed of macromolecules (chitosan, chitin, lipid, glucan, phospholipids), which contain amino, carboxyl, melanin, phosphate, hydroxides and sulphate groups (Fogarty and Tobin 1996; Bowman and Free 2006; Hardison et al. 2012).

Fungi are typically heterotrophic organisms, nonmotile, and have quite simple nutritional requirements for their growth. Most of the fungi are a saprophytic organism in origin, in which they get the nutrient by breaking down of dead organic material by hydrolyses enzymes such as lipase, cellulase, amylase, chitinase, pectinase and protease. Other fungi have pathogenicity for human and animals such as Histoplasma, Blastomyces, Paracoccidioide and Coccidioides which have dimorphic or 'endemic' form. These fungi grow as filamentous mycelium at ambient temperature and as yeasts form or spherules within the host tissue (35–37 °C). Some of the fungi such as *Aspergillus* spp. are opportunistic organisms that may have a potential to cause secondary infection for human (Noman et al. 2016).

Fungi are the supreme examples of spore-producing organisms (Deacon 2005). They have sexually and asexually reproduction, but the majority of them reproduce by the generation of asexual spores. In asexual reproduction, the fungi generate many of the conidia (spores) mounted on a special structure called conidiophore. These spores are distributed into the environment and germinate to fungal mycelium.

The taxonomy of fungi has been subjected to series of developments since nineteenth century. A brief history of the fungi taxonomy is illustrated in Table 9.1.

The first taxonomy was suggested by Haeckel, who classified fungi along with bacteria, protozoa and algae in the Kingdom of Protists. However, the development of the electron microscope created a revolution in the field of taxonomy. The electron microscope technique revealed the microstructure differences between prokaryotic and eukaryotic cells based on nuclear membrane of the cell nucleus. Fungi appeared to have a true cell nucleus and classified as eukaryotic cells in a separate kingdom called fungi (Whittaker 1969).

The main characteristics which are used for the identification and the classification of fungi include cell structure (unicellular or filamentous), mycelia (septate or aseptate), nutrition (saprophytic, parasitic) and reproduction spores (sexual, asexual, motile, non-motile). However, it has been mentioned that some of these characteristics such as nutrition are common in several fungi, so the dependence on one parameter was not accurate to identify fungal isolates to the species and strains level. For example, *Aspergillus* spp. and *Penicillium* spp. are saprophytic but they may also be opportunistic depending on the environmental condition. Therefore, nutrition may be useful as a general characteristic to distinguish fungi among others organisms, but not between the species within the kingdom of fungi.

Several characteristics need to be considered simultaneously in the identification of fungi by the phenotypic method. Fungi have high diversity in their culture characteristics such as colony size (diameter, mm), texture, surface, zonation and sporulation on different culture media including Potato Dextrose Agar (PDA), Sabouraud Dextrose Agar (SDA), Czapek-Dox Agar (CZ), Czapek Yeast Extract Agar (CYA) and Malt Extract Agar (MEA) (Promputtha et al. 2005).The microscope morphologies which include the characteristics of vegetative and reproductive structures such as hyphal colour and structures, as well as spore shape, surface ornamentation and size, are very important for the identification of fungi (Efaq et al. 2017). The morphological characteristics are more useful to identify some of the fungal isolates such as *Aspergillus* sp. *Penicillium* sp., *Trichoderma* sp., *Curvularia* sp. and *Rhizopus* sp. to species level (Kumara and Rawal 2008; Emine et al. 2010). For instance, in *Aspergillus* sp. the structure of conidiophore is very important for their identification (Robert et al. 2011; Silva et al. 2011).

The phialides (spore-producing cells) are located on the vesicles and may partially (columnar vesicles) or entirely (radiate head) cover the surface of the vesicles. It may also be attached to the vesicle directly (uniseriate) or via metula (biseriate) (Jensen et al. 2013). Fungi have different spore shape and size as well as spore surface ornamentations. These represent an important key for identification by phenotype method. However, it has to be mentioned that the phenotypic characteristics of fungi depend on environmental conditions and culture media (Guarro et al. 1999). Hence, several culture media are recommended to be used in the phenotypic identification.

Table 9.1 Brief history of the fungi taxonomy

References	Fungal Taxonomy
Haeckel (1866)	The fungi were classified alongside bacteria, protozoa and algae in kingdom Protists
Copeland (1956)	The living organisms were classified into four kingdoms. Fungi were within Kingdom of Protista, while bacteria and algae were classified in Kingdom of Monera. This classification relied on the presence of nuclear membrane of organism cell nucleus as determined by Electron Microscope
Nolan and Margoliash (1968)	The fungi were classified in a separate kingdom based on cytochrome. Fungi form phylogeneticl ine differ from animals and plants kingdoms
Alexopoulos (1962, 68)	The fungi were classified in Division of Mycota. This division has been divided into two subdivisions (Myxomycotina and Eumycotina) and nine classes (Myxomycetes, Chytridiomycetes, Hyphochytridiomycetes, Zygomycetes, Trichomycetes, Oomycetes, Ascomycetes, Basidiomycetes and Plasmodiophoromycetes) based on the presence or absence of cell wall as well as reproduction spores. The problematic with this classification was the Deuteromycetes only reproduce by asexual spores. Therefore, Deuteromycetes have placed in a form class included fungi with septate mycelia and asexual spores
Whittaker (1969)	The fungi were classified in a separate kingdom (Fungi) included fungi and slime mould groups
Ainsworth (1971) Ainsworth et al. (1973)	The fungi were classified in Kingdom of Mycota has 2 divisions, 5 subdivisions and 21 classes. This taxonomy has relied on cell structure (unicellular or filamentous), mycelia (septate or aseptate), nutrition (saprophytic, parasitic), reproduction spores (sexual, asexual, motile, non-motile)
Woese and Fox (1977)	A new taxonomy for all living organism by phylogenetic methods was suggested based on 16S rRNA sequences into three domains called Archaea, Bacteria and Eucarya. This classification has named phylogenetic tree. Fungi have been placed in the domain of Eucarya in a septate kingdom has named Fungi
Alexopoulos and Mims (1979)	All fungi including the slime fungi were placed in Kingdom of Myceteae. This kingdom has 3 divisions, 8 subdivision and 11 classes. The taxonomy has depends on nutrition and presence or absence of motile structure, while reproduction spores (sexual or asexual) which have been used to classify fungi before has not used here. Therefore, Deuteromycotina has been classified as a separate subdivision like Ascomycotina

(continued)

Table 9.1 (continued)

References	Fungal Taxonomy
Alexopoulos et al. (1996)	The fungal classification was developed the previous fungal taxonomy suggested by Alexopoulos and Mims (1979). The fungi were classified based on cell wall structure, mycelia structure (septate and aseptate), reproduction spores (sexual and asexual), motile or non-motile (number and position of flagella, nutrition (saprophytic, parasitic) and habitat (solid, water). They have divided fungi into three kingdoms, 11 phyla
Kirk et al. (2008)	The fungi classified into 10 phyla. The highest number of fungal species were within phylum of Ascomycota (64,163 species) and Basidiomycota (31,515 species) which represent ecological, industrial and medical importance

The developments in the fluorescent microscopy, scanning electronic microscopy (SEM) and flow cytometry have improved the recognition of several microstructures of fungal conidiophore and spores (Cole and Samson 1979; Guarro et al. 1999). The SEM analysis for fungal spores has the higher efficiency to show the microstructure in the spore shape and surface ornamentation which enhance the accurate identification of the fungi to the varieties level. The utilisation of SEM might be equivalent to the molecular technique, for instance: *Curvularia lunata* has two varieties included *C. lunata* var. *aeria* with smooth conidia and stromata in culture; and *C. lunata* var. *lunata* with smooth to roughly conidia but without stromata in culture (Ellis 1971; Sivanesan 1987). A study conducted by Nakada et al. (1994) and Cunha et al. (2013) revealed that the molecular analysis of 16S rRNA sequencing in these varieties identified as *C. lunata* and *C. area*.

The main limitations of phenotype methodologies lie in the fungal species which have no spores in the culture media due to the unavailability of the growth factors which induce the sporulation (Timnick et al. 1951). It has been demonstrated that the specific carbon and nitrogen sources are required for the sporulation in some of the fungi (Gao et al. 2007). In order to overcome this matter, Leslie and Summerell (2006) have described a sub-cultured method on Carnation Leaf Agar (CLA) for induction of sporulation. In this method, the fungi were incubated at 25 °C under UV lamp for 1 week on PDA. In the absence of sporulation, the targeted fungal culture colonies were transported into Petri dishes containing small pieces of autoclaved and dried fresh carnation leaves placed on moist filter paper and distributed on the edge of 1.8% water agar plate. Subsequently, the plates were sealed and incubated under UV lamp for 3 weeks. Schwarz et al. (2007) had suggested the carbon assimilation profiles by using API CH50 as a tool for identification of Zygomycetes. The largest differences have been recorded between the genera and species of this class. However, the phenotyping methods need longer time not only for isolation but also for searching the different references, which may require days or weeks. Fungal characteristics of non-culturable fungi are difficult to be determined in the laboratory. On the other hand, some morphological characteristics are useful for the identification of/at genus levels such as *Aspergillus* sp., *Penicillium*, *Trichoderma* sp., *Curvularia* sp. and *Rhizopus* sp., but these are not enough for the identification process of/at the species levels and require accurate techniques.

Alongside, the developments in the phenotypic taxonomy of fungi, Woese and Fox (1977) investigated the phylogenetic methods based on 16S rRNA sequences for the identification of all organisms. Fungi were placed in the domain of Eucarya in a separate Kingdom named Fungi. The application of polymerase chain reaction (PCR) with universal primers that have been developed since the 1990s had provided the mycologist with the information necessary to design a new classification of fungi especially for non-culturable species (White et al. 1990).

The identification of fungi by molecular techniques has improved the traditional classification for some fungi that was relied on phenotypic characteristics. For example, although Oomycetes and Chytridiomycota have motile spores and while Basidiomycota and Ascomycota have non-motile spores, Alexopoulos et al. (1996) have classified Chytridiomycota along with Basidiomycota and Ascomycota within the kingdom of fungi based on their cell wall structure. Meanwhile, Oomycota has been classified in the kingdom of Stramenopila. The analysis of 18S rRNA sequences of these fungi has revealed that Chytridiomycota were differing neither from Basidiomycota, Ascomycota nor Oomycota (Hendriks et al. 1991; Guarro et al. 1999). Kirk et al. (2008) have classified the fungi into 10 phyla based on characteristics determined by sequencing analysis. This classification includes a high number of fungal species within phylum of Ascomycota (64,163 species) and Basidiomycota (31,515 species) which represent ecological, industrial and medical importance.

Currently, the identification of fungi is relied on 18S rRNA sequences by using the universal primer. The selection of universal oligonucleotide primers specific to fungi has provided easy access to nucleotide sequences and to the identification of several fungi (White et al. 1990; Vilgalys and Hester 1990; Sandhu et al. 1995; Gyaurgieva et al. 1996). The technique is fast and has large Database of 18S rDNA sequences in GenBank belongs to several fungi. However, Bhadury et al. (2009) have discovered one important note: they detected 18S rRNA sequences of fungi in conjunction with nematode 18S rRNA sequences. The fungi have belonged to Chaetothyriales and Hypocreales. The study has revealed that the ecological interactions between nematodes and fungi in the marine environment might play an important role in the taxonomy of fungi and others organisms.

According to the above-aforementioned studies, the identification of fungi by the phenotyping method is easy but it needs more skills and well-trained personnel to look for the small details which might make a big difference in the accuracy of identification. According to the American Society for Microbiology, 89% of laboratories identified the fungi based on morphological characteristics (Diba et al. 2007). The morphological characteristics can be used to determine the genus, but physiological, chemical or molecular technique is required for species identification especially for uncommon fungi.

9.3   Fungi in the Peatland

Peat soil is a type of soil which has higher organic matter ranging from 50 to 75% of weight and more than 1500% of moisture. The peat forms as a result of the accumulation of partially decomposed plant biomass in the different natural environment due to the slow decay of materials and dependent on the temperature and tropical locations (Kalantari and Prasad 2014). In some definitions, the peat soil is called organic soil or histosols, where this term is not only related to the thickness of the soil layers and their high contents of organic matter, but also to their origin, underlying material, clay content and water saturation. Therefore, the histosols term is used to represent the peat and peaty soils with 40 cm total thickness which is located under 10 cm of the surface layer of the soil. The definition differs from most of the European definitions of peat, where the slightly thicker layer and slightly lower organic matter content are not included (Joosten and Clarke 2002).

The peat deposits cover large areas of northern Europe, northern North America, western Siberia, Indonesia and Southeast Asia (Glaser et al. 2004) (Fig. 9.1). It can be noted that the dominant area of the peatlands is located in Canada by 1,500,000 km^2 followed by the US which has 600,000 km^2. In Indonesia, the total area of the peatlands is 170,000 km^2, while is about 26,000 km^2 in Malaysia. This value appears to be very low in comparison to Canada, US and Indonesia but it represents 8% of the total area of Malaysia (Huat 2004).

The distribution of peatland in Malaysia is presented in Fig. 9.2. It seems that the largest area of the peatland is located in Sarawak, which represents 63% of the total area in Malaysia. This might explain the fewer development levels in the Sarawak

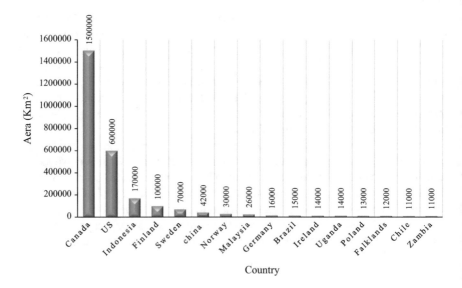

Fig. 9.1 Peatland area in different countries (adopted from Mesri and Ajlouni 2007)

■ Johor   ■ Selangor   ■ Perak   ■ Pahang   ■ Terengganu   ■ kelantan   ■ Sarawak

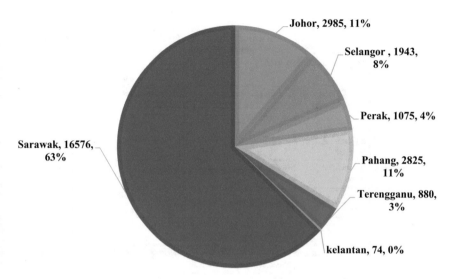

Fig. 9.2   Distribution of peatland in Malaysia

in comparison with others states of Malaysia. In addition, the area of the peatland in Johor and Pahang represents a large portion by 22%. Peat is considered a problematic soil due to its engineering characteristics such as poor strength, large deformation, high compressibility, and high magnitude and rates of creep which make it unsuitable for construction at its natural stage (Kazemian et al. 2011).

The peatland in Malaysia with more wood materials have lower CO_2 emissions in comparison to the temperate peatland (Veloo et al. 2014). The CO_2 emissions depends mainly on the decomposition level in the peat which also relies on the peatland layer (fabric, hemic and sapric). The maximum level of the decomposition takes place in the sapric due to the high level of water activities, while the fabric has the lowest decomposition level. The decomposition process is a biochemical or biological process that takes place by the microorganisms which inhabit the peatland. However, among the high diversity of these microorganisms, fungi are saprophytic in nature in that they get the energy and carbon source through the production of hydrolysis enzymes such as cellulase and lignocellulose enzymes, which degrade the organic matter in the peat. One of the main factors is that the peats provide the suitable conditions for fungal diversity that lie in the low pH. Peatland has pH values ranged from 4 to 5. This value is within the range of favourable for fungal growth, while the optimal pH values for bacterial growth is ranging between pH 6 and 8.

The micro-fungal diversity in the peatland has been investigated extensively in different regions around the world including Europe, North and South America and Asia. The studies presented in Table 9.2 confirm that the peatland have various

Table 9.2 Enzymatic degradation of XOCs

Fungal strain	XOCs	Medium	Enzyme	Degradation percentage (%)	References
Pycnoporus sanguineus	Nonylphenol (NP) and Triclosan (TCS)	Distilled water, 100 U/L, pH 5, 25 °C, treatment periods 4 and 6 h,	Laccase	94	Ramírez-Cavazos et al. (2014)
				92	
Irpex lacteus, Pleurotus ostreatus,	4-n-NP, technical 4-nonylphenol, 17α-ethinylestradiol, bisphenol A and triclosan	Culture medium	ligninolytic enzyme	80–90	Cajthaml et al. (2009)
Cerrena unicolor	Bisphenol A (BPA),	Aqueous solution, 50 μmol, pH 3	Laccase	80	Songulashvili et al. (2012)
	NP			40	
	TCS			60	
Chaetomiaceae	NP	1.1 g/L, 12 h of treatment, at 40 °C and pH 7	laccase	99	Saito et al. (2004)
T. versicolor	Triclosan (TCS)	5.8 mg/L, 25 °C and pH 5; after 2 h	Laccase	69	Kim and Nicell (2006)
Cirripectes polyzona	Nonylphenol (NP) and	pH 5; 50 °C, 4 h	Laccase	95	Cabana et al. (2007)
	Triclosan (TCS)	pH 5; 50 °C, 8 h		35	
Trametes versicolor	Diclofenac indomethacin naproxen	Sodium acetate buffer with ABTS and HBT (1-hyroxybenzotriazole) as inducers pH 4.5, 30 °C	Laccase	100	Tran et al. (2010)
P. chrysosporium	Azo Black Reactive 5 dye	3 days		92	Enayatizamir et al. (2011)
P. chrysosporium URM6181 and *Curvularia lunata* URM6179	Textile indigo dye	Textile effluents, 10 days		95	Miranda et al. (2013)

species of fungi and the explanation for the high diversity would be related to the characteristics of the soil in terms of organic matter, macro-elements and low pH. It estimated that 648 different fungal species have been recorded in the peatland soil including 408 species (63%) belonging to the anamorphic Ascomycetes fungi, which include *Acremonium* spp., *Aspergillus* spp., *Cladosporium* spp., *Fusarium* spp., *Oidiodendron* spp., *Penicillium* spp., *Trichoderma* spp. and *Verticillium* spp. (Thormann 2006). Some of these fungi are pathogenic and produce different toxins, while other has higher economic and environmental values.

Nevertheless, the aim of the exploration of peatland should be to focus on the presence of novel fungal strains which have environmental and industrial importance. The industrial application of fungi includes the species from *Fusarium* spp., *Chaetomium* spp., *Trichoderma* spp., and *Penicillium* spp. have higher ability to produce hydrolysis enzymes such as amylase, cellulase and pectinase enzymes. The applications of cellulase enzyme include the formulation of detergents, production of animal feed, the paper industry as well as juice clarification and wine production (Al-Gheethi 2015). The environmental application for fungi lies in their being bioremediation pollutants in the environment such as wastewater. Several fungi species include *Aspergillus niger, Humicola lanuginose, Penicillium* spp., *Phanerochaete chrysosporium, Pycnoporus cinnabarinus, Phlebia radiate, Rhizopus delemar, Trametes versicolor, Trichoderma harzianum* and *Yarrowia lipolytica* which have been used in the bioremediation process (Yamane 1989; Sayadi and Ellouz 1993; Wu et al. 2006; AbdulKarim et al. 2011). The advantages for the utilisation of fungi in the bioremediation process include low cost, eco-friendly, ecologically simple and as treating wastes on site (Nadeau et al. 1993).

9.4   Biodegradation of XOCs in the Greywater by Fungi

One of the most recent applications for fungi is to treat the contaminated environment with XOCs through bioremediation process. XOCs are complex organic compounds with high persistence for the natural degradation of the environment. Bioremediation is the optimisation of biodegradation. This process can be accelerated fertilising (adding nutrients) and/or seeding (adding microbes) in order to overcome certain environmental factors that might limit or prevent biodegradation. The import of specific microorganisms to enhance degradation of these compounds in the contaminated site is known as bioaugmentation. The reports indicate that the efficiency of biodegradation by soil fungi is more than (6–82%) that for soil bacteria (0.13–50%) (Pinholt et al. 1979; Hollaway et al. 1980). However, the criteria for choosing the microorganism in the environmental application include the non-pathogenicity of the microorganism, high efficiency and the longer period survival under stress conditions.

Bioremediation has been successfully applied in the treatment of petroleum hydrocarbon pollutants, refinery effluents, wastewaters and textile effluents (Ojumu et al., 2005; Okonko and Shittu 2007; Bako et al. 2008). The bioremediation is more effective in the environmental conditions, which permit microbial growth and activity (Vidali 2001). Hence, one of the simplest ways to get higher potential microorganisms with the fast degradation rate of XOCs is to isolate this organism from contaminated environments with XOCs, in which the microorganism naturally acclimatised to tolerate these compounds and developed a specific mechanism to degrade them into simple substances.

Many of the XOCs have been recorded in the soil. These compounds are accumulated in the soil due to the utilisation of pesticides and chemical fertilisers. However, since peatlands are less appropriate for the agriculture purpose, the concentrations of these compounds might be low. Moreover, the justifications for using the peatlands as a source for isolating fungal strains with high efficiency in the biodegradation of XOCs lies in the similarity of the degradation mechanism of XOCs and peatlands composition mainly the lignin. The lignin is a random phenylpropanoid polymer which has a heterogeneous chemical structure and thus is difficult to degrade (Kirk and Farrell 1987).

White Rot Fungi (WRF) are capable to degrade lignin by producing several extracellular enzymes which are called ligninolytic enzymes and include manganese peroxidase (MnP, EC 1.11.1.13), lignin peroxidase (LiP, EC 1.11.1.14), and laccase (LAC, EC 1.10.3.2). These enzymes are the key factor in the oxidation of the lignin and xenobiotic with a compound of aromatic structures (Lee et al. 2014). WRF are basidiomycetes which inhabit the forest ecology and they play an important role in the wood decomposition and xenobiotic biodegradation. Basidiomycetes are the group of fungi which have high exceptional adaptation and grow under detrimental environmental conditions and act as natural degraders of lignocellulose (Choi et al. 2009). The reasons which make WRF isolated from peatlands attractive in the bioremediation of XOCs include wide distribution in the forest soils and their role in the natural decay as well as the ability of these fungi to produce extracellular enzyme required for the degradation of XOCs (hydrophilic or hydrophobic compounds). Moreover, the ligninolytic enzymes produced by WRF obviating the need for internalising the substrates (Marco-Urrea and Reddy, 2012). The WRF does not use XOCs as carbon and energy. A study by Anastasi et al. (2009) has claimed that WRF uses XOCs as a sole carbon source and, therefore, the lignocellulose materials can be used for fungal colonisation as well as bio-carrier for WRF in the wastewater treatment.

There are fungi which have been reported for their ability to produce ligninolytic enzymes such as *Bjerkandera adusta, Ceriporiopsis subvermispora, Dichomitus squalens, Ganoderma lucidum, Mycoaciella bispora, Phanerochaete chrysosporium, Peniophora cinerea, Pleurotus eryngii, Peniophora incarnate, Pleurotus ostreatus, Physisporinus rivulosus, Phanerochaete sordida, Phlebia tremellosa, T. versicolor* and *Trichaptum abietinum* (Moreira et al. 2005; Hakala et al. 2006; Sugiura et al. 2009; Ruiz-Dueñas et al. 2009; Cañas and Camarero 2010; Hofrichter et al. 2010; Lee et al. 2014).

The fungal ligninolytic enzymes have recently become the focus of much attention due to their possible biotechnological applications (Lee et al. 2014). MnP and laccases enzymes have been reported as the main enzymes which degrade of XOCs. The peatlands in Malaysia is classified as a wood peatlands, therefore, it might represent an exciting medium for laccase and peroxides producing fungi. The applications of WRF which produce laccase and peroxidases in detoxification a vast range of xenobiotic environmental pollutants are gaining increasing importance (Michel et al. 1991).

The successful enzymatic bioremediation of polycyclic aromatic hydrocarbons (PAHs), which is a class of XOCs, by fungi requires the selection of species with desirable characteristics. Lee et al. (2014) tested the ability of 55 white rot fungi species, for their performance in dye decolorization and ligninolytic enzymes as well as tolerance for fluoranthene, anthracene, phenanthrene, and pyrene. The screening for the potential of fungal isolated for the degradation of PHAs was conducted by using Remazol brilliant blue R (RBBR) as a substrate. It is due to the decolorization of the dye that indicates the presence of lignin metabolism pathway in the fungal cells (Anastasi et al. 2009). However, in the bioassay for ligninolytic enzymes production, the XOCs should be used as a substrate to induce the enzyme production. A study conducted by Lee et al. (2014) indicated that six isolates identified as *P. incarnate, P. cinerea, P.sordida, T. abietinum, M. bispora* and *P. tremellosa* exhibited the high tolerance to polycyclic aromatic hydrocarbons (PAHs) and accelerated the decolourising of dye within ten days. *P. incarnata* and *P. brevispora* showed a significant degradation of PAHs. The degradation percentage of fluoranthene, phenanthrene, and pyrene by *P. incarnata* exceeded 90%, which indicated that these fungi have a metabolic pathway and can use the XOCs as the sole carbon source. The efficiency of this fungus to degrade PAHs is associated with the high activity of ligninolytic enzymes (MnP, LiP, and laccase). Therefore, *P. incarnata* was the most potent and applicable fungus for the degradation of xenobiotic compounds in the environment.

Martirani et al. (1996) revealed that *P. ostreatus* mycelium has the efficiency to detoxify phenols in the olive wastes without any externally added nutrient. Nevertheless, in some cases, where the fungal growth is weak the addition of simple or complex nutrient is essential in order to obtain an efficient decolorization (Sayadi and Ellouz 1993, 1995). The utilisation of WRF in the degradation of structurally diverse xenobiotic organic pollutants is enormous.

The degradation of XOCs in the soil and some of the industrial wastewater by several fungi species have been reported in the literature. The most common fungi which have used in the treatment of industrial wastewater with XOCs are *A. niger, B. adusta, Phanerochaete chyrosporium, T. versicolor*, and *Trametes hirsute* (Livernoche et al. 1983; Aust 1990; Cripps et al. 1990; Bhole et al. 2004; Mohorcic et al. 2004). However, very few studies were conducted on the utilisation of fungi as an advance treatment technology for the degradation of these compounds in the greywater. This might be due to the strategies used in the treatment of the greywater where these wastes are combined with the black water and then subject to the traditional treatment process in the wastewater treatment plants. Moreover, most of the international standards have not adopted a regulation for the XOCs in the greywater except for some European countries. Denmark has created regulations for the XOCs concentration in the treated greywater.

Eaton et al. (1980) used *P. chyrosporium* for treating pulp and paper industrial effluents. The results revealed high efficiency for *P. chyrosporium* in degrading polymeric lignin molecules and thus removing of colours. The high potential for *P. chyrosporium* in degrading XOCs and their derivatives lies in their ability to produce isoenzymes, including lignin peroxidases and Mn-dependent peroxidases. These isoenzymes are capable to degrade and chlorinate lignins (Lankinen et al. 1990). *P. chyrosporium* is among those wood rotting fungi which have a higher capability to degrade a wide range of xenobiotic compounds, including azo dyes (Aust 1990; Cripps et al. 1990). In a view for greywater compositions, the dyes such as p-phenylenediamine, toluene- 2,5-diamine and 2,5-diaminoanisole are among XOCs which resulted from hair colourants and food additives (Grčić et al. 2015). Therefore, the utilisation of fungi in the degradation of dyes compounds in the greywater is applicable.

Laccases (benzenediol: oxygen oxidoreductase) are extracellular glycoproteins with four copper blue oxidases, presented in wide plants and fungal species (Songulashvili et al. 2012). It was described and extracted by Yoshida in 1883 from the exudates of *Rhus vernicifera* (Thurston 1994). It is one of the very robust enzymes for the transformation and degradation of XOCs to generate free radicals and water, the oxygen being the only co-substrate required for the activity (Oberdörster and Cheek 2001). Laccase enzyme was reported in *Aspergillus* spp., *Penicillium* spp. *Curvularia* spp. (Scherer and Fischer 1998), while *Trametes* spp. and *Pleurotus* spp. are among WRF which recorded production of laccase.

Margot et al. (2013) revealed that *T. versicolor* produces laccase 20 times more than *Streptomyces cyaneus*, which was the best among four *Streptomyces* spp. investigated in the study and including *S. cyaneus, S. psammoticus S. griseus* and *S. ipomoea*. The study also compared between the laccase from *S. cyaneus* (LSc) and that from *T. versicolor* (LTv) in terms of the activity, XOCs oxidation efficiency and stability as a response to pH, temperatures and substrate range. The results revealed that LTv exhibited high activity within a wide range of pH (acidic to near-neutral pH) and temperature ranged from 10 to 25 °C. The laccase from *T. versicolor* occurred rapid degradation of bisphenol A, mefenamic acid and diclofenac in comparison to the one secreted by *S. cyaneus*. These findings indicated that the *T. versicolor* is more applicable for degradation of XOCs in the wastewater than *S. cyaneus*. However, different fungal strains have unique properties of the laccase in terms of catalytic efficiency, and tolerance to stress environmental conditions (Dantán-González et al. 2008).

Ramírez-Cavazos et al. (2014) purified two thermostable laccases (Lac I and Lac II) from *Pycnoporus sanguineus* and investigated the potential of these enzymes for the degradation of EDCs. Both enzymes have thermostability at 50 and 60 °C and act effectively on 2,6-dimethoxyphenol (DMP), 2,2-azino-bis (3-ethylbenzthiazoline-6-sulfonate (ABTS), and guaiacol. The enzyme exhibited high efficiency (95%) for the degradation of nonylphenol and triclosan at pH 5 and within 8 h of the treatment process. Since the optimal pH for this enzyme was pH 4, they are not suitable for the degradation of XOCs in the greywater with pH ranging from 6 to 8. Many of the others WRF strains have been used for the degradation of nonylphenol and triclosan such as *T. versicolor*, *Coriolopsis polyzona*, *Cerrena unicolor* and *C. gallica* (Cabana et al. 2007; Songulashvili et al. 2012; Torres-Duarte et al. 2012). Hence, there are a wide variety of the fungi which might be used for the degradation of XOCs, but the source of fungal isolates should be considered in order to be more applicable.

9.5   Bio-Carrier System

One of the main factors which play an important role in improving the biodegradability of XOCs lies in inducing the microorganism to produce the specific enzyme. In the biodegradation of XOCs, the manganese peroxidase and laccase are the specific enzymes for the degradation of XOCs. Some of the XOCs are inducible for fungi to produce the manganese peroxidase and laccase, while others have no inducible effect.

It has been described in the literature that the cultivation of WRF in the presence of lignocellulosic residues has a significant stimulation of the ligninolytic enzyme secretion (Elisashvili et al. 2009). Therefore, in order to achieve high biodegradation efficiency for XOCs, the fungi should be carried on a bio- carrier consisting of inducible substances for manganese peroxidase and laccase production.

In the peatland, the production of manganese peroxidase and laccase enzymes by WRF is induced by lignin components. Hence, the use of wood chips as a bio-carrier might be an option. In spite, the presence of Cu in the medium has also been reported as the inducer and acted as direct stimulators for laccase production (Songulashvili et al. 2007). It was confirmed by several authors that the greywater contains several macro-elements ions with low concentrations. The use of wood chips as a bio-carrier and loaded with laccase producing fungi for the degradation of XOCs in the greywater is possible, but it needs to be confirmed in the laboratory.

9.6   Mechanism of XOCs Degradation by Fungi

The mechanisms of XOCs by fungi are reported to include biosorption, biodegradation and enzymatic mineralisation (MnP, LiP, manganese independent peroxidase (MIP) and Laccase) (Wesenberg et al. 2003). However, one or more of these mechanisms could be involved in the colour removal, depending on the fungus used. It has been mentioned that *P. chyrosporium* degrades a wide variety of XOCs via a non-specific H_2O_2 dependent extracellular lignin degradation enzyme system (Cripps et al. 1990). A study by Abadulla et al. (2000) indicated that *T. hirsuta*is is highly potent to produce laccase in the degradation of textile dyes. The study revealed that the depredation of textile dyes took place by enzyme remediation with laccases.

The degradation levels of XOCs might be detected directly by determining XOCs studies in the treated wastewater by GC-MS or based on the reduction in COD which might reach the 97.8% (Jin et al. 1999). According to Al-Ahmad et al. (1999), the tested compound is classified as biodegradable if the percentage of oxygen consumed in the test vessel exceeds 60% of the highest theoretical consumption during 10 days.

The ability of WRF to produce extracellular ligninlytic enzymes make them attractive to degrade a broad array of XOCs such as polychlorinated biphenyls (PCBs), dioxins, trinitrotoluene, petroleum hydrocarbons, herbicides, industrial dye effluents and pesticides (Aust 1990; Reddy 1995; Pointing and Vrijmoed 2000). Laccase enzyme has a low specificity to the reducing substrate, which means that it has the ability to oxidise a wide spectrum of aromatic compounds such as ortho- and para-diphenols, aminophenols, polyphenols, and aromatic or aliphatic amines. It can also catalyse the oxidation from different types of XOCs such as pesticides and chlorinated phenolics (Torres et al. 2003; Pozdnyakova et al. 2004). The enzyme acts on the phenolic units, leading to cleavage of C oxidation, C-C and aryl-alkyl bonds. This reaction takes place by removing an electron and a proton of the OH group to form water (Thurston 1994; Viswanath et al. 2014). In polyphenols, aromatic amines and methoxy-substituted monophenols, laccases are able to reduce one molecule of dioxygen to generate free radical which is undergoing for the polymerization and partial precipitation (Bourbonnais et al. 1995; Archibald et al. 1997; Niku-Paavola and Viikari 2000).

9.7   Conclusion

It can be concluded that the peatlands contain several species of fungi that might have high environmental values. Biodegradation of XOCs by the fungi is found as one of the alternative green technology depending on the enzymatic reactions of peroxidase and laccase enzymes. The efficiency of peroxidase and laccase for degrading of XOCs has been reported in the literature. Therefore, it has high potential to be used as an alternative technology for treating greywater.

Acknowledgments The authors wish to thank the Ministry of Higher Education (MOHE) for supporting this research under FRGS vot 1574 and also the Research Management Centre (RMC) UTHM for providing grant IGSP U682 for this research.

References

Abadulla E, Tzanov T, Costa S, Robra KH, Paulo AC, Gubitz GM (2000) Decorization and deyoxification of textile dyes with a laccase from *Trametes hirsuta*. Appl Environ Microbiol 66(8):3357–3362
AbdulKarim MI, Daud NA, Alam MDZ (2011). Treatment of palm oil mill effluent using microorganisms. In: Alam MDZ, Jameel AT, Amid A (eds) Current research and development in biotechnology engineering at International Islamic University Malaysia (IIUM) Vol. III. IIUM Press, Kuala Lumpur, pp 269–275. ISBN 9789674181444
Ainsworth GC (1971) The fungi: an advanced treatise Vol. 4B: taxonomic review with keys, Basidiomycetes and lower fungi. Academic Press
Ainsworth GC, Sparrow FK, Sussman AS (1973) The Fungi, Vol. IVA. A: A taxonomic review with keys: ascomycetes and fungi imperfecti. Academic Press, New York, NY
Al-Ahmad A, Daschner FD, Kümmerer K (1999) Biodegradability of cefotiam, ciprofloxacin, meropenem, penicillin G, and sulfamethoxazole and inhibition of waste water bacteria. Arc Environ Cont Toxicol 37(2):158–163
Alexopoulos CJ (1962) Introductory mycology. Introductory mycology
Alexopoulos CJ, Mims CW (1979) Introducción a la Micología (No. QK603. A4318 3A ED) Eudeba
Alexopoulos CJ, Mims CW, Blackwell M (1996) Introductory mycology, 4th edn. Wiley, New York, USA, 868 p. ISBN: 978-0-471-52229-4
Al-Gheethi AAS (2015) Recycling of sewage sludge as production medium for cellulase by a Bacillus megaterium strain. Int J Rec Organic Waste Agr 4(2):105–119
Anastasi A, Prigione V, Cas L, Casieri L, Varese GC (2009) Decolourisation of model and industrial dyes by mitosporic fungi in different culture conditions. World J Microbiol Biotechnol 25:1363–1374
Archibald FS, Bourbonnais R, Jurasek L, Paice MG, Reid ID (1997) Kraft pulp bleaching and delignification by *Trametes versicolor*. J Biotechnol 53(2–3):215–236
Aust SD (1990) Degradation of environmental pollutants by *Phanerochaete chyrosporium*. Microb Ecol 20:197–209

Bako SP, Chukwunonso D, Adamu AK (2008) Bioremediation of refinery effluents by strains of *Pseudomonas aeruginosa* and *Penicillium janthinellum*. Appl Ecol Environ Res 6(3):49–60

Bhadury P, Bridge PD, Austen MC, Bilton DT, Smerdon GR (2009). Detection of fungal 18S rRNA sequences in conjunction with marine nematode 18S rRNA amplicons. Aquat Biol, 5:149–155

Bhole BD, Ganguly B, Madhuram A, Deshpande D, Joshi J (2004) Biosorption of methyl violet, basic fuchsin and their mixture using dead fungal biomass. Curr Sci 86(12):1641–1645

Bourbonnais R, Paice MG, Reid ID, Lanthier P, Yaguchi M (1995) Lignin oxidation by laccase isozymes from *Trametes versicolor* and role of the mediator 2,2′-azinobis(3-ethylbenzthiazoline-6-sulfonate) in kraft lignin depolymerisation. Appl Environ Microbiol 61(5):1876–1880

Bowman SM, Free SJ (2006) The structure and synthesis of the fungal cell wall. BioEssays 28(8):799–808

Cabana H, Jiwan JLH, Rozenberg R, Elisashvili V, Penninckx M, Agathos SN, Jones JP (2007) Elimination of endocrine disrupting chemicals nonylphenol and bisphenol A and personal care product ingredient triclosan using enzyme preparation from the white rot fungus *Coriolopsis polyzona*. Chemosphere 67(4):770–778

Cajthaml T, Křesinová Z, Svobodová K, Möder M (2009) Biodegradation of endocrine-disrupting compounds and suppression of estrogenic activity by ligninolytic fungi. Chemosphere 75(6):745–750

Cañas AI, Camarero S (2010) Laccases and their natural mediators: biotechnological tools for sustainable eco-friendly processes. Biotechnol Adv 28:694–705

Choi Y-S, Kim G-H, Lim YW, Kim SH, Imamura Y, Yoshimura T, Kim J-J (2009) Characterization of a strong CCA-treated wood degrader, unknown Crustoderma species. Antonie Van Leeuwenhoek 95(3):285–293

Cole GT, Samson RA (1979) Patterns of development in conidial fungi. Pittman, London, United Kingdom

Copeland HF (1956) The classification of lower organisms. Pacific Books, Palo Alto

Cripps C, Bumpus JA, Aust SD (1990) Biodegradation of azo and heterocyclic dyes by *Phanerochaete chyrosporium*. Appl Environ Microbiol 56:1114–1118

Cunha KC, Sutton DA, Fothergill AW, Gené J, Cano J, Madrid H, Hoog S, Crous PW, Guarro J (2013) In vitro antifungal susceptibility and molecular identity of 99 clinical isolates of the opportunistic fungal genus Curvularia. Diag Microbiol Inf Dis 76(2013):168–174

Dantán-González E, Vite-Vallejo O, Martínez-Anaya C, Méndez-Sánchez M, González MC, Palomares LA, Folch-Mallol J (2008) Production of two novel laccase isoforms by a thermotolerant strain of *Pycnoporus sanguineus* isolated from an oil-polluted tropical habitat. Int Microbiol 11(3):163–169

Deacon J (2005) Fungal Biology, 4th edn. Blackwell Publishing Ltd., Malden. https://doi.org/10.1002/9781118685068

Diba K, Kordbacheh P, Mirhendi SH, Rezaie S, Mahmoudi M (2007) Identification of *Aspergillus* species using morphological characteristics. Pak J Med Sci 23(6):867–872

Eaton D, Chang H, Kirk TK (1980) Fungal decolorization of kraft bleach plant effluent. Tappi J 63:103–106

Efaq AN, Rahman NNNA, Nagao H, Al-Gheethi AA, Kadir MA (2017) Inactivation of Aspergillus spores in clinical wastes by supercritical carbon dioxide. Arab J Sci Eng 42(1):39–51

Elisashvili V, Kachlishvili E, Tsiklauri N, Metreveli E, Khardziani T, Agathos SN (2009) Lignocellulose-degrading enzyme production by white-rot Basidiomycetes isolated from the forests of Georgia. World J Microbiol Biotechnol 25(2):331–339

Ellis MB (1971) Dematiaceous hyphomycetes. Commonwealth Mycological Institute, Kew

Emine S, Kambol, R, Zainol N (2010) Morphological characterization of soil *Penicillium* sp. Strains—Potential producers of statin. In: Biotechnology symposium IV, 01–03 Dec 2010, Universiti Malaysia Sabah, Sabah, Malaysia

Enayatizamir N, Tabandeh F, Rodriguez-Couto S, Yakhchali B, Alikhani HA, Mohammadi L (2011) Biodegradation pathway and detoxification of the diazo dye Reactive Black 5 by *Phanerochaete chrysosporium*. Biores Technol 102(22):10359–10362

Fogarty RV, Tobin JM (1996) Fungal melanins and their interactions with metals. Enz Microb Technol 19(4):311–317

Gao L, Sun MH, Liu XZ, Che YS (2007) Effects of carbon concentration and carbon to nitrogen ratio on the growth and sporulation of several biocontrol fungi. Mycol Res 111(1):87–92

Glaser PH, Chanton JP, Morin P, Rosenberry DO, Siegel DI, Ruud O, Reeve AS (2004) Surface deformations as indicators of deep ebullition fluxes in a large northern peatland. Global Biogeochem Cycles 18(1)

Grčić I, Vrsaljko D, Katančić Z, Papić S (2015) Purification of household greywater loaded with hair colorants by solar photocatalysis using TiO_2-coated textile fibers coupled flocculation with chitosan. J Water Process Eng 5:15–27

Guarro J, Gene J, Stchigel AM (1999) Developments in fungal taxonomy. Clin Microbiol Rev 12(3):454–500

Gyaurgieva OH, Bogomolova TS, Gorshkova GI (1996) Meningitis caused by *Rhodotorula rubra* in an HIV infected patient. J Med Vet Mycol 34:357–359

Hakala TK, Hildén K, Maijala P, Olsson C, Hatakka A (2006) Differential regulation of manganese peroxidases and characterization of two variable MnP encoding genes in the white-rot fungus *Physisporinus rivulosus*. Appl Microbiol Biotechnol 73:839–849

Hardison MT, Brown MD, Snelgrove RJ, Blalock JE, Jackson P (2012) Cigarette smoke enhances chemotaxis via acetylation of proline-glycine-proline. Front Biosci (Elite edition) 4:2402–2409

Hendriks L, De Baere R, Van De Peer Y, Neefs J, Goris A, De Wachter R (1991) The evolutionary position of the rhodophyte *Porphyra umbilicalis* and the basidiomycete *Leucosporidium scottii* among other eukaryotes as deduced from complete sequences of small ribosomal subunit RNA. J. Mol. Evol 32:167–177

Hofrichter M, Ullrich R, Pecyna MJ, Liers C, Lundell T (2010) New and classic families of secreted fungal heme peroxidases. Appl Microbiol Biotechnol 87:871–897

Hollaway Stephen L, Faw Gary M, Sizemore Ronald K (1980) The bacterial community composition of an active oil field in the Northwestern Gulf of Mexico. Marine Poll Bull 11(6):153–156

Huat BK (2004) Organic and peat soils engineering. University Putra Malaysia Press, Serdang

Jahangeer S, Khan N, Jahangeer S, Sohail M, Shahzad S, Ahmad A, Khan SA (2005) Screening and characterization of fungal cellulases isolated from the native environmental source. Pakistan J Bot 37(3):739

Jensen AB, Aronstein K, Flores JM, Vojvodic S, Palacio MA, Spivak M (2013) Standard methods for fungal brood disease research. J Apic Res 52(1):1–20

Jin B, van Leeuwen J, Patel B (1999) Production of fungal protein and glucoamylase by *Rhizopus oligosporus* from starch processing wastewater. Process Biochem 34:59–65

Joosten H, Clarke D (2002). Wise use of mires and peatlands—background and principles including a framework for decision-making. International Mire Conservation Group/International Peat Society, 304 pp

Kalantari B, Prasad A (2014) A study of the effect of various curing techniques on the strength of stabilized peat. Transp Geotech 1(3):119–128

Kazemian S, Prasad A, Huat BB, Barghchi M (2011) A state of art review of peat: geotechnical engineering perspective. Int J Phy Sci 6(8):1974–1981

Kim YJ, Nicell JA (2006) Laccase-catalysed oxidation of aqueous triclosan. J Chem Technol Biotechnol 81(8):1344–1352

Kirk TK, Farrell RL (1987) Enzymatic combustion: the microbial degradation of lignin. Annu Rev Microbiol 41:465–505

Kirk PM, Cannon PF, Minter DW, Stalpers JA (2008) Dictionary of the fungi, 10th edn. CABI, Wallingford, UK

Kumara KLW, Rawal RD (2008) Influence of carbon, nitrogen, temperature and pH on the growth and sporulation of some Indian isolates of Colletotrichum gloeosporioides causing anthracnose disease of papaya (Carrica papaya L). Trop Agric Res Ext 11:7–12

Lankinen VP, Inkeroinen MM, Pellien J, Hatakka AI (1990) The onset of lignin modifying enzyme, decrease of AOX and colour removal by white rot fungi: growth on bleach plant effluent. Water Sci Technol 24:189–198

Lee H, Jang Y, Choi YS, Kim MJ, Lee J, Lee H, Kim JJ (2014) Biotechnological procedures to select white rot fungi for the degradation of PAHs. J Microbiol Methods 97:56–62

Leslie JF, Summerell BA (2006) The Fusarium Laboratory manual. Blackwell Publishing Ltd., Iowa

Livernoche D, Jurasek L, Desrochers M, Dorica J (1983) Removal of colour from kraft mill wastewater with cultures of white rot fungi ad with immobilized mycelium of *Coriolus versicolor*. Biotechnol Bioeng 25:2055–2065

Marco-Urrea E, Reddy CA (2012) Degradation of chloro-organic pollutants by white rot fungi. In: Microbial degradation of xenobiotics. Springer, Berlin, pp 31–66

Margot J, Bennati-Granier C, Maillard J, Blánquez P, Barry DA, Holliger C (2013) Bacterial versus fungal laccase: potential for micropollutant degradation. AMB Express 3(1):63

Martirani L, Giardina P, Marzullo L, Sannia G (1996) Reduction of phenol content and toxicity in olive oil mill wastewater with the linolytic fungus *Pleurotus ostreatus*. Water Res 30:1914–1918

Mesri G, Ajlouni M (2007) Engineering Properties of fibrous peats. J Geotech Geoenviron Eng 133(7):850–866

Michel FC, Dass SB, Gulkcand EA, Reddy CA (1991) Role of manganese peroxidase and lignin peroxidase of *Phanerochaete chrysosporium* in the decolorization of kraft bleach plant effluent. Appl Environ Microbiol 57:2368–2375

Miranda RC, Gomes EB, Pereira NJ, Marin-Morales MA, Machado KM, Gusmao NB (2013) "Biotreatment of textile effluent in static bioreactor by *Curvularia lunata* URM 6179 and *Phanerochaete chrysosporium* URM 6181. Biores Technol 142:361–367

Mohorcic M, Friedrich J, Pavko A (2004) Decoloration of the diazo dye reactive black 5 by immobilized *Bjerkandera adusta* in a stirred tank bioreactor. Acta Chim Slov 51:619–628

Moreira PR, Duez C, Dehareng D, Antunes A, Almeida-Vara E, Frère JM, Malcata FX, Duarte JC (2005) Molecular characterization of a versatile peroxidase from a Bjerkandera strain. J Biotechnol 118:339–352

Nadeau RR, Singhvi J, Lin I, Syslo J (1993) Monitoring bioremediation for bioremediation efficiency: the marrow marsh experience, proceeding of the 1993 oil spill conference. Am petrol inst, Washington, DC, pp 477–485

Nakada M, Tanaka C, Tsunewaki K, Tsuda M (1994) RFLP analysis for species separation in the genera *Bipolaris* and *Curvularia*. Mycoscience 1994(35):271–278

Niku-Paavola M-L, Viikari L (2000) Enzymatic oxidation of alkenes. J Mol Catal B 10(4):435–444

Nolan C, Margoliash E (1968) Comparative aspects of primary structures of proteins. Ann Rev Biochem 37(1):727–791

Noman EA, Al-Gheethi AA, Rahman NNNA, Nagao H, Kadir MA (2016) Assessment of relevant fungal species in clinical solid wastes. Environ Sci Poll Res 23(19):19806–19824

Oberdörster E, Cheek AO (2001) Gender benders at the beach: endocrine disruption in marine and estuarine organisms. Environ Toxicol Chem 20(1):23–36

Ojumu TV, Bello OO, Sonibare JA, Solomon BO (2005) Evaluation of microbial systems for bioremediation of petroleum refinery effluents in Nigeria. Afr J Biotechnol 4(1):31–35

Okonko IO, Shittu OB (2007) Bioremediation of wastewater and municipal water treatment using latex from *Caloptropis procera* (Sodom apple). Electr J Environ, Agr Food Chem 6(3):1890–1904

Pinholt Y, Struwe S, Kjøller A (1979) Microbial changes during oil decomposition in soil. Ecography 2(3):195–200

Pointing SB, Vrijmoed LLP (2000) Decolorization of azo and triphenylmethane dyes by *Pycnoporus sanguineus* producing laccase as the sole phenoloxidase. World J Microbiol Biotechnol 16:317–318

Pozdnyakova NN, Rodakiewicz-Nowak J, Turkovskaya OV (2004) Catalytic properties of yellow laccase from *Pleurotus ostreatus* D1. J Mol Catal B 30(1):19–24

Promputtha I, Jeewon R, Lumyong S, McKenzie EHC, Hyde KD (2005) Ribosomal DNA fingerprinting in the identification of non sporulating endophytes from *Magnolia liliifera* (Magnoliaceae). Fungal Divers 20:167–186

Ramírez-Cavazos LI, Junghanns C, Ornelas-Soto N, Cárdenas-Chávez DL, Hernández-Luna C, Demarche P, Parra R (2014) Purification and characterization of two thermostable laccases from *Pycnoporus sanguineus* and potential role in degradation of endocrine disrupting chemicals. J Mol Catal B Enzym 108:32–42

Reddy CA (1995) The potential for white-rot fungi in the treatment of pollutants. Curr Opin Biotechnol 6:320–328

Robert AS, János V, Christian FJ (eds) (2011) Taxonomic studies on the genus *Aspergillus*-DTU Orbit. Studies in Mycology. Publication Research—peer-review. Book—Annual report year: 2011. CBS-KNAW Fungal Biodiversity Centre

Ruiz-Dueñas FJ, Morales M, García E, Miki Y, Martínez MJ, Martínez AT (2009) Substrate oxidation sites in versatile peroxidase and other basidiomycete peroxidases. J Exp Bot 60:441–452

Saito T, Kato K, Yokogawa Y, Nishida M, Yamashita N (2004) Detoxification of bisphenol A and nonylphenol by purified extracellular laccase from a fungus isolated from soil. J Biosci Bioeng 98(2004):64–66

Sandhu GS, Kline BC, Stockman L, Roberts GD (1995) Molecular probes for diagnosis of fungal infections. J Clin Microbiol 33:2913–2919

Sayadi S, Ellouz R (1993) Screening of white rot fungi for the treatment of olive mill waste waters. J Chem Tech Biotechnol 57:141–146

Sayadi S, Ellouz R (1995) Roles of lignin peroxidase and manganese peroxidase from *Phanerochaete chrysosporium* in the decolorization of olive mill wastewaters. Appl Environ Microbiol 61:1098–1103

Scherer M, Fischer R (1998) Purification and characterization of laccase II of *Aspergillus nidulans*. Arch Microbiol 170(2):78–84

Schwarz P, Lortholary O, Dromer F, Dannaoui E (2007) Carbon Assimilation Profiles as a Tool for Identification of Zygomycetes. J Clin Microbiol 45(5):1433–1439

Silva DM, Batista LR, Rezende EF, Fungaro MHP, Sartori D, Alves E (2011) Identification of fungi of the genus *Aspergillus* section Nigri using polyphasic taxonomy. Braz J Microbiol 42:761–773

Sivanesan A (1987) Graminicolous species of Bipolaris, Curvularia, Drechslera, Exserohilum, and their teleomorphs. Mycol Pap 1987(158):1–261

Songulashvili G, Elisashvili V, Wasser SP, Nevo E, Hadar Y (2007) Basidiomycetes laccase and manganese peroxidase activity in submerged fermentation of food industry wastes. Enzyme Microb Technol 41:57e61

Songulashvili G, Jimenéz-Tobón GA, Jaspers C, Penninckx MJ (2012) Immobilized laccase of Cerrena unicolor for elimination of endocrine disruptor micropollutants. Fungal biology 116(8):883–889

Sugiura T, Yamagishi K, Kimura T, Nishida T, Kawagishi H, Hirai H (2009) Cloning and homologous expression of novel lignin peroxidase genes in the white-rot fungus *Phanerochaete sordida* YK-624. Biosci Biotechnol Biochem 73:1793–1798

Thormann MN (2006) Diversity and function of fungi in peatlands: a carbon cycling perspective. Canadian J Soil Sci 86:281–293

Thurston CF (1994) The structure and function of fungal laccases. Microbiology 140:19e26

Timnick MB, Lilly VG, Barnett HL (1951) The effect of nutrition on the sporulation of *Melanconium fuligineum* in culture. Mycologia 43(6):625–634

Torres E, Bustos-Jaimes I, Le Borgne S (2003) Potential use of oxidative enzymes for the detoxification of organic pollutants. Appl Catal B 46(1):1–15

Torres-Duarte C, Viana MT, Vazquez-Duhalt R (2012) Laccase-mediated transformations of endocrine disrupting chemicals abolish binding affinities to estrogen receptors and their estrogenic activity in zebrafish. Appl Biochem Biotechnol, 1–13

Tran NH, Urase T, Kusakabe O (2010) Biodegradation characteristics of pharmaceutical substances by whole fungal culture *Trametes versicolor* and its laccase. J Water Environ Technol 8(2):125–140

Veloo R, Paramananthan S, van Ranst E (2014) Classification of tropical lowland peats revisited: The case of Sarawak. CATENA 118:179–185

Vidali M (2001) Bioremediation. An overview. Pure Appl Chem 73(7):1163–1172

Vilgalys R, Hester M (1990) Rapid genetic identification and mapping of enzymatically amplified ribosomal DNA from several Cryptococcus species. J Bacteriol 172:4238–4246

Viswanath B, Rajesh B, Janardhan A, Kumar AP, Narasimha G (2014) Fungal laccases and their applications in bioremediation. Enzyme Res, 2014

Wesenberg D, Kyriakides I, Agathos SN (2003) White-rot fungi and their enzymes for the treatment of industrial dye effluents. Biotechnol Adv 22:161–187

White T, Bruns T, Lee S, Taylor J (1990) Amplification and direct sequencing of fungal ribosomal RNA genes for phylogenetics. In: Innis M, Gelfand D, Sninsky J, White T (eds) PCR protocols. Academic Press Inc., New York, pp 315–322

Whittaker RH (1969) New concepts of kingdoms or organisms. Evolutionary relations are better represented by new classifications than by the traditional two kingdoms. Science 163(3863):150–160

Woese C, Fox G (1977) Phylogenetic structure of the prokaryotic domain: the primary kingdoms. Proc Natl Acad Sci USA 74(11):5088–5090

Wu L, Luo YP, Wan JB, Li SG (2006) Use of *Yarrowia lipolytica* for the treatment of oil/grease wastewater. Res Environ Sci (China) 19(5):122–125

Yamane T (1989) Enzyme technology for the lipid industry. An engineering overview. J Am Oil Chem Soc 64:1657–1662

Chapter 10
Disinfection Technologies for Household Greywater

**Adel Ali Saeed Al-Gheethi, Efaq Ali Noman,
Radin Maya Saphira Radin Mohamed, Balkis A. Talip,
Amir Hashim Mohd Kassim and Norli Ismail**

Abstract The treatment technologies for greywater are followed by the disinfection processes in order to achieve safe disposal into the environment. The disinfection technologies aim at reducing or minimising the concentrations of the pathogenic microorganism of greywater which have a high potential risk for humans and plants, and, thus, provide safe and aesthetically acceptable greywater that is appropriate for the purpose of irrigation. The disinfection processes include chemical (chlorination and ozonation), physical or mechanical (filtration process) and radiation disinfection (UV irradiation, solar disinfection (SODIS)). The degree of the disinfection process proposed must take into account the type of reuse and the risk of exposure to the population. In this chapter, the disinfection techniques of greywater are reviewed and discussed based on their efficiency to eliminate the pathogenic bacteria and other toxic by-products. The objective of this chapter was to discuss the advantages and disadvantages of disinfection processes. Among the several disinfectant technologies for greywater, SODIS appears to be the most potent technology which is widely applicable in most of the developing countries experiencing arid and semi-arid

A. A. S. Al-Gheethi (✉) · R. M. S. Radin Mohamed · A. H. Mohd Kassim
Micro-Pollutant Research Centre (MPRC), Department of Water and Environmental Engineering, Faculty of Civil and Environmental Engineering, Universiti Tun Hussein Onn Malaysia (UTHM), 86400 Parit Raja, Batu Pahat, Johor, Malaysia
e-mail: adel@uthm.edu.my

R. M. S. Radin Mohamed
e-mail: maya@uthm.edu.my

E. A. Noman
Faculty of Applied Sciences and Technology (FAST), Universiti Tun Hussein Onn Malaysia (UTHM), Pagoh, Johor, Malaysia

E. A. Noman
Department of Applied Microbiology, Faculty Applied Sciences, Taiz University, Taiz, Yemen

B. A. Talip
Faculty of Applied Sciences and Technology (FAST), Universiti Tun Hussein Onn Malaysia (UTHM), 84000 KM11, Jalan Panchor, Pagoh Muar, Johor, Malaysia

N. Ismail
Environmental Technology Division, School of Industrial Technology, Universiti Sains Malaysia (USM), 11800 George Town, Penang, Malaysia

© Springer International Publishing AG, part of Springer Nature 2019   185
R. M. S. Radin Mohamed et al. (eds.), *Management of Greywater in Developing Countries*,
Water Science and Technology Library 87,
https://doi.org/10.1007/978-3-319-90269-2_10

atmospheric conditions due to the high density of sunlight which is more effective for inactivating pathogenic microorganisms.

Keywords SODIS · AOPs · Pathogenic bacteria · PGP · Non-culture methods

10.1 Introduction

The disinfection of greywater is the last stage of the greywater treatment process. It should be conducted after the suspended solids and organic matter are removed to enhance the inactivation of pathogenic organisms. In the disinfection process, the greywater is subjected to the chemical and physical processes in which the concentrations of the infectious agents are reduced to less than the detection limits. However, this process exhibits different efficiency in the inactivation percentages which depends on the mechanism of the inactivation process as well as the characteristics of the greywater. The concept of the disinfection processes is quite different for the sterilisation processes which aimed to irreversible inactivation of the pathogens. The disinfection technologies are divided into chemical (chlorination and ozonation) physical or mechanical (filtration process, heat and pasteurisation) and radiation disinfection (UV irradiation, SODIS) or combination of these techniques. In the physical disinfection process, the pathogens are inactivated as a result of the destruction of the cells, where the target is to damage the cell morphology and then their functions. In the filtration techniques, the pathogens are adsorbed on the surface of the filter as live cells. In contrast, the chemical disinfection acts by inactivation of the metabolic and anabolic pathways by the inactivation of the main enzymes used in these processes. The radiation disinfection acts by the damage of the DNA of the cells which may lead to the inhibition of the protein synthesis and the cell growth. The degree of disinfection process proposed must consider the type of reuse and the risk the exposure will pose to the population. It has to be mentioned that the microbial loads in the greywater are fewer than that in the blackwater. Therefore, the disinfection processes provide the alternative resource for the water which might be used for toilet flushing or car washing (WHO 2006). Moreover, the effectiveness of the disinfection process depends on the occurrence of regrowth in the disinfected greywater. This chapter focuses on reviewing various disinfection techniques of greywater based on the efficiency of the treatment system to eliminate the pathogenic bacteria and other toxic by-products. The advantages and disadvantages of the techniques used for the inactivation of pathogens in the greywater as well as the detection of the potential of pathogenic growth in the disinfected greywater are discussed.

10.2   Disinfection Technologies of Greywater

The disinfection of the greywater is a treatment process that has been widely reported by several authors in the literature. Many of the technologies have been suggested for the reduction of infectious agents such as pathogenic bacteria, parasites and viruses. However, the reduction of bacteria in the treated greywater is the major challenge confronting researchers due to their ability to regrow in the disinfected greywater. This is because they have no intermediate hosts compared to the parasites and viruses. Therefore, the disinfection processes should have the potential to reduce the bacterial cells to be lower than the detection limits. There are some concepts which indicate that the pathogens should be reduced to less than their infective dose, but there is only limited dose–response information available for pathogenic bacteria and viruses, while the infective dose for most of the pathogens still remains unconfirmed because it depends on the immunity of the host (Rowe and Abdel-Magid 1995). The most common technologies used in the disinfection of greywater are presented in the next session.

10.2.1   Filtration Systems

The filtration system is a simple and inexpensive technology for the removal of pathogens from the water. One of the effective filtration systems is the slow sand filtration system. The principle of the system depends on the passing of the water or greywater slowly through a chamber or multi-chambers consisting of a bed of porous sand and gravel layers. The efficiency of this system in removing the pathogenic cells depends on the pores size and flow rate since the rapid filtration results in less removal efficiency. This slow sand filtration system is more applicable in the developing countries as a non-central system. The sand filtration removes the pathogens by decrease the pore size due to trapped of the pores by the particles presented in the greywater to be less than the diameter of the pathogens cell size. However, the increase in the level of suspended solids might lead to clogging of the pores and reduce the removal efficiency. The biofilter system has also been used as the alternative technology for the removal of pathogens and biodegradable organic matters simultaneously (Haarhoff and Cleasby 1991). In this system, the degradation of organic matter takes place in the natural microorganisms present in the sand, which might act as decomposers for the organic matter from the greywater.

The mechanical filtration system is effective for the removal of organic matter but not for the inactivation of pathogenic bacteria, which are active even after the separation from the greywater (Al-Gheethi et al. 2016). Therefore, the use of membrane bioreactors such as membrane chemical reactor and membrane bioreactor (MBR) is getting research attention by authors. In this system, two or more of the processes are combined such as aerobic reactor with UV and titanium dioxide. Winward et al. (2008) indicated that this system removed TC and *Pseudomonas aeruginosa* by 4.0 and 2.0 $\log_{10}$ CFU 100 mL^{-1}, respectively. Jong et al. (2010) revealed that this system removed *Escherichia coli* by 67.5, *Staphylococcus aureus* by 27.7 and *Salmonella typhimurium* by 20.4%. The submerged membrane bioreactor (SMBR) system which is another system of the bioreactors exhibited 99.99% of the TC and FC removal within 42 days of treatment (Bani-Melhem et al. 2015). These findings indicate that the filtration system has better efficiency for the removal of pathogens but the limitations are its need for prolonged periods to achieve high removal percentage and their less efficiency in the highly polluted wastes.

10.2.2 Chemical Disinfection

The chemical disinfection techniques include bromine chloride, chlorine, calcium hypochlorite, hydrogen peroxide and ozone. Chlorination is the most common process which is used extensively for the disinfection of water and wastewater due to the low cost and the simple usage as well as its effectiveness for the inactivation of most of the infectious agents (Ottosson 2003). The mechanism in which the chlorine deactivates the pathogenic cells is through the generation of chlorine radical which acts as the oxidation of organic components of the cells such as enzymes. The behaviour of the pathogens in the greywater for the disinfection through chlorination depends on the organic matter present in the water, since the high contents of the organic compounds might consume the chlorine by the oxidation reactions, and thus the chlorine residues are not enough to achieve high reduction on the microbial loads (Al-Gheethi et al. 2016). In order to overcome this challenge, some of the authors suggested that a preliminary treatment process for the removal of the organic content should be undertaken before chlorination (Santasmasas et al. 2013). The turbidity of the greywater might also affect negatively the disinfection efficiency, therefore a double dose of chlorination is needed (Mohamed et al. 2015).

The adverse effect of chlorination is the toxicity to aquatic life as a result of the presence of free and combined chlorine residues which are often classified as carcinogenic compounds. Some of these compounds include nitrosodimethylamine (NDMA), trihalomethane (TTHM) and haloacetic acid (HAA) (Cantor et al. 1987; Pehlivanoglu-Mantas et al. 2006). Therefore, in the USA, chlorination is not used for the disinfection of drinking water, although it is still in use in most of the developing countries due to its low cost. Other limitations for the utilisation of chlorination are the microbial resistance and the regrowth which is associated with the highly reactive characteristics that may lead to accelerating the chlorine decay process (Tal et al. 2011). This limitation occurs more in the antibiotic-resistant bacteria since the resistance to the antibiotics is correlated with that for chlorine (Shi et al. 2013). In greywater, the potential of the pathogenic bacteria to resist for chlorine might be more than the one for the bacteria in the drinking water because the chlorine is available in the greywater generated from the washing machines and kitchens. As a result, the bacteria have developed a mechanism for the resistance of chlorine residues and became more resistant during the disinfection process by the chlorine (Al-Gheethi et al. 2016).

The chlorine has less efficiency against parasites and protozoa organism and therefore, the ozonation is the alternative most effective method for inactivating protozoan cysts (Robertson et al. 1994). Ozonation is one of the most efficient methods for the disinfection of waters due to the high oxidative potential of ozone which could lead to the destruction of semipermeable membrane of the cells and, as a result, bacterial cell death (Facile et al. 2000). The extensive use of the ozonation is due to the cheap and low energy needed for the inactivation as well as the fast reactions which can reduce the pathogenic bacteria cells by 98% in less than 5 min (Tripathi et al. 2011). In the greywater, the factors which might affect negatively on the efficiency of ozonation are the chemical oxygen demand (COD) and total suspended solids (TSS). Both parameters might also induce the bacterial resistance and the regrowth in the disinfected greywater (Janex et al. 2000; Xu et al. 2002). Nevertheless, the occurrence of toxic by-products by ozonation such as aldehydes, bromate ions (BrO_3) and peroxides has also been reported (Vital et al. 2010).

10.2.3   UV Irradiation

The application of UV irradiation is one of the radiation disinfection processes which is used for the greywater reuse excluding the toilet flush water due to the high contents of the suspended solids which prevent the penetration of UV. This technology is more appropriate for the greywater treated with constructed wetland (Lindgren and Grette 1998). UV irradiation is an advanced disinfection method, in comparison with chlorination, and has several advantages including the applicability for small-scale treatment plants without the need for dosing apparatus or a storage process of disinfected greywater as well as the absence of toxic by-products (USEPA 2003).

The mechanism in which UV deactivate the pathogen cells lies in the localisation lesions in DNA as a result of the mutation caused by the formation of pyrimidine dimers particularly thymine (Smith and Hanawalt 1969). The selection of UV with 260 nm for the disinfection of water and inactivation of microbial cells was based on the laboratory results which revealed that the high absorption spectrum of DNA by the spectrophotometer is 260 nm (Setlow 1968). UV disinfection occurred due to the high reduction of many of the pathogenic bacteria, however, other bacterial species have also exhibited a resistance for the UV action, due to their ability to repair the damage caused in the DNA by UV. The disinfection of the greywater using UV might be more applicable in terms of the absence of residual disinfection by-products as in the case of chlorination. In the drinking water, the presence of residual chlorine in the disinfected water might be necessary in order to face any possible contamination. Conversely, these precautions might also be necessary for the disinfected greywater to prevent the regrowth of inactivated pathogens since the chlorination would not totally destroy them, and some might remain available in the dormant state and might grow back in the absence of residual chlorine as in the disinfected greywater with UV. Therefore, the UV technology is an alternative for the chlorination in terms of the absence of toxic by-products, but the chlorination might be the best option for the microbial regrowth (Chang et al. 1985).

In fact, the authors have demonstrated regrowth of pathogens inactivated by UV or chemical disinfections. In the UV disinfection, the regrowth appears as a result of dark repair of damages caused by the UV in the DNA structure. In contrast, in the case of the chemical disinfection, the regrowth occur due to the oxidation process of organic matter which could lead to producing AOC which are nutrients inducible for the microbial growth (Al-Gheethi et al. 2015). Gilboa and Friedler (2008) studied the UV disinfection kinetics and the efficiency for the inactivation of HPC, FC, *S. aureus* and *P. aeruginosa* as well as the survival and regrowth of these pathogens in the disinfected greywater. The greywater samples were first treated with rotating biological contactor (RBC) followed by sedimentation. The inactivation rate coefficient of UV with 69 mWS cm^{-2} of the dose was 0.0687, 0.201 cm^2 mW^{-1}S^{-1} for FC, 0.113 for HPC, 0.129 for *P. aeruginosa* and 0.201 cm^2 mW^{-1}S^{-1} for *S. aureus*. Among these pathogens, FC exhibited the highest resistance for the UV, the microscopic examination indicated that was due to the FC self-aggregate in the greywater. FC, *S. aureus* and *P. aeruginosa* have no regrowth in the next 6 months of the disinfection process with UV doses (19–439 mWS cm^{-2}). In contrast, the regrowth of HPC in the disinfected greywater was explained because of the absence of the competition with other bacteria which were eliminated by the irradiation.

10.2.4 Solar Disinfection (SODIS)

The utilisation of SODIS appeared to be a promising disinfection technology for the greywater when these waters are used for toilet flushing or car washing. More attention and research should be conducted to evaluate this technology in the reduction of pathogens in the greywater if these waters will be used for irrigation of disposal for the natural water systems. Since some of the studies indicated that the efficiency of SODIS might be low against some pathogens due to the ability of these infectious agents to regrow after the disinfection process. SODIS is more applicable for the developing countries especially those located in the arid and semi-arid region because the solar radiation might reach more than at least 500 W/m^2 which is the minimum solar radiation required to achieve an acceptable reduction in the microbial load of the greywater. In SODIS, two or more factors are the keys for the inactivation of the pathogens which include the temperature and UV radiation. The photocatalysis by the visible light might also play an important role in the reduction of pathogens because this process could lead to the oxidation of organic matter and then the release of the inhibitory substances for pathogen growth.

The temperature is the main factor in the inactivation of pathogens cells because it leads to deactivation or destruction of cell enzymes, but it should be more than the optimal temperature required for the microbial growth. Since the ambient temperature has no significant role in the destruction of the pathogenic cell. One of the solutions used for increasing the temperature to be between 45 and 50 °C is to use transparent polyethylene terephthalate (PET). However, the use of PET is not applicable for the large-scale treatment system and therefore the plastic bag SODIS reactors are the alternative method capable of absorbing a greater quantity of solar radiation and then more reduction of the microbes can be achieved in the water (Walker et al. 2004). The selection of plastic bag polymers depends on the potential of the polymers to transmit high quantities of UV and the stability at significant low unitary cost, therefore, the low-density polyethylene (LDPE) is the most commonly used. In order to achieve high inactivation, the LDPE-SODIS reactors are performed by washing it with H_2O_2 or TiO_2 (Sciacca et al. 2010; Ciavola 2011). In this manner, Harding and Schwab (2012) investigated the efficiency of SODIS with limes and psoralens to enhance the inactivation of *E. coli*. The study revealed that the limes and psoralens exhibited a synergistic effect with UV radiation to accelerate the reduction of microbes. The reduction of *E. coli* was >6.1 logs by SODIS coupled with lime slurry and 5.6 logs by SODIS coupled with lime juice within 30 min of solar exposure, while it was 1.5 log reduction by SODIS alone.

The UV of the sunlight leads to the destruction of the pathogen genome by the destruction of DNA bases. It has been reported that half of the lethal effect for the SODIS is attributed to UV wavelengths which are below 370 nm and the UV wavelengths between 370 and 400 nm as well as the blue–green visible spectrum between 400 and 500 nm (Gameson and Gould 1985). Moreover, it has been noticed by the authors that the SODIS is more effective against the actively metabolising cells, which means that the presence of bacterial cells in the dormant state might make it more resistant for the SODIS.

The effectiveness of SODIS in the inactivation of pathogenic cells belongs to the formation of the reactive oxygen, which leads to the destruction of cell membrane permeability, metabolic and anabolic pathways due to the irreversible destruction of the specific enzymes required in these pathways. Although some reports indicated that the harmful wavelength of the sunlight against the cells is lower than 280 nm, others have claimed that the sunlight has a photodynamic action (Al-Gheethi et al. 2015). This process acts through the induction of the oxidation reactions for the hydroxyl groups on the cell-wall and cell-membrane to generate hydroxyl radicals which may lead to damages of the functional group of the bacterial cells wall such as an absorption of nutrients and transport system through the cell membrane into the cytoplasm. Moreover, the bacterial cells usually grew as colonies and in the water, they grew as biofilm which means they are attached together by the slim layers, and the presence of sunlight might lead to oxidation of the polymers of these slim layers. The bacterial cells grown as a biofilm are more resistant to many of the stress environmental conditions while they are very sensitive as individual cells. Furthermore, unlike the damages caused by UV, the destruction of the DNA structure by the visible light cannot be repaired by the bacterial cells (Eisenstark 1971).

The efficiency of SODIS for the reduction of the indicator and pathogenic bacteria in different water and wastewater has been reported. Al-Gheethi et al. (2013) investigated the reduction of FC, *Enterococcus faecalis*, *Salmonella* spp. and *S. aureus* in the lake water and secondary effluents using SODIS. The samples were placed in the PET bottles for a period of 8 h. The study found that the FC, *Salmonella* spp. and

S. aureus were reduced by 4 $\log_{10}$ CFU/100 mL within 6 h of the SODIS process. However, this period was not enough to eliminate completely these pathogens, since the regrowth assay for these bacteria in the laboratory revealed the ability of the inactivated pathogens to grow in the culture media after the incubation of disinfected samples for 24 h at 37 °C. In contrast, for 8 h the SODIS reduced the level of pathogens to below the detection limits without regrowing after the assay on the culture media. Among the investigated bacteria, *E. faecalis* was not totally eliminated by SODIS even after 8 h. Moreover, the bacteria was more sensitive for the storage system where it has reduced to less than the detection limits (1 CFU/100 mL) after 16 days of the storage at room temperature. These findings indicate that the efficiency of the SODIS depends on the bacterial species. In some cases, the SODIS should be followed by the storage system in order to achieve the high reduction in the bacterial loads.

Bosshard et al. (2009) examined the effect of SODIS on *S. typhimurium* and *Shigella flexneri*, and the inactivation level of SODIS on the pathogens was evaluated based on efflux pump activity, cellular ATP levels, polarisation and integrity of the cytoplasmic membrane, as well as glucose uptake ability. The study revealed that the respiratory chain was the main target of sunlight and UVA irradiation. In order to study the behaviour of the inactivated bacteria after the SODIS, the pathogens were stored in the dark and the results indicated that the physiological state of the cells continued to deteriorate even in the absence of irradiation. These findings concluded that the investigated pathogens are very sensitive for SODIS and a small light dose (700 W m^2) might be enough to irreversibly damage the cells without the need for the chemical additives.

10.2.5 Advanced Oxidation Processes (AOPs)

Advanced oxidation processes (AOPs) are a combination of chemical and irradiation disinfection techniques and more specifically by photocatalytic processes. This technology has been used mainly for the degradation of non-degradable compounds in the wastewater such as XOCs, as discussed in Chap. 9. However, the AOPs are also efficient in the disinfection processes. Examples for the AOPs include the combination of titanium dioxide (TiO_2) and UV (TiO_2-UV). TiO_2 alone is effective against bacterial cells due to the formation of highly reactive hydroxyl radicals (Joo et al. 2005). Recently, TiO_2 is frequently used in the nanotechnology treatment of the water, due to the large surface area which can adsorb many of the pollutants from the water (Khalaphallah et al. 2012). However, one of the limitations of the application of TiO_2 in the large-scale water treatment process is the particle size and morphology of TiO2 practices (Yu et al. 2002).

The examination of different AOPs in the disinfection of water has been reported in the literature. Teodoro et al. (2014) investigated the efficiency of the photo-Fenton system (Fe^{2+}/H_2O_2 coupled with UV) as an advanced oxidation process for disinfection of greywater and inactivation of *P. aeruginosa*. The greywater was treated by a treatment system consisting of an evapotranspiration tank and constructed wetland with the horizontal flow and *Heliconia psittacorum* L.f. In one experiment, the authors examined the efficiency of H_2O_2 150 mg L^{-1} coupled with UV (4.3 mW cm^{-2}) (H_2O_2/UV), while in another experiment Fe^{2+} was used with 10 mg L^{-1} and adjusted to pH at 3 and then used in coupled with UV (Fe/UV). The study found that both systems were most efficient in the total inactivation of *P. aeruginosa*. The authors concluded that the system of H_2O_2/ UV was the main factor for the inactivation of bacteria and the exertion of greater influence compared to the system with Fe/UV at low pH.

The mechanism in which photo-Fenton system deactivates pathogenic growth is explained based on the ability of UV to regenerate of Fe^{2+} ions by the photoreduction of $Fe(OH)^{2+}$, this reaction resulting to the hydroxyl radical (OH^-).

10.3   Detection of Inactivated Pathogens in Disinfected Greywater

The concentration of pathogenic bacteria in the disinfected greywater needs a critical and accurate method because these pathogens are variable with a lower concentration than the one that can be detected by the culture-dependent method, or that is available in a dormant state which failed to grow in the culture media. Therefore, the use of the enrichment methods might be the alternative option to detect the presence or absence of these pathogens. However, the overtime required for the detection and identification of pathogens might be extended for several days. Many of the authors have suggested the use of the molecular technique such as the quantitative polymerase chain reaction (qPCR). This technique is fast, but the limitations lie in the principle of the molecular analysis, which depends on the detection of the pathogens based on the nucleic acids. The molecular analysis might be affected by the free DNA fragments of the non-viable cells, resulting in over- or underestimation of bacterial densities (Bae and Wuertz 2009; Orlofsky et al. 2015). The combination of molecular and culture-dependent methods such as non-specific enrichment and the most probable number (MPN) is a useful tool to increase the likelihood of pathogen detection with low concentrations as in the case of disinfected greywater (Krämer et al. 2011; Russo et al. 2014). In this section, the techniques used for the detection of pathogenic bacteria have been reviewed.

The isolation of pathogenic bacteria from the environment is quite difficult and different from that of clinical samples. Many factors should be considered during the isolation procedure such as the selection of the best dilution, and enrichment media. However, in many cases, conventional techniques fail to isolate these pathogens. It has been reported that 99.9% of bacteria in the environment are still uncultured (Zubair et al. 2010). Moreover, the failure to isolate these bacteria from the environment on the culture medium might be due to the failure in replicating essential aspects of their environment such as the factors necessary for their growth (Stewart 2012). However, Straškrabová (1983) claimed that the bacterial cells died due to high nutrient shock.

Culture-based techniques are more effective in determining available pathogenic microorganisms in water and wastewater samples. Utilisation of direct plating, MPN or membrane filtration depends on the density of the pathogens in the samples. In contrast, molecular-based methods (PCR, antibody-based and metabolic-based) are quite useful in the identification of pathogenic species. However, they are not efficient in distinguishing between viable and non-viable pathogens. Moreover, culture-based methods can be used for the determination of health risks associated with the exposure to pathogens due to the presence of human virulence genes while molecular methods might be able to discern cell viability. Although the presence of virulence genes can also be determined by molecular methods, the deficiency lies in the absence of entire microorganisms which makes further characterisation of pathogens limited (Center and Warrenton 2007).

In a comparison study between cultures based method and the DNA-based methods, Benami et al. (2015) used both techniques to assess the presence of pathogenic bacteria in the disinfected greywater with chlorine and UV. The culture-dependent method indicated that the concentrations of *E. coli*, FC, *Enterococcus* sp. *S. aureus*, *Salmonella enterica* and *P. aeruginosa* have differed significantly in the disinfected greywater compared with the disinfected samples. Conversely, the culture-independent DNA-based method recorded no significant differences in the concentrations of these pathogens before and after the disinfection process. The study concluded that the inactivation efficiency of the disinfection processes angst of pathogens could not be estimated by DNA-based qPCR.

Bedrina et al. (2013) examined the efficiency of a combined magnetic immuno-capture and enzyme immunoassay for the detection of *Legionella pneumophila* in water samples in comparison to the culture-based method. The method depends on the anti-*L. pneumophila* antibodies immobilised on magnetic microspheres. The results revealed that the method was more applicable for the fast detection of *L. pneumophila* in water samples without the need for bacterial growth on the culture medium.

PCR technique is the best and most successful technique for the identification of pathogenic bacteria since it depends on the nucleotide sequence of a DNA strand. The efficiency of this identification technique may reach up to 99.99% (Nissen and Sloots 2002). The main challenge lies in the determination of pathogenic bacteria or fungi concentrations. The determination of bacterial concentrations represents a very serious point in terms of pathogenicity which depends on the dosage of bacteria at which the bacteria might cause infections in humans.

Pathmanathan et al. (2003) investigated the potential of the PCR procedure based on the use of hilA primers for detecting *Salmonella* spp. In the study, 33 *Salmonella* strains and 15 non-*Salmonella* strains were used. The results revealed that PCR produced 784 bp DNA fragments in *Salmonella* strains but none in non-*Salmonella* strains. The detection limit of PCR was 100 pg based on the genomic DNA which is equivalent to 3×10^4 CFU mL^{-1} based on serial dilutions of bacterial culture. The use of the enrichment-PCR method has increased the sensitivity of the hilA primers' efficiency for a concentration of 3×10^2 CFU mL^{-1}. The study concluded that hilA primers are selective and specific for detecting *Salmonella* spp. in faeces through the PCR method. However, the presence of *Salmonella* spp. with s concentration of 300 CFU mL^{-1} which was detected as a detection limit can be achieved through the direct culture method on a selective medium without the need for other enrichment methods or molecular techniques. Moreover, one of the issues in the detection of potent pathogenic bacteria such as *Salmonella* spp. and *Shigella* spp. in wastewater is when they are available in concentrations less than the detection limits of direct isolation. It is also an issue when there are high concentrations of other bacteria such as *E. coli*, FC, *Proteus* sp., *Pseudomonas* sp., *Klebsiella* sp. which might be similar to *Salmonella* spp., *Shigella* spp. colonies in terms of some morphological characteristics in their grown colonies on the selective medium. This is because even though the medium is selective for isolation specific pathogens, some other bacteria can also be grown there. In this case, the use of a more specific method is required to detect *Salmonella* spp. and *Shigella* spp.

Silbert et al. (2006) proposed a new technique for detecting bacteria based on the interaction between membrane-active compounds which are secreted by the bacterial cell and nanoparticles embedded in an agar medium which leads to the formation of phospholipids and the chromatic polymer polydiacetylene (PDA). The results revealed that the PDA changed from visible blue-to-red transformations alongside an intense fluorescence emission due to the induction process caused by the molecules released from the bacterial cells. The generated colour can be detected by the naked eye. Moreover, the time required for fluorescence change is less than that required for the bacteria to form a colony on the culture medium. This method that is called chromatic technology is acceptable in terms of its simple procedure and the short time required to obtain the results. Therefore, it might be more useful for the detection of bacterial contamination of foods as well as antibiotic-resistant bacteria.

The culture-based method depends on the measurement of metabolic activities of the cells or the level of growth in the culture medium. However, these measurements need an incubation period of 24 h. The molecular methods have the potential to measure cellular activities in less than an hour (Tanchou 2014). Noble and Weisberg (2005) reviewed advanced technologies for the rapid detection of bacteria in the water. These methods include immunoassay techniques, molecule-specific probes, quantitative PCR (qPCR), nucleic acid sequence-based amplification (NASBA) and microarrays as well as the enzyme/substrate methods which depend on the utilisation of chromogenic or fluorogenic substrates. These techniques are also effective for the detection of pathogenic bacterial cell concentration in water.

The use of electrokinetic methods such as electrophoresis (EP) and AC dielec-trophoresis (DEP) to determine the bacterial concentration in samples has been reported in the literature. The DEP technique depends on the polarisation of a particle in a non-uniform electric field which will be attracted to the regions with high field or low field, based on the polarisability of these particles relative to the medium. The characteristics of particles in terms of size, shape and the conductivity play an important role in detecting the living and dead cells by DEP (Camacho-Alanis and Ros 2015).

Another technique for the detection of microbial cells in the environment is the use of spectroscopy such as ultraviolet and visible UV–Vis, infrared IR and Raman spectroscopy. These techniques have high sensitivity towards molecular differences and complex structures of bacterial cells as well as DNA and chemical compounds. Both IR and Raman spectroscopy depends on measuring the bending, vibrating and stretching modes of the molecule bonds. Therefore, they have high sensitivity towards different molecular structures (Hou et al. 2007; Davis and Mauer 2010). The use of surface-enhanced Raman scattering (SERS) for rapid detection and identification of bacteria which is estimated to be the 20 s, but the bacteria supposed to be in log phase with the concentration of bacteria reach to 100 CFU mL^{-1}. Moreover, the bacteria need to be immobilised and this represents the main challenge for the process known as the on-chip diagnostics technique (Hou et al. 2007).

10.4   Pathogen Growth Potential (PGP)

The main concern in the treated wastewater lies in the ability of the inactivated pathogenic bacteria to grow in treated wastewater which is disposed into the environment or reused for irrigation. The disinfection process which includes chemical and physical disinfections has been demonstrated to reduce pathogenic bacteria in wastewater. However, the remaining bacteria even in very small populations can multiply rapidly under suitable conditions. This case is true for bacteria but not for viruses, helminths and protozoa which cannot regrow outside their specific host organism(s). Therefore, once these pathogens have been reduced through treatment, their populations cannot increase again in the environment (USEPA 2007).

The methods used for studying the potential of the inactivated pathogenic bacteria or fungi to regrow in treated wastewater include culture and non-culture techniques. Moreover, the selection of the culture media and method of isolation in the culture-based method plays an important role in detecting the presence or absence of these pathogens. Chun-ming (2007) claimed that the use of DCA or MLCB for the direct isolation of FC and *Salmonella* spp. from heat-treated cow dung is effective in detecting the presence or absence of these pathogens. The study depends on the storage of the treated samples at 30 °C for 7, 14 and 21 days, respectively, in which some of the damaged bacterial cells can repair themselves during the storage period. Besides, the study depends on optimal growth conditions for these pathogens such as the use of a selective media and an optimal temperature of 37 °C. This method might be effective if some of the pathogenic bacteria remain active after the heat treatment or if the pathogenic bacteria have not been totally damaged. However, the absence of growth on the selective media does not mean that the bacteria have been totally eliminated from the samples since in many cases, the bacteria will still be available in the samples in an inactivated state. Sugumar and Mariappan (2003) reported that *Salmonella* spp. were found to have survived for more than 3 months in water without supplemental nutrition and metabolic injury. However, they failed to grow on selective media. Besides, even after the bacteria undergo the treatment process, they have different colony characteristics compared to those of the typical colonies before the treatment. This might lead to misidentification of the grown bacteria. Markova et al. revealed that the bacterial cells inactivated using autoclave have the ability to regrow. However, they have grown in L-form where the bacterial cells have grown without cell walls. This case also shows that bacterial cells can be inactivated by using antibiotics whose function is to prevent cell wall synthesis.

It has been demonstrated that bacterial or fungal cells under stressed conditions go into the dormant state which is called viable but non-culturable (VBNC) (Weaver et al. 2010; Al-Gheethi et al. 2016). VBNC means that the cells would not produce hydrolysis enzymes even if isolated on specific or enrichment media. The ability of microorganisms to be reactivated depends on many factors such as the surrounding incubation temperature and the availability of glucose and amino acids which represents the main source for energy and anabolism pathways (Choi et al. 1999). Therefore, the treatment process which does not lead to the reduction of nutrients in wastewater or the irreversible destruction of pathogenic cells is not considered an effective technique to produce high-quality wastewater. For instance, the use of chemical disinfectants such as chlorine or ozone has the potential to inactive the microorganism cells by inhibiting the enzymatic reactions of their energy pathway. However, some microorganism cells have alternative pathways for metabolic activities thus causing the inactivation process for these pathogens to be temporary. Using irradiation as a form of disinfection works because the cell is inactivated due to the ability of UV irradiation to form a dimer between thymine bases in the DNA nucleotide. Nevertheless, the cells have the ability to repair the damage caused by UV irradiation (Al-Gheethi et al. 2013). The possibility of pathogenic bacteria or fungi to regrow in the treated wastewater is a serious point which limits the safe handling and disposal of wastewater into the environment. However, the main challenge is to find an effective technique to determine the potential of ABNC pathogenic bacteria or fungi to regrow in the treated wastewater or the environment. Based on the physiological status of ABNC cells, the culture-based method is not efficient for recovering these pathogens on the culture medium. The absence of bacterial or fungal growth in the culture medium after inoculation with treated samples does not mean that these pathogens have been totally eliminated. Instead, it indicates that the pathogens might exist in concentrations less than the detection limits or that the cells might be in an inactivated state and need more time to regrow. Banana (2013) found that pathogenic bacteria in blood waste samples treated with supercritical carbon dioxide (SC-CO_2) were reduced to less than the detection limits. However, the sample stored at room temperature for 2 months revealed the regrowth of the pathogenic bacteria. The level of the growth was less than that in the raw samples, but the presence of the regrowth confirmed that the cells have the ability to regrow in the treated samples.

The alternative technique to evaluate the efficiency of the disinfected or sterilised waste samples is to use non-culture methods which might be able to determine the level of destruction caused by the treatment method or to determine the possibility of inactivated cells to regrow in treated waste. Noman et al. (2016) investigated the level of destruction caused by the fungal spores inactivated by autoclave and SC-CO_2. In the study, the inactivated fungal spores were scanned using the scanning electronic microscope (SEM) technique which revealed that the spore surface was totally damaged. The use of SEM for detecting the level of destruction in bacterial or fungal cells is a critical step but it is not enough to confirm that no regrowth will occur after the disposal of waste into the environment because the SEM technique cannot detect the validity of the cells.

One-dimensional SDS-PAGE analysis for the protein banding patterns of inactivated bacteria was used by Hossain (2013) in order to detect the validity of bacterial cells after the treatment process by SC-CO$_2$. The results revealed the absence of the protein banding patterns of inactivated bacteria. Kim et al. used two-dimensional electrophoresis (2-DE) and principal component analysis (PCA) to detect the protein profiling of *S. enterica* after the inactivation process. The study revealed that the cell fatty acids and proteins are alternated. The analysis of cell fatty acids by GC-MS noted that the total fatty acid quantity reduced in comparison to the control. Conducting a test on the cell membrane permeability can also be used to determine the destruction level in the inactivated cells. The test is carried out using flow cytometry. Two types of stains are used including ethidium bromide (EB) and propiumiodide (PI) to evaluate the permeability of the cell membrane and efflux pump system of bacteria. The ability of EB to stain the bacterial genomic (DNA) and the ability of PI to stain cell cytoplasm confirm that the treatment process has damaged the membrane cell and the efflux pump (Humphreys et al. 1994; Ericsson et al. 2000). In untreated cells, EB is transported into the cytoplasm, but the validity of the efflux pump leads to the stain being pumped out of the cell (Jernaes and Steen 1994). These studies have not been conducted on pathogenic bacteria inactivated in wastewater by different disinfection methods. Moreover, the analysis tools can also be applied to the inactivated bacteria or fungi regardless of the sample type.

Physical treatment such as the use of high temperature can cause denaturation of proteins and enzymes necessary for metabolic and anabolic pathways. The destruction by thermal treatment might be irreversible. In contrast, the chemical treatment might lead to change in the protein structure, but these changes are reversible. For example, the change of pH to extreme acidic or alkaline conditions may cause the destabilisation of enzymes by dissociation of the enzyme subunits or loss of correct assembly structure which may be reversible or irreversible (Kamihira et al. 1987; Poltorak et al. 1999; Fernandez-Lafuente 2009). Nevertheless, the main challenge here is to detect if these changes in the bacterial or fungal cell protein and lipids are enough for irreversible inactivation and preventing regrowth. In other words, it is important to detect if these changes can lead to the death of bacterial or fungal cells.

The detection of proteins available in disinfected samples based on the PCR technique might be unsuitable because the presence of DNA fragments in treated or disinfected samples does not mean that the bacterial or fungal cells are active. The viability of bacterial or fungal cells in the treated sample might be accessed based on metabolic activities, RNA transcripts, a positive energy status and responsiveness. Some authors have combined culture and molecular methods to detect the validity of inactivated cells or those which are present in VBNC state. Jiang et al. (2013) used MPN method and reverse transcription quantitative PCR (RT-qPCR) for the quantification of *S. typhimurium*, *E. coli* and *S. flexneri* in the VBNC state. The study stated that this procedure provided an improved evaluation of pathogen inactivation efficiency.

In addition, the toxicity of the bacterial and fungal protein structure are important factors to be considered. The treatment process might inactivate the microorganism cells and prevent their regrowth in the environment. However, it does not mean that the health risks for these pathogens have been totally eliminated. For example, even though viruses are invalid or nonliving outside host cells, they can become more pathogenic in the cells. The microorganism protein with high molecular weight is considered toxic for other organisms even in the absence of the cells. The toxins produced by the bacterial and fungal cells consist of protein and polysaccharides which might exhibit high resistance towards chemical or physical treatment. The proteins might also pose a high risk for humans if they are not removed from the samples such as prion proteins (PrPs) which are infectious agents and consist of protein materials (Bartelt-Hunt et al. 2013).

10.5   Conclusions

The evaluation of the treatment efficiency of the inactivation of pathogenic bacteria or fungi in wastewater should include the determination of the validity of pathogens to regrow and the toxicity of proteins and other fragment structures of the cell. The best treatment method should have the ability to kill pathogens by causing irreversible destruction of the cell, energy pathways, toxicity of the proteins, polysaccharides and lipid fragments released from the cells as well as the irreversible destruction of DNA and RNA fragments to prevent transmission to other microorganisms. In contrast, the evaluation procedure should be conducted using different techniques to ensure that the cell and their components have been totally damaged by the selected treatment process.

Acknowledgements The authors wish to thank the Ministry of Higher Education (MOHE) for supporting this research under FRGS vot 1574 and also the Research Management Centre (RMC) UTHM for providing grant IGSP U682 for this research.

References

Al-Gheethi AA, Norli I, Lalung J, Azieda T, Kadir MOA (2013) Reduction of faecal indicators and elimination of pathogens from sewage treated effluents by heat treatment. Caspian J Appl Sci Res 2(2):29–45

Al-Gheethi AA, Ismail N, Efaq AN, Bala JD, Al-Amery RM (2015) Solar disinfection and lime stabilization processes for reduction of pathogenic bacteria in sewage effluents and biosolids for agricultural purposes in Yemen. J Water Reuse Desalin 5(3):419–429

Al-Gheethi AA, Mohamed RM, Efaq AN, Amir HK (2016) Reduction of microbial risk associated with greywater utilized for irrigation. Water health J 14(3):379–398

Bae S, Wuertz S (2009) Discrimination of viable and dead fecal *Bacteroidales* bacteria by quantitative PCR with propidium monoazide. Appl Environ Microbiol 75(9):2940–2944

Banana AAS (2013) Inactivation of pathogenic bacteria in human body fluids by steam autoclave, microwave and supercritical carbon dioxide. Ph.D. thesis, Environmental Technology Division, School of Industrial Technology, Universiti Sains Malaysia (USM), Penang, Malaysia

Bani-Melhem K, Al-Qodah Z, Al-Shannag M, Qasaimeh A, Qtaishat MR, Alkasrawi M (2015) On the performance of real grey water treatment using a submerged membrane bioreactor system. J. Membrane Sci 476:40–49

Bartelt-Hunt SL, Bartz JC, Saunders SE (2013) Prions in the environment. In: Prions and diseases. Springer, New York, pp 89–101

Bedrina B, Macián S, Solís I, Fernández-Lafuente R, Baldrich E, Rodríguez G (2013) Fast immuno sensing technique to detect *Legionella pneumophila* in different natural and anthropogenic environments: comparative and collaborative trials. BMC Microbiol 13(1):88

Benami M, Gillor O, Gross A (2015) The question of pathogen quantification in disinfected greywater. Sci Total Environ 506:496–504

Bosshard F, Berney M, Scheifele M, Weilenmann HU, Egli T (2009) Solar disinfection (SODIS) and subsequent dark storage of *Salmonella typhimurium* and *Shigella flexneri* monitored by flow cytometry. Microbiology 155(4):1310–1317

Camacho-Alanis F, Ros A (2015) Protein dielectrophoresis and the link to dielectric properties. Bioanalysis 7(3):353–371

Cantor KP, Hoover R, Hartge P, Mason TJ, Silverman DT, Altman R, Austin DF, Child MA, Key CR, Marrett LD (1987) Bladder cancer, drinking water source and tap water consumption: a case control study. J Nat Cancer Ins 79:1269–1279

Center A, Warrenton V (2007) Report of the experts scientific workshop On critical research needs for the development of new or revised recreational water quality criteria

Chang JC, Ossoff SF, Lobe DC, Dorfman MH, Dumais CM, Qualls RG, Johnson JD (1985) UV inactivation of pathogenic and indicator microorganisms. Appl Environ Microbiol 49(6):1361–1365

Choi JW, Sherr BF, Sherr EB (1999) Dead or alive? A large fraction of ETS-inactive marine bacterioplankton cells, as assessed by reduction of CTC, can become ETS-active with incubation and substrate addition. Aquat Microb Ecol 18(9):105–115

Chun-ming GONG (2007) Microbial safety control of compost material with cow dung by heat treatment. J Environ Sci 19:1014–1019

Ciavola M (2011) Water disinfection in developing countries: design of a new household solar disinfection (SODIS) system. University of Salerno (IT), Tattarillo Award 2011 Appropriate Technologies for sustainable development in any South of the World

Davis R, Mauer LJ (2010) Fourier transform infrared (FT-IR) spectroscopy: a rapid tool for detection and analysis of foodborne pathogenic bacteria. Curr Res Technol Educ Top Appl Microbiol Microb Biotechnol 2:1582–1594

Eisenstark A (1971) Mutagenic and lethal effects of visible and near-ultraviolet light on bacterial cells. Adv Genet 1971(16):167–198

Ericsson M, Hanstorp D, Hagberg P, Enger J, Nystrom T (2000) Sorting out bacterial viability with optical tweezers. J Bacteriol 182:5551–5555

Facile N, Barbeau B, Prevost M, Koudjonou B (2000) Evaluating bacterial aerobic spores as a surrogate for *Giardia* and *Cryptosporidium* inactivation by ozone. Water Res 34(12):3238–3246

Fernandez-Lafuente R (2009) Stabilization of multimeric enzymes: strategies to prevent subunit dissociation. Enzyme Microb Technol 45:405–418

Gameson ALH, Gould JD (1985) Bacterial mortality, Part 2. In: Investigations of sewage discharges to some British coastal waters. WRcTechn. Rep. TR 222. WRc Environment, Medmenham, UK

Gilboa Y, Friedler E (2008) UV disinfection of RBC-treated light greywater effluent: kinetics, survival and regrowth of selected microorganisms. Water Res 42(4):1043–1050

Haarhoff J, Cleasby LJ (1991) Biological and physical mechanisms in slow sand filtration. In: Logsdon (ed) Slow sand filtration. ASCE, New York

Harding AS, Schwab KJ (2012) Using limes and synthetic psoralens to enhance solar disinfection of water (SODIS): a laboratory evaluation with norovirus, *Escherichia coli*, and MS2. Am J Trop Med Hyg 86(4):566–572

Hossain S (2013) Supercritical carbon dioxide sterilization of clinical solid waste. Ph.D. thesis, Environmental Technology Division, School of Industrial Technology, University Science Malaysia, Penang, Malaysia

Hou D, Maheshwari S, Chang HC (2007) Rapid bioparticle concentration and detection by combining a discharge driven vortex with surface enhanced Raman scattering. Biomicrofluidics 1(1):014106

Humphreys MJ, Allman R, Lloyd D (1994) Determination of the viability of *Trichomonas vaginalis* using flow cytometry. Cytometry 15:343–348

Janex ML, Savoye P, Xu P, Rodriguez J, Lazarova V (2000) Ozone for urban wastewater disinfection: a new efficient alternative solution. In: Proceedings of the specialized conference on fundamental and engineering concepts for ozone reactor design, Toulouse, France. International Ozone Association, Stamford, Connecticut, pp 95–98

Jernaes MW, Steen HB (1994) Staining of *Escherichia coli* for flow cytometry: influx and efflux of ethidium bromide. Cytometry 17:302–309

Jiang Q, Fu B, Chen Y, Wang Y, Liu H (2013) Quantification of viable but nonculturable bacterial pathogens in anaerobic digested sludge. Appl Microbiol Biotechnol 97(13):6043–6050

Jong J, Lee J, Kim J, Hyun K, Hwang T, Park J, Choung Y (2010) The study of pathogenic microbial communities in greywater using membrane bioreactor. Desalination 250:568–572

Joo JH, Wang SY, Chen JG, Jones AM, Fedoroff NV (2005) Different signaling and cell death roles of heterotrimeric G protein alpha and beta subunits in the Arabidopsis oxidative stress response to ozone. Plant Cell 17:957–970

Kamihira M, Taniguchi M, Kobayashi T (1987) Sterilization of microorganisms with supercritical and liquid carbon dioxide. Agricul Biol Chem 51:407–412

Khalaphallah R, Maroga-Mboula V, Pelaez M, Hequet V, Dionysiou DD, Andres Y (2012) Inactivation of *E. coli* and *P. aeruginosa* in greywater by NF-TiO2 photocatalyst under visible light. In: Conference WWPR 2012, Water Reclamation & Reuse, Heraklion, Crete, Greece, 28–30 March

Krämer N, Löfström C, Vigre H, Hoorfar J, Bunge C, Malorny B (2011) A novel strategy to obtain quantitative data for modelling: combined enrichment and real-time PCR for enumeration of salmonellae from pig carcasses. Int J Food Microbiol 145:S86–S95

Lindgren S, Grette S (1998) Vatten-och avloppssystem. EkoportenNorrköping [Water and sewerage system. Ekoporten in Norrköping]. SABO Utveckling. Trycksak 13303/1998-06.500

Machulek Jr A, Moraes JEF, Okano LT, Silverio CA, Quina FH (2009) Photolysis of ferric ion in the presence of sulfate or chloride ions: implications for the photo-Fenton process. Photochem Photobiol Sci 8(2009):985–991

Mohamed H, Brown J, Njee RM, Clasen T, Malebo HM, Mbuligwe S (2015) Point-of-use chlorination of turbid water: results from a field study in Tanzania. J Water Health 13(2):544–552

Nissen MD, Sloots TP (2002) Rapid diagnosis in pediatric infectious diseases: the past, the present and the future. Pediatr Infect Dis J 21(6):605–612

Noble RT, Weisberg SB (2005) A review of technologies for rapid detection of bacteria in recreational waters. J Water Health 3(4):381–392

Noman EA, Rahman NN, Shahadat M, Nagao H, Al-Karkhi AF, Al-Gheethi A, Omar AK (2016) Supercritical fluid CO2 technique for destruction of pathogenic fungal spores in solid clinical wastes. Clean—Soil, Air, Water 44(12):1700–1708

Orlofsky E, Benami M, Gross A, Dutt M, Gillor O (2015) Rapid MPN-Qpcr screening for pathogens in air, soil, water, and agricultural produce. Water Air Soil Pollut 226(9):1

Ottosson J (2003) Hygiene aspects of greywater and greywater reuse. Doctoral dissertation, Mark och vatten. Royal Institute of Technology (KTH), Department of Land and Water Resources. Engineering Swedish Institute for Infectious Disease Control (SMI), Department of Water and Environmental Microbiology

Pathmanathan SG, Cardona-Castro N, Sanchez-Jimenez MM, Correa-Ochoa MM, Puthucheary SD, Thong KL (2003) Simple and rapid detection of Salmonella strains by direct PCR amplification of the hilA gene. J Med Microbiol 52(9):773–776

Pehlivanoglu-Mantas E, Elisabeth L, Hawley R, Deeb A, Sedlak DL (2006) Formation of nitrosodimethylamine (NDMA) during chlorine disinfection of wastewater effluents prior to use in irrigation systems. Water Res 40(2):341–347

Poltorak OM, Chukhrai ES, Kozlenkov AA, Chaplin MF, Trevan MD (1999) The putative common mechanism for inactivation of alkaline phosphatase isoenzymes. J Mol Cata B: Enzymatic 7:157–163

Robertson LJ, Smith HV, Ongerth JE (1994) *Cryptosporidium* and cryptosporidiosis. Part III: development of water treatment technologies to remove and inactivate oocysts. Microbiol Eur(Jan/Feb)

Rowe DR, Abdel-Magid IM (1995) Handbook of wastewater reclamation and reuse. CRC Press, CRC Lewis, London

Russo P, Botticella G, Capozzi V, Massa S, Spano G, Beneduce L (2014) A fast, reliable, and sensitive method for detection and quantification of *Listeria monocytogenes* and *Escherichia coli* O_{157}: H_7 in ready-to-eat fresh-cut products by MPN-qPCR. BioMed Res Int

Santasmasas C, Rovira M, Clarens F, Valderrama C (2013) Greywater reclamation by decentralized MBR prototype. Res Conser Rec 72:102–107

Sciacca F, Rengifo-Herrera J, Wethe J, Pulgarin C (2010) Dramatic enhancement of solar disinfection (SODIS) of wild *Salmonella* spp. in PET bottles by H_2O_2 addition on natural water in Burkina Faso containing dissolved iron. Chemosphere 78:1186–1191

Setlow RB (1968) The photochemistry, photobiology, and repair of polynucleotides. Prog Nucleic Acid Res Mol Biol 8:257–295

Shi P, Jia S, Zhang XX, Zhang T, Cheng S, Li A (2013) Metagenomic insights into chlorination effects on microbial antibiotic resistance in drinking water. Water Res 45(1):111–120

Silbert LE, Liu AJ, Nagel SR (2006) Structural signatures of the unjamming transition at zero temperature. Physical Rev E 73(4):041304

Smith WD, Hanawalt CP (1969) Repair of DNA in UV irradiated mycoplasma laidlawii B. J Mol Biol 46(1):57–77

Stewart EJ (2012) Growing unculturable bacteria, mini review. J Bacteriol 194(16):4151–4160

Straškrabová V (1983) The effect of substrate shock on populations of starving aquatic bacteria. J Appl Bacteriol 54:217–224

Sugumar G, Mariappan S (2003) Survival of *Salmonella* sp. in freshwater and seawater microcosms under starvation. Asian Fish Sci 16(3/4):247–256

Tal T, Sathasivan A, Bal Krishna KB (2011) Effect of different disinfectants on grey water quality during storage. J Water Sustain 1:127–137

Tanchou V (2014) Review of methods for the rapid identification of pathogens in water samples—ERNCIP Thematic Area Chemical & Biological Risks in the Water Sector Task 7. Publications Office of the European Union

Teodoro A, Boncz MÁ, Júnior AM, Paulo PL (2014) Disinfection of greywater pre-treated by constructed wetlands using photo-Fenton: influence of pH on the decay of *Pseudomonas aeruginosa*. J Environ ChemEng 2(2):958–962

Tripathi S, Pathak V, Tripathi DM, Tripathi BD (2011) Application of ozone based treatments of secondary effluents. J Biores Technol 102(3):2481–2486

USEPA J (2003) Ultraviolet disinfection guidance manual, pp 1–556. EPA-815-D-03-007

USEPA (2007) Pathogens, pathogen indicators and indicators of fecal contamination. Airlie Center, Warrenton, Virginiam U.S. Environmental Protection Agency, Office of Water, Office of Research and Development. EPA 823-R-07-006

Vital M, Stucki D, Egli T, Hammes F (2010) Evaluating the growth potential of pathogenic bacteria in water. Appl Environ Microbiol 67(19):6477–6484

Walker DC, Len SV, Sheehan B (2004) Development and evaluation of a reflective solar disinfection pouch for treatment of drinking water. Appl Environ Microbiol 70:545–2550

Weaver L, Michels HT, Keevil CW (2010) Potential for preventing spread of fungi in air-conditioning systems constructed using copper instead of aluminium. Lett Appl Microbiol 50(1):18–23

Winward GP, Stephenson ALM, Jefferson B (2008) Chlorine disinfection of grey water for reuse: effect of organics and particles. Water Res 42:483–491

World Health Organization (2006) Overview of greywater management health considerations. Regional Office for the Eastern Mediterranean Centre for Environmental Health Activities Amman, Jordan

Xu P, Savoye P, Cockx A, Lazarova V (2002) Wastewater disinfection by ozone: main parameters for process design. Water Res 36(4):1043–1055

Yu JC, Yu JG, Ho WK, Jiang ZT, Zhang LZ (2002) Effects of F-doping on the photocatalytic activity and microstructures of nanocrystalline TiO$_2$ powders. Chem Mater 2002(14):3808–3816

Zubair A, Yasir M, Khaliq A, Matsui K, Chung YR (2010) Mini Review: too much bacteria still unculturable. Crop Environ 1:59–60

Chapter 11
Recycle of Greywater for Microalgae Biomass Production

Adel Ali Saeed Al-Gheethi, Efaq Ali Noman,
Radin Maya Saphira Radin Mohamed, Najeeha Mohd Apandi,
Maizatul Azrina Yaakob, Fadzilah Pahazri and Amir Hashim Mohd Kassim

Abstract The potential of greywater to be used as a production medium for biomass lie in the high concentrations of nitrogen and phosphorus as well as the organic matter necessary for microalgae growth. Microalgae have high potential to adapt and utilise nitrogen, phosphate and other nutrients available in wastewater. Other factors which affect the production of biomass in microalgae include light, temperature, aeration and mixing. The effect of pH might also contribute to the quality and quantity of the produced biomass. The critical step in the production of biomass lies in the harvesting of microalgae cells, extraction of the lipids, proteins and carbohydrates. The objective of this review was to identify the criteria required for selecting greywater as a production medium and microalgae species. The harvesting and extractions techniques used in this process are also discussed and also the quality of the produced biomass and the further utilisation based on the toxicity, nutrients values and microbiological aspects.

Keywords Greywater · Microalgae biomass · Quality · Harvesting process
Application

A. A. S. Al-Gheethi (✉) · R. M. S. Radin Mohamed (✉) · N. M. Apandi · M. A. Yaakob
F. Pahazri · A. H. Mohd Kassim
Micro-Pollutant Research Centre (MPRC), Department of Water and Environmental Engineering,
Faculty of Civil and Environmental Engineering, Universiti Tun Hussein Onn Malaysia (UTHM),
86400 Parit Raja, Batu Pahat, Johor, Malaysia
e-mail: adel@uthm.edu.my

R. M. S. Radin Mohamed
e-mail: maya@uthm.edu.my

E. A. Noman
Faculty of Applied Sciences and Technology (FAST), Universiti Tun Hussein Onn Malaysia
(UTHM), Pagoh, Johor, Malaysia

E. A. Noman
Department of Applied Microbiology, Faculty Applied Sciences, Taiz University, Taiz, Yemen

© Springer International Publishing AG, part of Springer Nature 2019   205
R. M. S. Radin Mohamed et al. (eds.), *Management of Greywater in Developing Countries*,
Water Science and Technology Library 87,
https://doi.org/10.1007/978-3-319-90269-2_11

11.1   Introduction

The improper management of the greywater is a major challenge in most of the developing countries. The discharge of these wastes into the environment constitutes many adverse effects on the natural biodiversity. Conversely, the characteristics of the greywater in terms of the nutrients and elements make these wastes a suitable production medium for the generation of biomass. Among several types of the microorganism which might be cultured in the greywater, the microalgae are the most appropriate organisms because they have chlorophyll which can obtain the required energy from the light by the process of photosynthesis. This process is used by the microalgae cell to convert light and CO_2 into glucose which is the main substrate in the anabolic pathways and production of biomass (Shekhawat et al. 2012).

The generation of microalgae biomasses and their application in the different sectors of the life have been started since the 1970s. However, the applications such as biofuel production and bio-generation of bio-products such as a source of valuable chemicals, food additives and pharmaceuticals have increased significantly since 2008 (Pahazri et al. 2016). In the recent years, several companies are working on producing microalgae biomass in the marine and freshwater (Jais et al. 2017). The most common microalgae species used are *Botryococcus sudeticus*, *Dunaliella* sp., *Chlorella vulgaris*, *Haematococcus pluvialis*, *Nannochloropsis oculata* and *Spirulina platensis*. It has been estimated that the total amount of the *Haematococcus* sp. by 30 tonnes/years and that for *Spirulina* sp. by 20 tonnes/year. It is estimated that more than 6000 l of water are required to produce biomass yield enough to generate one litre of algal oil based on the conventional systems of cultivation, which indicates that the use of a large-scale algal cultivation in freshwater is not an economically suitable option due to the problems of water shortage in many of the developing countries (Ozkan et al. 2012). In this review, the wastewater might provide the alternative source of the water.

Many of the developing countries in the East Asia and Middle East countries have the environmental conditions suitable for the production of microalgae biomass. The microalgae biomass generated from the cultivation of microalgae species in the greywater have several applications in the industry, agriculture and medical activities. The commercial bio-products generated from the microalgae biomass are cosmetics, organic fish feed, human nutrition, pharmaceutical products and animal feed as well as biodiesel (Bala et al. 2016; Jais et al. 2017). However, there are several challenges which should be considered to ensure the recycling of the greywater as a production medium for microalgal biomass. The operating and harvesting process of biomass yield, as well as extraction of protein and lipids from the biomass, are the main points which are discussed in this chapter.

11.2 Recycle of Greywater for Microalgae Biomass Production

The recycling of greywater is consistent with the concept of zero discharge raised in 1980 for industrial wastewater and aims to recycle or reuse wastes (Efaq et al. 2015). The utilisation of effluents as a media for the production of enzymes such as β-lactamase and cellulase by bacteria has been investigated by Al-Gheethi (2015) and Al-Gheethi and Norli (2014). Many of the wastewater including municipal and industrial wastewater as well as dairy industry has been used as a culture medium for the microalgae biomass. Microalgae have high contents of carbohydrates, lipids and proteins. Therefore, it represents a good nutrition source as an animal feeds.

The recycling of greywater as a production medium for the microalgae biomass relied on the chemical, physical and biological characteristics of the greywater. The nutrients represent the key factors for the microalgae growth. The greywater has a content of nitrogen, phosphorus and trace elements in the range required for microalgae growth and production of biomass yield (Pahazri et al. 2016). Others parameters such as pH, temperature and light are also important, but these parameters are adjustable. The microbiological aspects of the greywater should be more investigated. Some of the microalgae could grow well in the greywater since it has the potential to compete with the indigenous organism where other microalgae have failed to grow. The relationship between microalgae and bacteria in the greywater has been investigated (Al-Gheethi et al. 2017). The secondary metabolic of the bacterial cells such as the release of CO_2 might induce the microalgae growth. Some of the microalgae have antibacterial activities which can inhibit the bacterial growth. In contrast, some of the bacterial species have algicidal activities. Therefore, the selection of microalgae species represents the bottleneck in the recycling of greywater as a culture medium. The other point is the presence of the pathogenic bacteria which might survive during the production process of the microalgae biomass, and they are harvested with the biomass yield and thus limit the application of biomass yields. One of the solutions to overcome this problem is to sterilise greywater before the recycling.

The choice of microalgae species is an important parameter to be considered, the ability of microalgae to survive under hard environmental conditions reflect its potential to grow in the greywater and thus results to overproduction of biomass yields. In this case, the best option to consider is to use indigenous microalgae species isolated from the surrounding environment. It is estimated that the microalgae species are more than 200,000 types; many of these species have the ability to survive in extreme conditions (Kalin et al. 2005). The ability of the microalgae to survive in different environmental conditions is attributed to their rapid rate of acclimatisation to the surrounding environment even with low concentrations of the nutrients. This process is called the natural selection process. However, the selected microalgae species in the biomass production should be non-pathogenic and should not have the potentials to produce toxins, since these toxins in the biomass yield might limit the application of biomass as fish or animal feeds or as soil fertilisers.

Microalgae are known to be autotrophic organisms; however, for overgrowth they require some of the nutrients in terms of nitrogen and phosphate as well as Fe, Mg, Na and Cu, as trace elements (Rahman et al. 2012). Therefore, the greywater might provide these requirements for the microalgae better than the freshwater. *Chlorella* sp., *Botryococcus* sp., *Euglena* sp. and *Scenedesmus* sp. are among the different species of the microalgae which have been investigated for their growth in the wastewater and exhibit good biomass production (Godos et al. 2012). The nutrients and trace elements in the greywater are the main factors which determine the potential of the greywater to act as a production medium for microalgae growth. Many of the other factors should be adjusted to improve the high quality and quantities of the biomass yield. In a view to examine the factors affecting microalgae growth and the characteristics of the greywater, it can be noted that pH, temperature, light, aeration, CO_2, salinity and mixing are the factors which need to be considered. These factors play a secondary effect on the microalgae growth but also contribute significantly to the amount of produced biomass (Abdel-Raouf et al. 2012).

11.2.1 Factors Affecting the Biomass Production in Greywater

The factors affecting the production of microalgae biomass are not independent variables as they interact together. The concentration of pH of the greywater ranges from 6 to 8, this range is within the optimal pH for microalgae growth which is between pH 7.5 and 11 for most of the microalgae species such as *Scenedesmus* sp. and *Chlorella vulgaris* (Sengar et al. 2011; Gong et al. 2014; Jais et al. 2017). The pH contributes to the microalgae growth which has direct effect or influence on the diffusion and transportation process of the nutrients through the cell membrane by the assimilation process of nutrients by microalgae cells as well as the activity of chlorophyll. The pH for CO_2 capturing from the atmosphere by the microalgae cells ranges from 7 to 9.5. However, pH during the growth of microalgae is changed. These changes depend on the photosynthesis activities, microalgae initial inoculums, aeration as well as the nature of nitrogen source. For instance, pH increases as a result of the photosynthetic CO_2 assimilation; this process takes place faster in the heavy inoculums and provides the high concentration of CO_2, HCO_3^- and H_2CO_3, while in the role to keep the pH within the optimal range for algal cultivation (Yaakob et al. 2014). In terms of CO_2, it should be noted that the maximum biomass of *Scenedesmus* sp. (0.2 g L^{-1}) and *Chlorella* sp. (0.12 g L^{-1}) was recorded with 24% of CO_2 concentrations (Makareviciene et al. 2014).

In contrast, pH is decreased when the concentration of ammonia is used as nitrogen source due to the release of H^+ ions which reduces the pH value below 4 (Sengar et al. 2011). In this case, the pH of the culture should be adjusted by using buffer solutions such as K_2HPO_4 to keep the pH value constant during the biomass production. Nevertheless, the authors have decided against the application of chemical additives into wastewater used as a production medium, since it might impact on the quality of biomass yield. Therefore, this problem can be overcome by using the continuous culturing system in which a balance between input and output greywater is adjusted (Brar et al. 2017). In contrast, the presence of nitrate as a secondary nitrogen source in the greywater medium contributes to the increase of pH value (Arumugam et al. 2013). However, in some of the microalgae species such as *Scenedesmus bijugatus*, nitrate is the preferred nitrogen source for the growth and production of biomass in the wastewater. The nitrogen contents in the microalgae cells range from 25 to 40% (Riano et al. 2016). Some of the microalgae species have the ability to fix nitrogen from the atmosphere but this pathway might be used by the microalgae cells as the alternative source if the concentration of nitrogen sources in the greywater is insufficient. The nitrogen contributes more than 4% in the production of microalgae biomass and approximately 10% in the production of lipids. However, it has been reported that some of the microalgae species such as *Neochloris* sp., *Tetraselmis* sp., *Nannochloropsis* sp. and *Scenedesmus* sp. produce maximum lipids and carbohydrate in the medium with low concentrations of nitrogen substances (Minhas et al. 2016).

Phosphorus concentration in the greywater is one the main factors which might induce or inhibit the microalgae growth. The microalgae need these elements between 0.03 and 0.06%, but it is necessary for the cell metabolism, since the phosphorus plays important elements for saving of the energy as adenosine triphosphate ATP required for the metabolic and anabolic pathways (Yaakob et al. 2014). Orthophosphate (PO_4^{3-}) is the superior form of most of the algae; the microalgae cells store the phosphorus in the form of polyphosphate to face the deficiency during the growth (Rasala and Mayfield 2015). As mentioned above, the interaction between the parameters in the greywater occurs in the case of phosphorus which is easy binds to Fe ions. Therefore, in some case, the concentrations of Fe ions in the greywater should be reduced to provide a favourable medium for biomass production (Yaakob et al. 2014).

Light is the sole source of energy in the microalgae; therefore, it represents the backbone for the production of microalgae biomass in the greywater. Even in the presence of high concentrations of the nutrients, the absence of the light makes the greywater not useful for microalgae growth. The limitation for the penetration of light through the greywater is that the turbidity, low density of the sunlight required for the microalgae growth which. For this reasons, the turbidity should be reduced before the recycling of the greywater as a culture medium. Meanwhile, the high density of the light might affect negatively the microalgae growth and the optimum light density recommended for high microalgae growth which is estimated to be 600 ft. candles (McKinney 2004). The optimum light density for *Spirulina platensis* was 232.26 fc, while it was 225 fc for *Chlorella kessleri* and 0.657–1.34 fc for *Botryococcus braunii* (Lee and Lee 2001; Qin and Li 2006; Fagiri et al. 2013). Moreover, the microalgae have developed the potentials using special mechanism to survive in the high density of the light; this takes place by the moving deeper or by releasing internal gases which are allowed to sink to desired level. In contrast, in the weak light, the microalgae are grown on the surface of the waters. Therefore, to understand the microalgae behaviour for the light might improve the overproduction of the biomass, since one of the problems which was recorded in the production of *Botryococcus* sp. in the greywater incubated under the direct sunlight was the participation of the microalgae cells. These observations were noted for several studies conducted in Malaysia which have 10,000 fc of the sunlight intensity. The effect of the sunlight on the microalgae growth also depends on the type of the greywater. The laboratory experiments found that *Botryococcus* sp. failed to grow in some greywater samples while grown more on the others. But in the view of greywater characterises, the inhibition of the microalgae growth might be due to the presence of xenobiotics organic compounds (XOCs). The toxicity of these is discussed more in Chap. 5.

The photoperiod of the microalgae growth is in the range of 12/12 to 18/6 h of light/dark (Mahale and Chaugule 2013). This cycle period is due to the nature or mechanism of the photosynthesis which is carried out in light and dark stages. In the light stage, the light is converted in the chloroplast into energy and then to $NADPH_2$ and ATP which are used to synthesise lipids, starch and sugar while in the dark cycle, the stored energy is used for the metabolic and anabolic reactions to synthesise amino acids (Jacob-lopes et al. 2009; Pérez-Pazos and Fernández-Izquierdo 2011). Therefore, the microalgae also exhibit a detectable growth in the dark. From the studies reported in the literature, the period of 12/12 h of light/dark is not standard for all the microalgae species. In some cases, the increase of the light periods is associated with maximum biomass yield and lipid contents, where the long period of the light cycle might improve the quality and quantity of the microalgae biomass by 100% (Rai et al. 2015). It is worthy of note that the light might induce the production of some of the specific compounds as in the case of *S. platensis* which requires 3500 lx for the production of carotenoid which requires only 2000 lx to achieve high biomass production (Kumar et al. 2015). There should be a clear understanding of the differences between the biomass yield and production of bio-products. The biomass production is a process that takes place during the log phase of the microalgae growth in which the cells used up the energy in the anabolic pathways and build more amino acids and thus increasing the microalgae cells which imply that more biomass is produced. In contrast, the production of specific compounds is a process that takes place during the stationary phase in which the cells have stopped the reproductions and the mature cells have high metabolic activity and thus more secondary metabolic bio-products are generated. Therefore, the cultivation of microalgae is proposed to be carried out in two stages. The first stage should be conducted with 12/12 light/dark to allow for the microalgae cells to grow and multiply and then produce more biomass. In the second stage, the period of lighting should be increased to induce the microalgae for producing the specific compounds.

The availability of trace elements in the greywater plays the very remarkable role in improving the microalgae growth. The elements such Ca, K, Mg, Ni, Cu and Mn act as cofactors for the many metabolic enzymes in the microalgae cell (Jais et al. 2017). For instance, Fe ions act as the electron, and Mg is required for the chlorophyll, while others such as Ca, K, Zn, Cu and Mn are required by algae to sustain the living cells. The greywater is rich with these elements which resulted from the detergents and others chemical compounds. The heavy metals might also be present in the greywater, but the reports indicated that they occur with a concentration below the sublethal levels (Wurochekke et al. 2016).

Temperature is a critical factor which has a real role in enhancing or inhibiting the microalgae growth. This role belongs to their effect on the diffusion and transportation of nutrients, protein and chlorophyll contents, metabolic activities, respiration intensity enzymatic reaction, specific affinity for nitrogen and phosphorus, CO_2 fixation as well as the cellular chemical composition and growth rate (Xin et al. 2011; Jais et al. 2017). Unlike the other factors which the highest effect might lead to inhibit or inactivate the microalgae cells, the high temperature might kill the cells by destroying the enzyme structure and function. Most of the microalgae grow at the ambient temperature (15 and 25 °C), some of them such as *C. vulgaris* have a maximum temperature growth reach of 30 and 35 °C, and *Spirulina* sp. has temperature growth range between 20 and 40 °C, while the temperature growth range of *Scenedesmus* sp. is between 10 and 40 °C (Cassidy 2011). Moreover, both light and temperature changes might be overcome by using the indigenous microalgae strains obtained from the surrounded and local environments.

The microalgae species which have been reported to produce biomass yield in different wastewater samples are listed in Table 11.1.

Based on the aforementioned, it can be concluded that the interactions between the factors affecting the microalgae growth and biomass production need to be optimised to detect the best operating parameters required for overproduction of microalgae. One of the best software programs to study the optimisation process is the response surface methodology which has been used by several authors to optimise the biotechnology applications including the biomass production (Efaq et al. 2016; Adeleke et al 2017; Hauwa et al. 2017b).

11.3   Harvesting Techniques of Microalgae Biomasses from the Culturing Media

The qualities and quantities of the biomass yield in the greywater depend mainly on the harvesting methods efficiency and recovery percentage from these wastes. Table 11.2 presents the percentage efficiency of the microalgae biomass using different harvesting techniques.

The harvesting microalgae biomass might be performed by physical, chemical and biological techniques or a hybrid system between them. The mechanical methods are superior due to the possible recycling of the culture media of the microalgae, since no chemical substances have been added, but the selection of the favourable method is based on the efficiency of the method with respect to other methods. The considerations for choosing the appropriate method depend on several factors which are related to the final utilisation and the economic and commercial points, since it is estimated that the harvesting process cost represents 20–30% of the total production cost (Pahazri et al. 2016). The factors affecting the harvesting processes include types

Table 11.1 Microalgae species grown in various types of wastewater

Microalgae	Type of wastewater	References
Botryococcus braunii	Greywater	Gokulan et al. (2013)
B. braunii	Household greywater	Gani et al. (2015)
Botryococcus sp.	Swine wastewater	Liu et al. (2013)
C. minutissimum	Sewage wastewater	Azarpira et al. (2014)
C. pyrenoidosa	Textile wastewater	Pathak et al. (2014)
C. sorokiniana Desmodesmus communis	Wastewater	Yao et al. (2015)
C. vulgaris	Chemical manufacturing wastewater	Rao et al. (2011)
C. vulgaris	Rubber latex concentrate processing wastewater	Bich et al. (1999)
Chlorella sp.	Poultry wastewater	Agwa and Abu (2014)
Chlorella sp.	Centrate Municipal wastewater	Min et al. (2011)
Chlorella saccharophila, Chlamydomonas pseudococcum, Scenedesmus sp., *Neochloris oleoabundans*	Dairy farm wastewater	Hena et al. (2015)
Chlorella vulgaris	Wastewater	Sengar et al. (2011)
- *Gloeocapsa gelatinosa* - *Euglena viridis* - *Synedra affinis*	Drain water	
Nostoc sp.	Dairy effluent	Kotteswari et al. (2012)
Pithophora sp.	Dairy wastewater	Silambarasan et al. (2012)
Phormidium sp.	Cattles laughter house wastewater	Maroneze et al. (2014)
S. obliquus *C. sorokiniana*	Raw sewage	Gupta et al. (2016)
Scenedesmus dimorphus	Anaerobically digested palm oil mill effluent	Kamarudin et al. (2013)
Scenedesmus sp.	Swine wastewater	Kim et al. (2007)
	Aqueous Solution	Xin et al. (2010)
	Swine wastewater	Michel et al. (2016)
	Municipal wastewater	Alva et al. (2013)
	Artificial wastewater	Song et al. (2014)
	Secondary wastewater	Kim et al. (2015)
	Sewage wastewater	Lekshmi et al. (2015)
	Tannery wastewater	Ajayan et al. (2015)
Spirulina platensis	Sago starchy wastewater	Phang et al. (2000)

Table 11.2 Harvesting methods of microalgae by different techniques

Algae species	Harvesting method	Efficiency percentage (%)	Reference
S. quadricauda	Flotation methods (SDS + Chitosan)	95	Chen et al. (1998)
Chlorella sp.	Flotation method (SDS + Chitosan)	85–90	Liu et al. (1999)
Scenedesmus quadricauda	Ultrafiltration	92	Zhang et al. (2010)
Parachlorella spp., *Scenedesmus* spp., *Phaeodactylum* spp., *Nannochloropsis* spp.	Cationic starch	90	Vandamme et al. (2010)
C. sorokiniana, Scenedesmus obliquus, Chlorococcum sp.	$FeCl_3$ and $Fe_2(SO_4)_3$	66–98	De Godos et al. (2011)
Chlorella sp.	Sedimentation and filtration	90.8	Li et al. (2011)
Chlorella sp.	Chitosan	99	Ahmad et al. (2011)
Chlorella vulgaris	Magnetic filtration	90	Ruiz-Martinez et al (2012)
Chlorella vulgaris	Submerged microfiltration	98	Bilad et al. (2012)
Chlorella vulgaris	*Maringa oleifera, Aluminium sulphate*	85	Teixeira et al. (2012)
Dunaliella salina	Aluminium sulphate	95	Hanotu et al. (2012)
	Ferric sulphate	98	
	Ferric Chloride	98.7	
Chlorella vulgaris	Chitosan	99	Rashid et al. (2013)
Nannochloris sp.	Centrifugation	>90	Dassey and Theegala (2013)
S. obliquus	Filtration	99	Ji et al. (2013)
Nannochloropsis oculata	$FeCl_3$ and $Fe_2(SO_4)_3$	93.80	Surendhiran and Vijay (2013)
		87.33	
Chlorella sp., *Chlamydomonas* sp.	Cationic guar gum (CGG)	94.5	Banerjee et al. (2013)
		92.15	
C. vulgaris, S. obliquus	Saponin and chitosan	93	Kurniawati et al. (2014)
C. protothecoides	Cationic starch	84–90	Letelier-Gordo et al. (2014)
Chlorella sp.	*Moringa oleifera*	90	Hamid et al. (2014)
S. dimorphus, S. minutum	Centrifugation	96	Gentili (2014)
		91	
Botryococcus sp.	*Maringa oleifera*	90	Hauwa et al. (2017a)

of culture media as wastewater or freshwater as well as the size of the microalgae cells (Barros et al. 2015). Microalgae cells grow in the culture medium in two layers, the surface layer (biofilm) which represents 2–7% of the total microalgae biomass and can be harvested either by the flotation or coagulant additives to enhance the participation process. In contrast, more than 90% of the biomass yield is suspended in the production medium and this quantity needs a critical selection for a suitable method for achieving the maximum recovering percentage (Atiku et al. 2016). In this section, the harvesting methods used for recovering of microalgae biomass from different culturing medium are discussed.

11.3.1 Centrifugation

The centrifugation method is the suitable methods used for recovering the microalgae biomass with small sizes. Centrifugation exhibit is effective and fast for harvesting the biomass without the need for chemical additives, it can achieve 80–95 of the recovery percentage within 2–5 min at $13,000 \times g$ for *Tetraselmis* sp. and *Chaetoceros calcitrans* and 30% at $6000 \times g$ (Dassey and Theegala 2013). The absence of chemical additive might assist in the storage of the biomass for a long time without any negative effect on the quality. Moreover, the process supposed to be carried out at appropriate speeds to avoid the damage of the cells by the high gravitational and shear forces as in the cases recorded for *Tahitian Isochrysis*, *Chaetoceros muelleri* and *Pavlova lutheri*, which are totally restricted due to the absence of hard cell walls (Caixeta et al. 2002; Knuckey et al. 2006). Nevertheless, the destruction of the microalgae cells might be an advantage for the centrifugation process, since no more extraction process is required for the biomass, but this case is good if the microalgae biomass yield will be used for biodiesel production. Among different designs of the centrifuge, the tubular bowl centrifugation is the more suitable for the small laboratory scale (Yaakob et al. 2014). Other limitation of the method includes the cost and increase in the temperature during the centrifugation. The temperature might be adjusted by providing the centrifugation with ice bath, but the cost is still the main challenge.

11.3.2 Membrane Filtration

The filtration techniques are the best alternative way in order to avoid the destruction of the microalgae cells as a result of high speed on the centrifugation (Atiku et al. 2016). However, this technique is more appropriate for the microalgae biomass with large sizes such as *Arthrospira* sp. but it is also dependent on the pore size of the used filter (Park et al. 2011). One of the challenges for using the membrane filtration is due to the microbial and microalgae growth on the surface of the membrane filter which leads to reduction in the effectiveness of this process in achieving the high recovering percentage of biomass from the culture medium as well as contamination of the harvested biomass (Bohdziewicz et al. 2003; Barros et al., 2015). Tangential flow filtration is a promising technique for the recovery from microalgae cells from the culture media with the efficiency ranging from 70 to 89% without any changes in the cell morphology or structure (Petrusevski et al. 1995). Micro-strainers is another type of the harvesting process which is dependent on the filtration mechanism; this process has simple implement and operation as well as can be used in both directions; therefore, no microalgae growth is accumulated on the surface (Chen et al. 2011).

11.3.3 Sedimentation

Very few studies have been reported on the use of sedimentation method for harvesting of the microalgae biomass. This process takes place in nature as a response to the changes in the pH of water, which leads to the settlement of the microalgae cells as a function of the gravity without the need for chemical additives. The auto-flocculation and bio-flocculation are one of the natural sedimentation methods which take place due to the extracellular polymeric substances (EPS) produced by some microalgae species such as *Scenedesmus obliquus*, *Micractinium* sp. and *Chlorella* sp. or by bacteria such as *Solibacillus silvestris*. In this case, the microalgae cell are agglomerated and then precipitated by the sedimentation (Guo et al. 2012; Wan et al. 2013; Ndikubwimana et al. 2014). The efficiency of this method is acceptable and inexpensive for the microalgae species with large cell size such as *Arthrospira* sp. and *Nannochloropsis* sp. In some cases, it can be accelerated by coagulants additives which might achieve 99% of the recovery (Barros et al. 2015).

11.3.4 Flotation

The flotation method contributes effectively to the harvesting of microalgae with cells size ranging from 10 to 500 μm (Hanotu et al. 2012). The air injection system produces air bubbles between 700 and 1500 lm and enhances in the process the efficiency of the recovery percentage of the biomass from the culture media (Rubio et al. 2002). In the dispersed air flotation (DAF), the air bubbles are between 10 and 100 lm, which make it more efficient than the settling process (Uduman et al. 2010). The combination between DAF and electro-flocculation method has achieved 98.9% of the recovery percentage of *Botryococcus braunii* within 14 min (Xu et al. 2010). The modification of the hydrophobicity of bubble surfaces by algogenic organic matter (AOM) or with the surfactant cetyltrimethylammonium bromide (CTAB) recorded high improvements for the recovering methods (Cheng et al. 2011). DAF has one limitation which lies in the oversized bubbles that could lead to the breakup of the flocs (Pragya et al. 2013). Flotation technique is used extensively in wastewater treatment processes after the removing of suspended solids by the coagulation process (Sim et al. 1988). It is more suitable for recovering of *Anabaena* sp., *Microcystis* sp. and *Arthrospira* sp., which have gas vesicles and thus have low density and are not harvested by centrifugation. Dispersed air flotation (DiAF) is similar to DAF, but the bubbles are generated by passing air continuously through a porous material. This process needs less energy. However, it needs more expensive equipment for the generation of the air bubble. The combination between DiAF and bio-surfactant saponin enhanced the recovery percentage of *S. obliquus* and *C. vulgaris* to more than 90% (Kurniawati et al. 2014). In ozonation-spread flotation (ODF), the recovering of microalgae biomass take place as a function of the interaction between negative functional groups on the surface of microalgae cell and positively charged bubbles. The method contributed in the extraction of lipids by 24% from *C. vulgaris* (Barros et al. 2015). However, the concern in the use of ozone is the formation of secondary products which might have carcinogenic effects (Rawat et al. 2011).

11.3.5 Chemical Flocculation and Coagulation

The flocculation and coagulation as a function of chemical coagulants are the most common technology used for the recovering of microalgae biomass from the culturing system. The coagulants used include polymers (organic and inorganic) or chelators (metal salts and alum) (Atiku et al. 2016). The polymers are more efficient than metal salts for many of microalgae species such as *C. vulgaris*, *Muriellopsis* sp., *Scenedesmus subspicatus*, *Chaperina fusca* and *Scenedesmus* sp. (De Godos et al. 2011). The hybrid flocculation system consists of metal salts and polymers might achieve more than 90% of the *C. vulgaris* (Gorin et al. 2015). The mechanism in which the flocculation by the multivalent metal ions acts to the harvest of microalgae biomass is explained based on the negative charge of the functional groups on the microalgae cell wall and the hydrolysed metals with positive charges. This process acts mainly as a function of pH where the optimal pH is 7, at which the flocculation achieves high efficiency, the functional group interacted with the ions to become unstable and then aggregated to form the flocs. The efficiency of flocculation to achieve high recovery percentage is a response for the type of coagulant (dosage of the coagulant, charge density and molecular weight), microalgae cell concentrations, operating parameters (pH, mixing speed, retention time and temperature (Hauwa et al. 2017a).

The concerns related to the use of flocculation method with the chemical coagulants are the health risks associated with the toxic by-products. The alum and acrylamide have been recorded by the authors as disease-causing agents such as Alzheimer's as well as carcinogenic substances (Ahmad et al. 2011). The concerns should be considered in the further application of harvested biomass as animal or fish feeds or fertilisers (Hamid et al. 2014). In this case, the alternative compounds which have been suggested by the authors are the polyelectrolytes compounds as well as the biodegradable organic polymers such as starch and chitosan. Both starch and chitosan recorded increasing efficiency (80–96–99%) of the microalgae biomass from the culture media, respectively (Vandamme et al. 2010; Rashid et al. 2013). Porcelanite is another alternative coagulant for the recovering of microalgae biomass, it has not been investigated before, but the chemical composition of porcelanite in terms of presence Al_2O_3, Fe_2O_3, CaO, MgO, CaO and TiO_2 indicated that it will have high harvesting efficiency (Atiku et al. 2016).

11.3.6   Natural Flocculation and Coagulation

The new directions in the field of harvesting method for microalgae biomass from the culturing process are by using the natural coagulants, which exhibited higher effectiveness in comparison with the chemical coagulants. One of the best advantages of using natural coagulants is the absence of chemical additives; rather than this, the natural coagulants might have more protein substance which might improve the quality of microalgae biomass. Among different natural coagulations, *Moringa oleifera* and *Strychnos potatorum* are the most studied; *M. oleifera* has recorded between 85 and 99% of the *Chlorella* sp. and *Botryococcus* sp. recovery from different culture media (Hamid et al. 2014; Hauwa et al. 2017a). The mechanism which makes the natural coagulants to have high efficiency for the harvesting of microalgae biomass is the presence of active polyelectrolytes with positive charges as well as their potential to dissolve in the water (Imtiazuddin et al. 2012). *M. oleifera* as a natural coagulant is non-toxic and has high efficiency with the low dosage, as well as lipid and protein contents with specific functional groups which make it one of the best alternative and natural coagulants for the chemical substances (Hauwa et al. 2017a). *S. potatorum* has also similar composition with *M. oleifera* such as carbohydrates, lipids and alkaloids so it plays an important role in the recovering of the microalgae biomass from the production media (Pahazri et al. 2016). The use of γ-glutamic acid in the harvesting of *N. oculata*, *Chlorella protothecoides*, *C. vulgaris*, *Phaeodactylum tricornutum* and *B. braunii* has recorded 90% of the harvesting efficiency (Zheng et al. 2012). Ecotan and Tanfloc coagulants have achieved 90% of biomass recovery from the culture medium (Gutiérrez et al. 2015), while cationic guar gum (CGG) recovered 94.5% of *Chlorella* sp. and 92.15% of *Chlamydomonas* sp. (Banerjee et al. 2013). In contrast, the use of nanotechnology in the preparation of coagulants such as cellulose nanocrystals (CNCs) increased the efficiency of the harvesting of microalgae biomass to 100% (Vandamme et al. 2015). The natural coagulants are biodegradable substances which are not toxic to human and biodiversity in nature.

11.3.7 Advance Technologies for Harvesting of Microalgae Biomass

The electrocoagulation is one of the most promising technologies for the harvesting of the microalgae biomass and is used as an alternative technology for the traditional methods which have several limitations (Matos et al. 2013). This technology is also used for the removal of xenobiotic organic compounds (XOCs). It depends on the release of metal ions by electrolytic oxidation of the anode which reacted with the microalgae cells to form the flocs. The flocs are separated by the sedimentation with 97% of the recovery percentage for some of the microalgae species such as *Nannochloropsis* sp. (Uduman et al. 2011).

The harvesting of the microalgae biomass from the culture medium by the magnetic separation acts based on the functionalized magnetic particles such as cationic polyelectrolytes (Fe_3O_4) and an external magnetic field with negative charges such as microalgae cells (Toh et al. 2012). This technology has achieved 95% of the recovery of *Chlorella ellipsoidea* (Hu et al. 2014). The magnetic particles are adsorbed on the functional groups of the microalgae cell wall by electrostatic bonds (Lim et al. 2012). Many of the magnetic particles have been used for recovery of microalgae from the production medium; the silica-coated magnetic particles exhibited more than 95% of the recovery of *Chlamydomonas reinhardtii, C. vulgaris, P. tricornutum* and *Nannochloropsis salina* (Cerff et al. 2012). Moreover, the developments in the magnetic separation have included the use of nanotechnology for the preparation of Fe_3O_4 nanoparticles, which exhibited more than 98% of *B. braunii, Corymbia ellipsoidea* and *Nannochloropsis maritima* (Xu et al. 2011). The low-gradient magnetophoretic separation coupled with iron oxide nanoparticles (NPs) and functionalized with cationic polyelectrolyte (diallyldimethylammonium chloride) (PDDA) harvested *Chlorella* sp. by 99% (Lim et al. 2012), while 95% of *C. vulgaris* recovery was achieved with iron oxide magnetic microparticles (IOMMs) (Prochazkova et al. 2013).

The milking technology is common for the harvesting of microalgae biomass as live cells from the low productive medium used for the extraction of high-value compounds such as carotenoids from *Arthrospira platensis* (Liu et al. 2009). In the milking process, the microalgal biomass is not harvested, rather than it performed like a continues culture system, where new cells are generated and grown in the log phase while others are in the stationary phase which is induced by the addition of the chemical substance to produce specific compounds. The addition of dodecane into the bioreactor of microalgae growth has improved the production of β-carotene (Hejazi et al. 2004).

11.4   Drying and Applications of Microalgae Biomass

In order to facilitate the management of the harvested microalgae biomass, it supposed to dehydrate the water contents and reduce the level of the moisture by the drying methods. However, the selection of the drying methods should be considered based on the future utilisation of this biomass. The drying process should have no effect on the quality of the biomass and the nutrients value as well as the chemical composition. The chemical composition of different microalgae biomass is presented in Table 11.3.

In some cases whereby the microalgae biomass is stored before the final utilisation, the conditions of the storage system are expected to have no negative effect on the biomass. The considerations should also include the selection of extraction method for lipids and protein from the biomass. The main challenge of the drying, stored and final utilisation is the destruction in the chemical composition and loss of the nutrients values of the biomass as well as the contamination by the bacteria or fungi which might grow on this biomass due to the high contents of the organic matter (Jais et al. 2017). Therefore, the dying methods need to have more than one function such as dehydration of water and disinfection of the biomass. Among several methods of the drying methods, the solar radiation might be the best promising technique due to their ability to remove the water and inactivates the pathogens as well as the degradation of some of the micro-pollutant organic compounds from the greywater during the harvesting process (Al-Gheethi et al. 2013; Atiku et al. 2016). However, the solar radiation is more applicable in the arid and semi-arid countries (Al-Gheethi et al. 2015).

Table 11.3 Protein and lipid contents in the microalgae biomass generated in different types of wastewater

Type of wastewater	Microalgae species	Protein content (%)	Lipid content (%)	References
Olive oil mill wastewater	*Spirulina platensis*	38.13	16.91	Wang et al. (2010)
Meat processing wastewater	*Chlorella* sp.	68.65	17.54	Lu et al. (2015)
Piggery wastewater		NA	21	Kuo et al. (2015)
Piggery wastewater	*Scenedesmus* sp.	NA	31	Hamid et al. (2014)
Dairy farm wastewater		45.09	21.82	Michels et al. (2014)
Aquaculture wastewater	*Spirulina plantesis*	48.5	4.7	Guerrero-Cabrera et al. (2014)
Secondary effluents	*Chlorella sorokiniana*	22.36 mg L^{-1}	24.91 mg L^{-1}	Ramsundar et al. (2017)

In contrast, the freeze-drying method might be recognised as an alternative method for drying of biomass; this process is not very common but it has high potential to remove water without destruction of nutrients values in the biomass. However, the technique would not deactivate the microbial contents; therefore, it might be more suitable for the biomass used in the biodiesel production (Munir et al. 2013).

The final utilisation of the microalgae biomass might include whole algal products or compounds extracted from this biomass. Unlike the microalgae biomass harvested from the water medium, there many of aspects should be conserved as well in the biomass recovered from the greywater due to the complex structure of these wastes. In a view of the chemical and biological composition of the greywater, it can be found that there are many of the available pollutants such as heavy metals, XOCs and pathogens. These parameters should be removed if the biomass will be used as dietary supplements for humans and animals (Kang et al. 2013). Microalgae have many of the bio-products such as carotenoids, pigments, vitamin, polyunsaturated fatty acids (PUFAs) and antioxidants which of course have more advantages in comparison with those synthesised in the laboratory (Maizatul et al. 2017).

Another aspect which needs to be considered is the microbial contents which might also be harvested with the microalgae biomass. However, the microbial contents might have no more concerns when the biomass is subjected to the extraction methods since some of the extraction processes such as supercritical carbon dioxide (SC-CO_2) have dual role to extract the compounds from the biomass and to inactivate the pathogens (Efaq et al. 2017).

11.5 Conclusion

The potential of greywater for recycling as a production medium for microalgae biomass depends mainly on the purpose and final utilisation of the biomass. Nevertheless, the chemical composition of the greywater in terms of nutrients and growth factors available in these wastes makes it an alternative source for the waters. However, there are many of the considerations which need to be evaluated as well to recycle the greywater as a production medium. On the other hand, the harvesting methods represent the bottleneck in the production of microalgae biomass which might increase or decrease the quality of the biomass. Among most of the harvesting techniques, the advanced technology such as milking methods appeared to be the best option in terms of high quality of the extracted compounds and regeneration of the biomass for several times.

Acknowledgements The authors also wish to thank The Ministry of Science, Technology and Innovation (MOSTI) for supporting this research under E-Science Fund (02-01-13-SF0135) and also the Research Management Centre (RMC) UTHM under grant IGSP U682 for this research.

References

Abdel-Raouf N, Al-Homaidan AA, Ibraheem IB (2012) Microalgae and wastewater treatment. Saudi J Biol Sci 19(3):257–275

Adeleke AO, Latiff AAA, Al-Gheethi AA, Daud Z (2017) Optimization of operating parameters of novel composite adsorbent for organic pollutants removal from POME using response surface methodology. Chemosphere 174:232–242

Agwa OK, Abu GO (2014) Utilization of poultry waste for the cultivation of *Chlorella* sp. for biomass and lipid production. Int J Curr Microbiol App Sci 3(8):1036–1047

Ahmad A, Yasin N, Derek C, Lim J (2011) Microalgae as sustainable energy source for biodiesel production: a review. Renew Sustain Energy Rev 15:584–593

Ajayan KV, Selvaraju M, Unnikannan P, Sruthi P (2015) Phycoremediation of tannery wastewater using microalgae Scenedesmus species. Int. J. Phytoremed. 17(10):907–916

Al-Gheethi AA (2015) Recycling of sewage sludge as production medium for cellulase enzyme by a *Bacillus megaterium* strain. International J Rec Org Waste Agri 4(2):105–119

Al-Gheethi AA, Norli I (2014) Biodegradation of pharmaceutical residues in sewage treated effluents by *Bacillus subtilis* 1556WTNC. J Environ Processes 1(4):459–489

Al-Gheethi AA, Norli I, Kadir MOA (2013) Elimination of enteric indicators and pathogenic bacteria in secondary effluents and lake water by solar disinfection (SODIS). J Water Reuse Des. 3(1):39–46

Al-Gheethi AA, Norli I, Efaq AN, Bala JD, Al-Amery Ramzy M A (2015) Solar disinfection and lime treatment processes for reduction of pathogenic bacteria in sewage treated effluents and biosolids before reuse for agriculture in Yemen. Water Reuse Des. 5(3):419–429

Al-Gheethi AA, Mohamed RM, Jais NM, Efaq AN, Wurochekke AA, Amir-Hashim MK (2017) Influence of pathogenic bacterial activity on removal of nutrients from wet market wastewater by *Scenedesmus* sp. Water and Health Journal (Online)

Alva MS, De Luna-pabello VM, Cadena E, Ortíz E (2013) Green microalga *Scenedesmus actus* grown on municipal wastewater to couple nutrient removal with lipid accumulation for biodiesel production. Biores Technol 146:744–748

Arumugam M, Agarwal A, Arya AC, Ahmed Z (2013) Influence of nitrogen sources on biomass productivity of microalgae *Scenedesmus bijugatus*. Biores Technol 131:246–249

Atiku A, Mohamed RMSR, Al-Gheethi AA, Wurochekke AA, Kassim Amir H (2016) Harvesting microalgae biomass from the phycoremediation process of greywater. Environ Sci Poll Res 23(24):24624–24641

Azarpira H, Dhumal K, Pondhe G (2014) Application of phycoremediation technology in the treatment of sewage water to reduce pollution load. Adv Environ Biol 2419–2424

Bala JD, Lalung J, Al-Gheethi AA, Norli I (2016) A Review on biofuel and bioresources for environmental applications. In: Renewable energy and sustainable technologies for building and environmental applications. Springer publishing, New York, pp 205–225

Banerjee C, Ghosh S, Sen G, Mishra S, Shukla P, Bandopadhyay R (2013) Study of algal biomass harvesting using cationic guar gum from the natural plant source as flocculant. Carbohydrate Poly 92(1):675–681

Barros AI, Gonçalves AL, Simões M, Pires JC (2015) Harvesting techniques applied to microalgae: A review. Renew Sustain Energy Rev 41:1489–1500

Bich NN, Yaziz MI, Bakti NA (1999) Combination of *Chlorella vulgaris* and *Eichhornia crassipes* for wastewater nitrogen removal. Water Res 33(10):2357–2362

Bilad MR, Vandamme D, Foubert I, Muylaert K, Vankelecom IF (2012) Harvesting microalgal biomass using submerged microfiltration membranes. Bioresoure Technol 111:343–352

Bohdziewicz J, Sroka E, Korus I (2003) Application of ultrafiltration and reverse osmosis to the treatment of the wastewater produced by the meat industry. Polish J Environ Stud 12(3):269–274

Brar A, Kumar M, Vivekanand V, Pareek N (2017) Photoautotrophic microorganisms and bioremediation of industrial effluents: current status and future prospects. 3 Biotech 7(1):18

Caixeta C, Cammarota M, Xavier A (2002) Slaughterhouse house wastewater treatment: evaluation of a new three-phase separation system in a UASB reactor. Biores Technol 81:61–69

Cassidy KO (2011). Evaluating algal growth at different temperatures. MSc Theses and Dissertations, Biosystems and Agricultural Engineering. University of Kentucky, United States

Cerff M, Morweiser M, Dillschneider R, Michel A, Menzel K, Posten C (2012) Harvesting fresh water and marine algae by magnetic separation: screening of separation parameters and high gradient magnetic filtration. Biores Technol 118:289–295

Chen YM, Liu JC, Ju YH (1998) Flotation removal of algae from water. Colloid Surface B 12(1):49–55

Chen CY, Yeh KL, Aisyah R, Lee DJ, Chang JS (2011) Cultivation, photobioreactor design and harvesting of microalgae for biodiesel production: a critical review. Biores Technol 102(1):71–81

Cheng YL, Juang YC, Liao GY, Tsai PW, Ho SH, Yeh KL, Lee DJ (2011) Harvesting of *Scenedesmus obliquus* FSP-3 using dispersed ozone flotation. Biores Technol 102(1):82–87

Dassey AJ, Theegala CS (2013) Harvesting economics and strategies using centrifugation for cost effective separation of microalgae cells for biodiesel applications. Biores Technol 128:241–245

De Godos I, Guzman HO, Soto R, García-Encina PA, Becares E, Muñoz R, Vargas VA (2011) Coagulation/flocculation-based removal of algal–bacterial biomass from piggery wastewater treatment. Biores Technol 102(2):923–927

Efaq AN, Nagao NNNA, Rahman H, Al-Gheethi AA, Shahadat M, Kadir MOA (2015) Supercritical Carbon dioxide as non-thermal alternative technology for safe handling of clinical wastes. J Environ Process 2:797–822

Efaq AN, Rahman NNNA, Nagao H, Alkarkhi AM, Al-Gheethi AA, Tengku NTL, Kadir MOA (2016). Supercritical fluid CO_2 technique for destruction of pathogenic fungal spores in solid clinical wastes. CLEAN—Soil, Air, Water 44(12):1700–1708

Efaq AN, Rahman NNA, Nagao H, Al-Gheethi AA, Kadir MOA (2017) Inactivation of *Aspergillus* Spores in Clinical Wastes by Supercritical Carbon Dioxide. Arab J Sci Eng (AJSE) 42(1):39–51

Fagiri YMA, Salleh A, El-Nagerabi SA (2013) Influence of chemical and environmental factors on the growth performance of *Spirulina platensis* strain SZ100. J. Algal Biomass Utln 4(2):7–15

Gani P, Sunar NM, Matias-Peralta HM, Latiff A, Aziz A, Kamaludin NS, Er CM (2015) Experimental study for phycoremediation of *Botryococcus* sp. on greywater. Appl. Mech. Mat. 773:1312–1317

Gentili FG (2014) Microalgal biomass and lipid production in mixed municipal, dairy, pulp and paper wastewater together with added flue gases. Biores Technol 169:27–32

Godos I, Vargas VA, Blanco S, Gonzalez MCG, Soto R, Garcia-Encina PA, Becares E, Munoz R (2012) A comparative evaluation of microalgae for the degradation of piggery wastewater under photosynthetic oxygenation. Biores Technol 101:5150–5158

Gokulan R, Sathish N, Kumar RP (2013) Treatment of grey water using hydrocarbon producing *Botryococcus braunii*. Int J Chem Tech Res 5(3):1390–1392

Gong Q, Feng Y, Kang L, Luo M, Yang J (2014) Effects of light and pH on cell density of *Chlorella Vulgaris*. Energy Procedia 64:2012–2015

Gorin KV, Sergeeva YE, Butylin VV, Komova AV, Pojidaev VM, Badranova GU, Shapovalova AA, Konova IA, Gotovtsev BM (2015) Methods coagulation/flocculation and flocculation with ballast agent for effective harvesting of microalgae. Biores Technol 193:178–184

Guerrero-Cabrera L, Rueda JA, García-Lozano H, Navarro AK (2014) Cultivation of *Monoraphidium* sp., *Chlorella* sp. and *Scenedesmus* sp. algae in batch culture using *Nile tilapia* effluent. Biores Technol 161:455–460

Guo SL, Zhao XQ, Wan C, Huang ZY, Yang YL, Alam A, Ho SH, Bai FW, Chang GS (2012) Characterization of flocculating agent from the self-flocculating microalga *Scenedesmus obliquus* AS-6-1 for efficient biomass harvest. Biores Technol 145:285–289

Gupta SK, Ansari FA, Shriwastav A, Sahoo NK, Rawat I, Bux F (2016) Dual role of *Chlorella sorokiniana* and *Scenedesmus obliquus* for comprehensive wastewater treatment and biomass production for bio-fuels. J. Cleaner Prod. 115:255–264

Gutiérrez R, Ferrer I, García J, Uggetti E (2015) Influence of starch on microalgal biomass recovery, settleability and biogas production. Biores Technol 185:341–345

Hamid ASH, Lananan F, Din WNS, Su SL (2014) Harvesting microalgae, *Chlorella* sp. by bioflocculation of *Moringa oleifera* seed derivatives from aquaculture wastewater phytoremediation. Int Biodeterioration Biodegrad, 270–275

Hanotu J, Bandulasena H, Zimmerman W (2012) Microflotation performance for algal separation. Biotechnol Bioengineer 109:1663–1673

Hauwa A, Mohamed RM, Al-Gheethi AA, Wurochekke AA, Amir HK (2017a) Optimizing *Botryococcus* sp. biomass harvesting from greywater by natural coagulants. Waste and Biomass Valorization (online)

Hauwa A, Mohamed RMSR, Al-Gheethi AA, Wurochekke AA, Hashim MA (2017b). Harvesting of Botryococcus sp. biomass from greywater by natural coagulants. Waste Biomass Valorization 1–13

Hejazi M, Wijeffels R, Holwerda E (2004) Milking microalga *Dunaliella salina* for β-carotene production in two-phase bioreactors. Biotechnol Bioeng 85:475–481

Hena S, Fatimah S, Tabassum S (2015) Cultivation of algae consortium in a dairy farm wastewater for biodiesel production. Water Res Ind 10:1–14

Hu YR, Xu L, Feng W (2014) A magnetic separator for efficient microalgae harvesting. Biores Technol 158:388–391

Imtiazuddin S, Mumtaz M, Mallick KA (2012) Pollutants of wastewater characteristics in textile industries. J Basic Appl Sci 8:554–556

Jacob-Lopes E, Scoparo CHG, Lacerda LMCF, Franco TT (2009) Effect of light cycles (night/day) on CO_2 fixation and biomass production by microalgae in photobioreactors. Chem Eng Process: Process Intensification 48(1):306–310

Jais NM, Mohamed RMSR, Al-Gheethi AA, Hashim Amir (2017) Dual role of phycoremediation of wet market wastewater for nutrients and heavy metals removal and microalgae biomass production. Clean Technol Environ Policy 19(1):37–52

Ji MK, Kim HC, Sapireddy VR, Yun HS, Abou-Shanab RA, Choi J, Jeon BH (2013) Simultaneous nutrient removal and lipid production from pre-treated piggery wastewater by *Chlorella vulgaris* YSW-04. Appl Microbiol Biotechnol 97(6):2701–2710

Kalin M, Wheeler WN, Meinrath G (2005) The removal of uranium from mining waste water using algal/microbial biomass. J Environ Radioact 78:151–177

Kamarudin KF, Yaakob Z, Rajkumar R, Takriff MS, Tasirin S (2013) Bioremediation of palm oil mill effluents (POME) using *Scenedesmus dimorphus* and *Chlorella vulgaris*. Adv Sci Lett 19(10):2914–2918

Kang HK, Salim HM, Akter N, Kim DW, Kim JH, Bang HT, Suh OS (2013) Effect of various forms of dietary Chlorella supplementation on growth performance, immune characteristics, and intestinal microflora population of broiler chickens. J Appl Poultry Res 22(1):100–108

Kim MK, Park JW, Park CS, Kim SJ, Jeune KH, Chang MU, Acreman J (2007) Enhanced production of *Scenedesmus* spp. (green microalgae) using a new medium containing fermented swine wastewater. Biores Technol 98(11):2220–2228

Kim G, Yun Y, Shin H, Kim H, Han J (2015) Scenedesmus-based treatment of nitrogen and phosphorus from effluent of anaerobic digester and bio-oil production. Biores Technol 196:235–240

Knuckey RM, Brown MR, Robert R, Frampton DMF (2006) Production of microalgal concentrates by flocculation and their assessment as aquaculture feeds. Aquacult Eng 35:300–313

Kotteswari M, Murugesan S, Kumar R (2012) Phycoremediation of dairy effluent by using the microalgae Nostoc sp. Int J Environ Res Dev 2(1):35–43

Kumar RR, Rao PH, Arumugam M (2015) Lipid extraction methods from microalgae: a comprehensive review. Front Energy Res 2:61. https://doi.org/10.3389/fenrg.2014.00061

Kuo CM, Chen TY, Lin TH, Kao CY, Lai JT, Chang JS, Lin CS (2015) Cultivation of Chlorella sp. GD using piggery wastewater for biomass and lipid production. Biores Technol 194:32

Kurniawati H, Ismadji S, Liu J (2014) Microalgae harvesting by flotation using natural saponin and chitosan. Biores Technol 166:429–434

Lee K, Lee CG (2001) Effect of light/dark cycles on wastewater treatments by microalgae. Biotechnol Bioprocess Eng 6(3):194–199

Lekshmi B, Joseph R, Jose A, Abinandan S, Shanthakumar S (2015) Studies on reduction of inorganic pollutants from wastewater by Chlorella pyrenoidosa and Scenedesmus abundans. Alex Eng J 54(4):1291–1296

Letelier-Gordo CO, Holdt SL, De Francisci D, Karakashev DB, Angelidaki I (2014) Effective harvesting of the microalgae Chlorella protothecoides via bioflocculation with cationic starch. Biores Technol 167:214–218

Li Y, Chen YF, Chen P, Min M, Zhou W, Martinez B, Ruan R (2011) Characterization of a microalga Chlorella sp. well adapted to highly concentrated municipal wastewater for nutrient removal and biodiesel production. Biores Technol 102(8):5138–5144

Lim JK, Chieh DCJ, Jalak SA, Toh PY, Yasin NHM, Ng BW, Ahmad AL (2012) Rapid magnetophoretic separation of microalgae. Small 8(11):1683–1692

Liu JC, Chen YM, Ju YH (1999) Separation of algal cells from water by column flotation. Separ Sci Technol 34(11):2259–2272

Liu D, Li F, Zhang B (2009) Removal of algal blooms in freshwater using magnetic polymer. Water Sci Technol 59(6):1085–1091

Liu J, Zhu Y, Tao Y, Zhang Y, Li A, Li T, Zhang C (2013) Freshwater microalgae harvested via flocculation induced by pH decrease. Biotechnol Biofuels 6(1):1

Lu Q, Zhou W, Min M, Ma X, Chandra C, Doan YT (2015) Growing Chlorella sp. on meat processing wastewater for nutrient removal and biomass production. Biores Technol 198:189–197

Mahale VE, Chaugule BB (2013) Optimization of freshwater green alga Scenedesmus incrassatulus for biomass production and augmentation of fatty acids under abiotic stress conditions. Phykos 43(1):22–31

Maizatul AY, Mohamed RMSR, Al-Gheethi AA, Hashim MA (2017) An overview of the utilisation of microalgae biomass derived from nutrient recycling of wet market wastewater and slaughterhouse wastewater. Int Aquat Res:1–17

Makareviciene V, Skorupskaite V, Levisauskas D, Andruleviciute V, Kazancev K (2014) The optimization of biodiesel fuel production from microalgae oil using response surface methodology. Int J Green Energy 11(5):527–541

Maroneze MM, Barin JS, Menezes CRD, Queiroz MI, Zepka LQ, Jacob-Lopes E (2014) Treatment of cattle-slaughterhouse wastewater and the reuse of sludge for biodiesel production by microalgal heterotrophic bioreactors. Scientia Agricola 71(6):521–524

Matos C, Santos M, Nobre B, Gouveia L (2013) Nannochloropsis sp. biomass recovery by Electro-Coagulation for biodiesel and pigment production. Biores Technol 134:219–226

McKinney RE (2004) Environmental pollution control microbiology: a fifty-year perspective. CRC Press, Boca Raton

Michel J, Luís M, Paola M, Pirolli M, Michelon W, Moreira H (2016) Enhancement of nutrient removal from swine wastewater digestate coupled to biogas purification by microalgae Scenedesmus spp. Biores Technol 202:67–75

Michels MH, Vaskoska M, Vermuë MH, Wijffels RH (2014) Growth of Tetraselmis suecica in a tubular photobioreactor on wastewater from a fish farm. Water Res 2014(65):290–296

Min M, Wang L, Li Y, Mohr MJ, Hu B, Zhou W, Ruan R (2011) Cultivating Chlorella sp. in a pilot-scale photobioreactor using centrate wastewater for microalgae biomass production and wastewater nutrient removal. Appl Biochem Biotechnol 165(1):123–137

Minhas AK, Hodgson P, Barrow CJ, Adholeya A (2016) A Review on the assessment of stress conditions for simultaneous production of microalgal lipids and carotenoids. Front Microbiol 7:546. https://doi.org/10.3389/fmicb.2016.00546

Munir NE, Sharif NA, Shagufta N, Saleem FA, Manzoor FA (2013) Harvesting and processing of microalgae biomass fractions for biodiesel production (A review). Sci Tech Dev 32:235–243

Ndikubwimana T, Zeng X, Liu Y, Chang JS, Lu Y (2014) Harvesting of microalgae Desmodesmus sp. F51 by bioflocculation with bacterial bioflocculant. Algal Res 6:186–193

Ozkan A, Kinney K, Katz L, Berberoglu H (2012) Reduction of water and energy requirement of algae cultivation using an algae biofilm photobioreactor. Biores Technol 114:542–548

Pahazri N, Mohamed RMS, Al-Gheethi AA, Amir Hashim (2016) Production and harvesting of microalgae biomass from wastewater, a critical review. Environ Technol Rev 5(1):39–56. Online

Park J, Craggs R, Shilton AN (2011) Wastewater treatment high rate algal ponds for biofuel production. Biores Technol 102:35–42

Pathak VV, Singh DP, Kothari R, Chopra AK (2014) Phycoremediation of textile wastewater by unicellular microalga Chlorella pyrenoidosa. Cell Mol Biol 60(5):35–40

Pérez-Pazos JV, Fernández-Izquierdo P (2011) Synthesis of neutral lipids in Chlorella sp. under different light and carbonate conditions. CT&F-Ciencia, Tecnología y Futuro 4(4):47–58

Petrusevski B, Bolier G, Van Breemen AN, Alaerts GJ (1995) Tangential flow filtration: a method to concentrate freshwater algae. Water Res 29:1419–1424

Phang SM, Miah MS, Yeoh BG, Hashim MA (2000) Spirulina cultivation in digested sago starch factory wastewater. J Appl Phycol 12(3–5):395–400

Pragya N, Pandey KK, Sahoo P (2013) A review on harvesting, oil extraction and biofuels production technologies from microalgae. Renew Sustain Energy Rev 24:159–171

Prochazkova G, Safarik I, Branyik T (2013) Harvesting microalgae with microwave synthesized magnetic microparticles. Biores Technol 130:472–477

Qin JG, Li Y (2006) Optimization of the growth environment of Botryococcus braunii strain CHN 357. J Freshwater Ecol 21(1):169–176

Rahman A, Ellis JT, Miller CD (2012) Bioremediation of domestic wastewater and production of bioproducts from microalgae using waste stabilization ponds. J Bioremed Biodeg 3:113

Rai MP, Gautom T, Sharma N (2015) Effect of salinity, pH, light intensity on growth and lipid production of microalgae for bioenergy application. J Biol Sci 15(4):260–267

Ramsundar P, Guldhe A, Singh P, Bux F (2017) Assessment of municipal wastewaters at various stages of treatment process as potential growth media for Chlorella sorokiniana under different modes of cultivation. Biores Technol 227:82–92

Rao HP, Kumar R, Raghavan BG, Subramanian VV, Sivasubramanian V (2011) Application of phycoremediation technology in the treatment of wastewater from a leather-processing chemical manufacturing facility. Water SA 37(1):07–14

Rasala B, Mayfield S (2015) Photosynthetic biomanufacturing in green algae; production of recombinant proteins for industrial, nutritional, and medical uses. Photosynth Res 123:227–239

Rashid N, Rehmana SU, Han JI (2013) Rapid harvesting of freshwater microalgae using chitosan. Process Biochem 48:1107–1110

Rawat I, Kumar RR, Mutanda T, Bux F (2011) Dual role of microalgae: phycoremediation of domestic wastewater and biomass production for sustainable production. Appl Energy 88:3411–3424

Riano B, Blanco S, Becares E, Garcia-Gonzalez MC (2016) Bioremediation and biomass harvesting of anaerobic digested cheese whey in microalgal-based systems for lipid production. Ecol Eng 97:40–45

Rubio J, Souza ML, Smith RW (2002) Overview of flotation as a wastewater treatment technique. Miner Eng 15:139–155

Ruiz-Martinez A, Garcia NM, Romero I, Seco A, Ferrer J (2012) Microalgae cultivation in wastewater: nutrient removal from anaerobic membrane bioreactor effluent. Biores Technol 126:247–253

Sengar RMS, Singh KK, Singh S (2011) Application of phycoremediation technology in the treatment of sewage water to reduce pollution load. Indian J Sci Res 2(4):33–39

Shekhawat K, Rathore SS, Premi OP, Kandpal BK, Chauhan JS (2012) Advances in agronomic management of Indian mustard (Brassica juncea (L.) Czernj. Cosson): an overview. Int J Agron

Silambarasan T, Vikramathithan M, Dhandapani R (2012) Biological treatment of dairy effluent by microalgae. World J Sci Techno. 2(7):132–134

Sim TS, Goh A, Becker EW (1988) Comparison of centrifugation, dissolved air flotation and drum filtration techniques for harvesting sewage-grown algae. Biomass 16:51–62

Song M, Pei H, Hu W, Zhang S, Ma G, Han L, Ji Y (2014) Identification and characterization of a freshwater microalga Scenedesmus SDEC-8 for nutrient removal and biodiesel production. Biores Technol 162:129–135

Surendhiran D, Vijay M (2013) Study on flocculation efficiency for harvesting nannochloropsis oculata for biodiesel production. Int J Chem Tech Res 5(4):1761–1769

Teixeira CM, Kirsten FV, Teixeira PCN (2012) Evaluation of *Moringa oleifera* seed flour as a flocculating agent for potential biodiesel producer microalgae. J Appl Phycol 24(3):557–563

Toh PY, Yeap SP, Kong LP, Ng BW, Chan DJC, Ahmad AL, Lim JK (2012) Magnetophoretic removal of microalgae from fishpond water: feasibility of high gradient and low gradient magnetic separation. Chem Eng J 211:22–30

Uduman N, Qi Y, Danquah MK, Forde GM, Hoadley A (2010) Dewatering of microalgal cultures: a major bottleneck to algae-based fuels. J Renew Sust Energy 2(1):012701

Uduman N, Bourniquel V, Danquah M, Hoadley A (2011) A parametric study of electrocoagulation as a recovery process of marine microalgae for biodiesel production. Chem Eng J 174:249–257

Vandamme D, Foubert I, Meesschaert B, Muylaert K (2010) Flocculation of microalgae using cationic starch. J Appl Phycol 22:525–530

Vandamme D, Eyley S, Van den Mooter G, Muylaert K, Thielemans W (2015) Highly charged cellulose-based nanocrystals as flocculants for harvesting *Chlorella vulgaris*. Biores Technol 194:270–275

Wan C, Zhao XQ, Guo SL, Alam MA, Bai FW (2013) Bioflocculant production from *Solibacillus silvestries* WO1 and its application in cost effective harvest of marine microalgae *Nannochloris oceanica* by flocculation. Biores Technol 135:207–212

Wang L, Li Y, Chen P, Min M, Chen Y, Zhu J, Ruan RR (2010) Anaerobic digested dairy manure as a nutrient supplement for cultivation of oil-rich green microalgae *Chlorella* sp. Biores Technol 101(8):2623–2628

Wurochekke AA, Mohamed RMS, Al-Gheethi AA, Amir HM, Matias-Peralta HM (2016) Household greywater treatment methods using natural materials and their hybrid system. J Water Health. Online

Xin L, Hong-Ying H, Ke G, Ying-Xue S (2010) Effects of different nitrogen and phosphorus concentrations on the growth, nutrient uptake, and lipid accumulation of a freshwater microalga *Scenedesmus* sp. Biores Technol 101:5494–5500

Xin L, Ying HH, Ping ZY (2011) Growth and lipid accumulation of a freshwater microalga *Scenedesmus* sp. under different cultivation temperature. Biores Technol 102:3098–3102

Xu L, Wang F, Li HZ, Hu ZM, Guo C, Liu CZ (2010) Development of an efficient electroflocculation technology integrated with dispersed-air flotation for harvesting microalgae. J Chem Technol Biotechnol 85(11):1504–1507

Xu L, Guo C, Wang F, Zheng S, Liu C (2011) A simple and rapid harvesting method for microalgae by in situ magnetic separation. Biores Technol 102:10047–10051

Yaakob Z, Ali E, Mohamad M, Takrif MS (2014) An overview: biomolecules from microalgae for animal feed and aquaculture. J Biol Res 21(6):1–10

Yao L, Shi J, Miao X (2015) Mixed wastewater coupled with CO_2 for microalgae culturing and nutrient removal. PLoS ONE 10(9):e0139117

Zhang X, Hu Q, Sommerfeld M, Puruhito E, Chen Y (2010) Harvesting algal biomass for biofuels using ultrafiltration membranes. Biores Technol 101(14):5297–5304

Zheng H, Gao Z, Yin J, Tang X, Ji X, Huang H (2012) Harvesting of microalgae by flocculation with poly (γ-glutamic acid). Biores Technol 112:212–220

Chapter 12
Centralised and Decentralised Transport Systems for Greywater and the Application of Nanotechnology for Treatment Processes

A. Athirah, Adel Ali Saeed Al-Gheethi, Efaq Ali Noman, Radin Maya Saphira Radin Mohamed and Amir Hashim Mohd Kassim

Abstract Centralised and decentralised treatment systems for greywater are used based on the available sites, economic conditions and facilities of the treatment system. However, the decentralised system is a priority for rural regions for the recycling of greywater because it is inexpensive and does not result in toxic byproducts. This system also does not require maintenance or chemical additives. This system reduces the amount of wastewater produced, reduces dependence on traditional water supply and saves cost especially for low-income populations. The utilisation of the decentralised treatment system facility for the treatment of wastewater may help to reduce organic matters in waste efficiently. Transport models are typically applied to simulate water and solute movement in different porous media in order to understand the movement of pollutants in the soils and nutrient uptake by plants. Moreover, the utilisation of nanotechnology in the treatment of wastewater has emerged recently. This technique appears to exhibit high efficiency in the removal of micro-pollutants from wastewater. However, recent research has indicated that certain types of nanomaterial used in wastewater treatment may be toxic. In this chapter, the advantages and disadvantages of centralised and decentralised systems, as well as transport models,

A. Athirah · A. A. S. Al-Gheethi · R. M. S. Radin Mohamed (✉) · A. H. Mohd Kassim
Micro-Pollutant Research Centre (MPRC), Department of Water and Environmental Engineering,
Faculty of Civil and Environmental Engineering, Universiti Tun Hussein Onn Malaysia (UTHM),
86400 Parit Raja, Batu Pahat, Johor, Malaysia
e-mail: maya@uthm.edu.my

E. A. Noman
Faculty of Applied Sciences and Technology (FAST), Universiti Tun Hussein Onn Malaysia
(UTHM), Pagoh, Johor, Malaysia
e-mail: adel@uthm.edu.my

E. A. Noman
Department of Applied Microbiology, Faculty Applied Sciences, Taiz University, Taiz, Yemen

© Springer International Publishing AG, part of Springer Nature 2019
R. M. S. Radin Mohamed et al. (eds.), *Management of Greywater in Developing Countries*,
Water Science and Technology Library 87,
https://doi.org/10.1007/978-3-319-90269-2_12

are reviewed. The challenges associated with the application of this technology in developing countries are discussed.

Keywords Transport model · Centralised and decentralised system Nanomaterials · Health risk

12.1 Introduction

The treatment systems applied in the greywater treatment process have built upon studies on water flow and the transport of pollutants in subsurface systems. Several models have been investigated in the literature. The DRAINMOD model is a multicomponent drainage and water-related management system used to describe the hydrology of poor and shallow water table soils (Skaggs et al. 2012). The aqua-cycle model is also a daily urban water balance model using spatial scales such as unit blocks, clusters and catchments (Pak et al. 2010). On the other hand, the Streeter–Phelps oxygen sag curve model is a water quality modelling tool to evaluate water pollution to solve the problem of untreated greywater discharged directly into water bodies or streams (Nas and Nas 2009). The use of a transport model can help engineers to simulate the hydrological process in the experimental assessment of pollutants and its impact on water quality (Haris et al. 2016). The efficiency of different measures will reduce the amount of pollutants discharged into receiving water bodies. Thus, accurate prediction is required for model evaluation and parameter measures will reflect the causes of errors in modelling. However, the weakness of these models includes the lack of user-friendly characteristics, large data requirements and the non-appearance of clear declarations of their boundaries. This chapter focused on the centralised and decentralised modelling systems for the transportation of greywater from a source to a discharge point in a drainage system.

12.2 Centralised and Decentralised Systems for Greywater

The main concern of human beings is the lack of clean water resources. Increasing development, especially in urban areas, requires a proper water management system. The implementation of strategies for urban water management is one of the main alternatives that has become a necessity. Reusing wastewater through resources from households such as greywater and black water is one of the most reliable options to reduce the water demand of urban areas. Both centralised and decentralised recycling systems are used based on available sites, economic conditions and facilities of the treatment system. Therefore, the decentralised system is a priority for recycling purposes because the centralised system involves high transportation costs (Ahmed and Arora 2012). Figure 12.1 shows the treatment process of wastewater collection using large diameter (centralised) and small diameter (decentralised) systems.

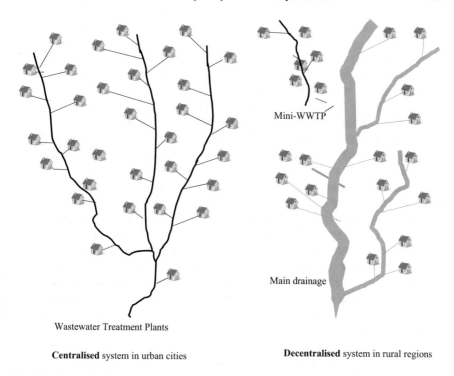

Mini-WWTP

Main drainage

Wastewater Treatment Plants

Centralised system in urban cities   **Decentralised** system in rural regions

Fig. 12.1   Centralised system in urban cities versus decentralised system in rural regions

According to Chirisa et al. (2016), the centralised system is no more sustainable for wastewater management especially for countries with high water consumption such as those in Latin America. Therefore, this system is associated with environmental pollution. In African countries such as Zimbabwe, the decentralised system is widely used due to the development of infrastructure which discourages sewer systems. In a comparison between centralised and decentralised systems, it can be noted that the centralised system includes different types of the wastewater and requires large diameters, area, long distances and well-trained technicians. In this system, all the water is collected in one area. Thus, its implementation is more complex and therefore represents high risk on a larger scale. In order to overcome all of these challenges, the decentralised system might be the best alternative system where treated wastewater can be reused at the same area, thus cutting down transportation costs (Chirisa et al. 2016).

Conventional centralised systems are considered as the main wastewater management systems in many developed and developing countries. In these systems, different types of wastewater which include domestic, commercial, industrial and hospital wastewater, as well as storm and runoff water, are connected to a central treatment plant which is planned, designed and operated by government agencies and then discharged into the nearest surface water bodies. The main challenge of the centralised treatment system is the cost of building a central treatment plant which is highly efficient in producing high-quality, treated effluents. This is due to the absence of funding and technical expertise needed to manage and control treatment plants in developing countries (Chirisa et al. 2016). However, the facilities to build and run central treatment plants have become more sustainable in terms of economic and nature conservation nowadays (Liang and Dijk 2012). The role of the centralised system is to collect wastewater from different sectors. The wastewater is then subjected to treatment and then redistributed to the respective sectors (industrial and agriculture sectors) which use the treated effluents for different purposes. Moreover, the centralised wastewater treatment plant (WWTP) ensures the sustainability of water use and improves public health by reducing the pollution of groundwater and surface water (Hendrawan et al. 2013).

One of the solutions to reduce the overload of WWTP is by the separating the types of wastewater connected to the plant. This means that it can be used for the treatment of sewage while rainwater, stormwater and greywater may be treated separately using wetlands, rainwater reuse and anaerobic digesters for generating energy on a small scale. In industrialised countries, companies should have a wastewater treatment plant to treat the industrial wastewater generated from their activities. These regulations are not applied in developing countries such as Yemen where the sewage treatment plants receive all the wastewater generated by different sectors. Thus, the treatment plant experiences capacity overloaded. This may reduce the efficiency of the treatment plant and result in low-quality treated effluents since the treatment period for each stage is shortened in order to receive more wastewater (Al-Gheethi et al. 2015). In developed countries, each housing cluster should have small treatment plants providing both primary and secondary treatment in order to reduce water pollutants. The partially treated effluents are connected to the main STP and subjected to the tertiary treatment process in order to produce treated effluents which comply with regulated standards for final disposal or reuse (Ma et al. 2015).

Decentralised treatment systems or on-site wastewater treatment systems are used in the regions which lack a centralised reticulated sewerage system (Gunady et al. 2015). The decentralised greywater treatment is usually performed in rural areas where people use this system for separating greywater from black water before discharging it into drainage systems (Capodaglio 2017). In contrast, the septic tank is a decentralised treatment system for domestic sewage in villages. The individual septic tank might contribute effectively to the improvement of the quality of effluents before disposal into drainage systems. In Malaysia, IST is an acceptable treatment method by Indah Water Konsortium (IWK) to meet the main standards regulated by EQA1974 in terms of COD, BOD and TSS. These systems are designed to contribute to the reduction of natural water via the direct discharge of untreated sewage. Therefore, the strategy implementation of public health and water quality can be controlled using decentralised rural sewage treatment technology (Guo et al. 2014).

The application of the decentralised system is an alternative approach for the proper management of wastewater by focusing on its development and practical implementation. This system is more applicable for the recycling of treated wastewater and reduces the dependency on traditional water supply. The decentralised system also reduces the cost of the treatment process as well as the cost of infrastructure for long-distance transport since wastewater is collected through distribution pipes in residential areas, stored and treated on-site (Ahmed and Arora 2012; Dhinadhayalan and Nema 2012). Greywater represents a good alternative resource for clean water. It might be used for toilet flushing, gardening, irrigation and construction.

Table 12.1 illustrated the most recent studies conducted on wastewater treatment using decentralised systems. Most of the studies were conducted in developing countries and the results have shown that these systems were significantly able to reduce pollutant load. Therefore, decentralised systems can be designed to overcome potential health risks and reduce the cost of long-distance transport. Both centralised and decentralised systems have advantages and disadvantages. The advantages of the centralised system include its applicability to high-density region and, its ability to protect the natural water system from the random discharge of wastewater as well as treat different types of the wastewater.

The disadvantages include the implementation, operation and maintenance costs. In contrast, the decentralised system is available for individual houses, has simple operation and maintenance requirements and is able to achieve high removal rates for most pollutants. However, the main disadvantages include its difficulty to be applied in any area, its reliance on the nature of soil permeability and density, as well as its potential threat to groundwater quality (Hendrawan et al. 2013; Ma et al. 2015).

12.3 Model Transport Simulation for Drainage Pollution

Computer model studies are mainly used to analyse the impact of wastewater on drainage systems. These models help researchers to understand the distribution or production of flows through pipes and ditches, overland flow and subsurface movement of water towards drain pipes and field ditches. Moreover, the model drainage is used to assess flood reduction and environmental benefits of subsurface drainage (Golmohammadi et al. 2016). The computer simulation model has used a simulation process for the flow within the water channel and drainage system, subject to lateral inflow owing to overland flow. Besides, it can also be used to investigate the effect of wastewater parameters in the water channel drainage via hydrographs and the accuracy of diffusion and kinematic equations in predicting flow in a drainage network (Pantelakis et al. 2013).

In the USA, the Environment Protection Agency (USEPA) uses mathematical relationship methods and simulation studies (computer model) for problem-solving (computer model) since 1970. Transport model simulation is used to assess the impacts of off-site transport in terms of economic and environmental outcomes based on pollutants. Some models have been effectively used to complete all the necessary information data for a more accurate analysis and to enhance the capabilities of the models. Model transport simulation needs to be improved for the experimental assessment of pollutants and its impact on water quality (Chung 1998).The multilevel model was used in a previous study on the transport of pollutants in the subsurface

Table 12.1 Recent studies on wastewater treatment using decentralised systems

Type of treatment	Wastewater/sources	Purpose	Treatment efficiency	References/Country
Septic tanks or pit latrines with up-flow filter	Blackwater (Toilet)	Waste stabilisation pond systems	COD, 95%; BOD, 98%; TSS, 99%	Dhinadhayalan and Nema (2012)/India
Microbial fuel cell (MFC)	Nitrogenous pollutants	Simultaneous removal of TN and organic matters in wastewater with electricity production	COD, 96%; NH4, 100%; TN, 60–68%	Feng et al. (2013)/China
Vermifiltration unit using the earthworm	Greywater (boy's hostel kitchen)	To know the efficiency of vermifilters and non-vermifilters	COD, 62%; BOD, 92%; TDS, 90%; TSS, 88%; Turbidity 93%	Lakshmi et al. (2014)/India
Aerobic biological using hydrogen peroxide	Greywater (Drains of bathroom sinks and showers greywater)	Treatment of greywater for non-potable usage	COD, 88%; TSS, 68%	Teh et al. (2015)/Malaysia
Constructed wetland pilot	Greywater (bathrooms, washing process, laundry greywater)	Wetland treatment	BOD, 75.99%; COD, 76.16%; TDS, 57.34%; TN, 62.08%; TP, 58.03%	Sudarsan et al. (2015)/India

as well as the water flow. The use of numerical models is an accurate technique to facilitate researchers in data analysis. Several studies have been using simulation models to simulate the process of pollutant transport and distribution (Pathirana et al. 2011). The benefit of the transport simulation model is to facilitate research work especially in terms of design and operational work (Kowalska et al. 2013).

Transport models are typically applied for simulating water and solute movement in different porous media to understand the movement of pollutants in soils as well as nutrient uptake by plants (Tsakiris and Alexakis 2012). Besides, the water quality model is one of the good examples in simulating the pollutants transported in a water environment. By doing the simulation, the pattern of the flow can be predicted thus saving labour cost if laboratory testing is done. This transport model has become an important tool to identify the behaviours of pollutants in the water environment (Wang et al. 2013). Peng et al. (2010) stated that model transport describing to simulate between point source and non-point source loads impacts also diffuse source pollution with analysing drainage impacts of wastewater and rainwater. Table 12.2 presents a summary of the previous studies conducted using different transport models.

There are numerous computerised models which have been used for simulating water transport systems as well as the hydrological transport model which have used in the experimental assessment of pollutants and impact on water quality such as DRAINMOD, aquacycle, Hydrus2D, QSWAT, Plugrisost and Streeter–Phelps. These models have been widely used by several researchers in studies on water quality (Longe and Omole 2008; Omole et al. 2012; Uzoigwe et al. 2015; Singh and Sharma 2015).

12.4 Model Development of Streeter–Phelps

Three conservative mass balances including DO, BOD and temperature are used to calculate the initial mixing of greywater and the stream. The theoretical mass balance calculations are performed based on DO present in greywater and streams. Water quality management is one of the major components in assessing the capability of a stream to absorb the waste load. DO is the most important characteristic in determining the quality of water. Most bacteria use oxygen in their metabolic processes. The lack of oxygen is directly related to the amount of organic material to measure the quantity of organic matter decomposed by bacteria. Waste discharge can be specified through the profile of DO concentration downstream. Oxygen demand for the oxidation process is measured as the concentration of low DO (Davis and Cornwell 2013). According to Cathey (2005), this profile which is called a DO sag curve represents how the DO concentration in a certain volume of wastewater changes over time or along a certain distance after oxygen-demanding materials are oxidised. This curve is calculated using the Streeter–Phelps equation.

A stream is able to self-purify as long as DO curve is going up the stream. It can still manage the waste discharge but if it goes below the critical point, this may cause changes in the living plants or animals in the stream. There are several factors

Table 12.2 Previous Studies conducted using different transport models

Software model	Application	Method	Scope	Findings	References
DRAINMOD model	To find an algorithm model to quantify the hydrologic components in the water	—By determining the effective dependent and independent value inputs	—Soil properties, —Site parameters —Weather data —Plant characteristics	—Water table depths with Nash–Sutcliffe modelling efficiency (EF = 0.68 and 0.72) —Daily drainage rates (EF = 0.73and 0.49) —Monthly drainage volumes (EF = 0.87 and 0.77)	Skaggs et al. (2012)
Aquacycle model	Analyse the water cycle in the urban region	Based on the sensitivity analysis for the Goonja drainage basin in the metropolitan Seoul region	—Base flow index (BI) was used as a sensitive parameter in the wet and dry season	—A valid model for rainfall and (R = 0.97) —The calibrated parameters significant for the validation of a run (r = 0.84)	Pak et al. (2010)
Hydrus 2D model	—Estimate the van Genuchten soil hydraulic parameters and heat transport parameters	—By examining the effect of water input and soil temperature on water flow	—Real-time sensors built to monitor soil temperatures at a depth between 40 and 160 m for 10 h, as well as the water temperature at the same location	—A favourable correlation was observed between simulated soil temperatures and corresponding observed values	Simunek et al. (2013)

(continued)

Table 12.2 (continued)

Software model	Application	Method	Scope	Findings	References
QSWAT model	Determine scanty rainfalls, land use changes, excess evaporation, deforestation and smaller amounts of groundwater recharge	QSWAT 1.2 was used between 1979 and 2012	The data collected daily included temperature, precipitation, radiation and wind velocity	The results revealed that the simulated data have reflected the hydrological parameters which included groundwater contribution, actual evapotranspiration, surface runoff and potential evapotranspiration	Bansode and Patil (2016)
PLUGRISOST model	—Evaluation the economic cost and environmental impact of rainwater and greywater for urban use	For the contribution in urban water planning for the development of smart cities	—Stimulation of rainwater harvesting (RWH) and greywater systems —Sizing of storage tanks, cost and quantitative analysis	—Infrastructure that incorporates the use of rainwater —Self-sufficiency analysis of water in cities	Gabarrell et al. (2014)
Streeter–Phelps model	—Assessment of water quality —Develop an oxygen sag curve	—The variations of DO were determined using the point source Streeter–Phelps model	—The concentration of DO, COD, temperature, pH TSS, TP, TN in the water was determined in response to time and distance	—DO was less than 4 mg/L —In the summer months, the consumption of DO was more than its production due to the increase in temperature and the low flow rate of the discharge	Nas and Nas (2009)

affecting self-purification in streams or rivers such as dilution (when wastewater is first being discharged into a water body, high DO level is available), current (the strong water current might prevent the sedimentation of solids in the discharged wastewater), temperature (the high activity of microorganisms in the stream with high temperature would accelerate self-purification), sunlight (the presence of algae in the stream might increase DO level due to photosynthesis which depends on the sunlight) and oxidation rate (based on the concentrations and type of organic matter in the discharged wastewater).

The dissolved oxygen in water bodies relies also on salinity, turbulence and atmospheric pressure. There are two factors which are the oxygen source and factors affecting oxygen shortage in developing a numerical expression for the DO sag curve. The oxygen shortage is caused by BOD which consists of CBOD (the oxygen demand resulting from the oxidation of organic carbon) and nitrogenous biochemical oxygen demand (NBOD) (resulting from the oxidation of nitrogen compounds) of the wastewater discharged (Cathey 2005; Haider et al. 2013). Next, non-point source pollution (respiration of living organisms and aquatic plants) of wastewater discharged (DO) into rivers is less than that in streams. Thus, the DO in rivers is lowered as soon as possible as wastewater is added even before any BOD is exerted. The parameter should consider only the initial condition such as DO reduction, carbonaceous BOD (CBOD) and re-aeration from the atmosphere to achieve the model development of Streeter–Phelps or DO sag equation (Qian 1999).

The deoxygenation phenomenon is used to define the deficiency in the DO of streams or rivers which receive the wastewater. This phenomenon depends on the quality of the organic matter (Lt) and temperature (T) at the receiving point where the reaction is high. However, the stream or river can replenish the oxygen supply at the interface between the river and the atmosphere in a process called reoxygenation or re-aeration. The rate of re-aeration is proportional to the DO deficit and a volumetric re-aeration coefficient. The re-aeration of oxygen is generated from the photosynthesis process of aquatic plants. Thus, the rate of re-aeration in streams or rivers depends on depth, velocity, oxygen deficit below saturation DO and the temperature of water (Cathey 2005).

12.5   Mathematical Analysis of the Streeter–Phelps Equation Model

The mathematical equation from Streeter–Phelps model was supported by Beck (1982) and Kumarasamy (2011). The equations of Streeter–Phelps have employed the form of a first-order partial differential equation model under steady-state conditions which included deoxygenation (K1) and reoxygenation (K2) (Kumarasamy 2011). According to the differential equation, the total change in oxygen deficit is equal to the difference between K1 and K2 at a specific time in the natural water system which receive the discharged greywater or wastewater. Thus, there is no specific method that might be used for determining the values which fit the reality of a given water body precisely. The equation methods used to estimate K_1 and K_2 provide reasonable approximations within predefined limits (Raymond et al. 2012; Haider et al. 2013; Benson et al. 2014). However, there is no special formula for each case because of the nonlinearity of these coefficients (Maroneze et al. 2014).

The Streeter–Phelps Eq. (1) is used to estimate the DO in the stream (Uzoigwe et al. 2015):

$$D_t = \frac{k_d L_o}{k_r - k_d} \left(e^{-k_d t} - e^{-k_r t} \right) + D_o \left(e^{-k_r t} \right) \tag{1}$$

where D_t = oxygen deficit in the stream after exertion of BOD for time, t, mg/L
L_o = initial ultimate BOD after stream and greywater have been mixed, mg/L
k_d = deoxygenation rate constant, d^{-1}
k_r = re-aeration rate constant, d^{-1}
t = time of travel of greywater discharged downstream, d
D_o = initial deficit after river and greywater have been mixed, mg/L

The equation for the determination of deoxygenation (k_d) and ultimate BOD_5 (L_o) uses the Thomas Slope method (Hendriarianti and Karnaningroem 2015). Furthermore, long-term BOD analysis results will be used to determine the value of carbon deoxygenation rate (k_d) and final BOD (BOD_u).

$$k_d = 2.61 \left(\frac{a}{b} \right) \tag{2}$$

$$L_o = \frac{1}{2.3 Kd\, a^3} \tag{3}$$

Note: Plotting $\left(\frac{t}{y} \right)^{1/3}$ as a function of t, the slope (b) and the intercept (a) of the line of best fit can be used to estimate the values of k_d and L_o.

The equation for the determination of re-aeration (K_r) uses the O'Conner Model method (Ugbebor et al. 2012). The developed standard table on a range of acceptable values for the reoxygenation coefficient f at 20 °C is shown below (Table 12.3).

Table 12.3 Values of the reoxygenation coefficient f at 20 °C (Ugbebor et al. 2012)

Types of water body	Value of f (day^{-1})
Small pond 0.05	1.0
Sluggish streams/lakes	1.0–1.5
Large stream with low velocity	1.5–2.0
Large streams with moderate velocity	2.0–3.0
Swift streams	3.0–5.0
Rapids	Rapids >5:0

The above values of the reoxygenation coefficient f imply that the higher the flow velocity, the higher the reoxygenation coefficient. Many research studies have been carried out on the mechanism of re-aeration as induced by temperature, river geometry and hydrodynamics factors.

$$k_r = \frac{3.9V^{0.5}}{H^{3/2}} \qquad (4)$$

$$k_r = \frac{3.9V^{0.5}\sqrt{1.037}^{T-20}}{H^{3/2}} \qquad (5)$$

where V = mean stream velocity m/s
H = average depth of stream
T = temperature (at 20 °C)

The equation for the determination of critical time of the deficit (Uzoigwe et al. 2015) shows the lowest point on the DO sag curve profile which is called the critical point. The critical point indicates the worst condition of a stream or a river. The critical distance can be calculated through Eq. (6).

$$t_c = \frac{x \ (m)}{U \left(\frac{m}{day} \right)} \qquad (6)$$

where x = distance (m)
U = average velocity of stream (m/s)
t_c = travel time from the basic relationship between time, distance and speed.

12.6 Advantages or Contribution of the Pollutant Transport Model

The pollutant transport model contributes effectively to the development of a modelling framework that enables a source-based pollution analysis of urban drainage systems. At present, an integrated approach in this model is crucial for the management of urban basins because these systems are special to various types of environmental problems (Maroneze et al. 2014). The Streeter–Phelps equation is useful for predicting the changes in surface water quality for environmental management around the world. Most developing countries are using this model to standardise water quality and it is widely used as a guide for environmental water quality assessment and surface water quality simulation (Wang et al. 2013). On the other hand, in cases for complete mixing starting from the point of wastewater discharge downstream, this model is important in terms of indicating the quality of time-based changes in dissolved oxygen deficits in rivers or streams. Thus, this model is capable of predicting the dissolved oxygen profile (Nas and Nas 2009).

The one-dimensional mathematical model is used with DO to predict the conditions when greywater is discharged into streams (Sinha et al. 2014). The DO simplified models can be used in many situations. This model is consistent with the experimental samples taken from streams to facilitate settlement in the Streeter–Phelps equation. The dissolved oxygen sag curve is able to help in the assessment of water quality in terms of oxygen content as well as the capability of stream waters to receive waste (Davis and Cornwell 2013). This model is related to the concentration of dissolved oxygen and the biological oxygen demand over time. This differential equation states that the rate of change in the DO deficit is the sum of two reactions between deoxygenation (Kd) and re-aeration (Kr) at any time (Cathey 2005).

This situation can be applied to the DO sag curve profile for the determination of DO concentration downstream. After passing a critical point, the rate of re-aeration is greater than the rate of deoxygenation for the process of self-purification in the stream (Jolankai 2000). This continuous addition of oxygen supply in water which is supplied by the atmosphere represents the capacity for self-purification because dissolved oxygen is gradually by the BOD load (Davis and Cornwell 2013).

The model in this study is used to estimate dissolved oxygen (DO) deficit, critical time after stream water and greywater are mixed and also to compare between measured and simulated DO. Analysis using the Streeter–Phelps equation helps to determine whether the quality of the river is affected when household greywater is disposed directly into water bodies near the river.

12.7   Nanotechnology Applications for Greywater Treatment

The application of nanomaterials in the treatment of wastewater is a new direction which has exhibited high efficiency in the removal of pollutants from wastewater as reported by several authors. Nanoplastics are loosely defined here as particles in the size range of 10–100 nm (Nolte et al. 2017). Engineered nanoparticles (ENPs) are being used as an alternative method to minimise the risks associated with chemical treatment (Hjorth et al. 2017).

However, there are some concerns pertaining to the application of nanomaterials in wastewater. There are consumer, occupational and environmental risks related to the use of nanomaterials in comparison to the use of traditional chemicals. The issue of the potential environmental toxicity of ENPs has been raised in 2003. Recently, a number of research studies by Hartmann and Baun (2010), Liguori et al. (2016) and others have confirmed the toxicity of nanomaterials. The concerns about nanotechnology are not only related to the risk of nanomaterials but also the methods used for the assessment of hazards and exposure. Many of the countries in the European Union, as well as the USA, have been focusing on the development of an accurate assessment method for the detection of hazardous chemicals (Cowan et al. 2014). In Malaysia, MOSTI has implemented the Safety Materials and Nano Products Risk

Benchmark. This step aims to reduce or eliminate the health risks associated with highly hazardous nanomaterials and subsequently replacing them with safer options. The alternative assessment of nanomaterials should include the identification of chemical substance with potential risks, the determination of alternative substances, which pose no health risks towards humans or the environment, the evaluation of alternative substances and finally, the determination of the economic cost of these alternatives (Jacobs et al. 2015).

The developments in this field are gradual. Therefore, before finding an efficient and accurate assessment method for the potential of nanoparticles, there is a growing concern from the view of scientists on the applications of nanotechnology. Some of the nanomaterials used such as silver, zirconia powder and TiO_2 nanomaterials are potentially hazardous nanomaterials. The preparation of nanomaterials from raw chemical substances which is risk-free might be an alternative way to avoid the threat of nanomaterials towards environmental biodiversity. The behaviours of these materials, as a result of the preparation process, chemical reactions with the other substances in wastewater as well as environmental conditions remain unknown. ENPs are defined as a complex mixture. Therefore, they are present in the environment as a complex structure in contrast to traditional chemical substances which are available as single contaminants (Hartmann and Baun 2010). A safe way to overcome the toxicity of nanomaterials is to use natural materials. Natural rubber latex (NRL) is non-toxic. Therefore, it might be used as a material to replace silver and zinc for the synthesis of nanomaterials (REF).

Baun et al. (2008) studied the potential of C_{60} nanoparticles (Buckminster fullerenes) for removing phenanthrene, atrazine, pentachlorophenol (PCP) and methyl parathion from aqueous solutions. The toxicity of these compounds in the presence or absence of C_{60}-nanoparticles against algae (*Pseudokirchneriellasubcapitata*) and crustaceans (*Daphnia magna*) was investigated. The results revealed that 85% of phenanthrene was sorbed to C_{60}-aggregates >200 nm. In contrast, only 10% of methyl parathion, atrazine and pentachlorophenol was sorbed. The toxicity of phenanthrene increased by 60% in the presence of C_{60}-aggregates while the toxicity of PCP reduced by 25% in tests with *D. magna*. There was no significant difference in the toxicity of atrazine and methyl parathion in the presence or absence C_{60}-aggregates. These findings indicated that the adsorption of phenanthrene to C_{60}-aggregates made it more available for uptake by living cells. Moreover, the study provided evidence on the toxicity of ENTs and their role in increasing the toxicity of some XOCs in water systems.

Nolte et al. (2017) investigated the ecological effects of plastic nanoparticles on green algae. The study focused mainly on the influence of polystyrene nanoparticles and coated (starch and PEG) gold nanoparticles on particle adsorption to *Pseudokirchneriellasubcapitata* cell walls based on depletion measurements and atomic force microscopy (AFM) analysis. The results revealed that positively charged and neutral plastic nanoparticles have a higher affinity to adsorb onto *P. subcapitata* cell walls compared to plastic nanoparticles with negative charge. This study introduces an accurate technique to evaluate toxicity effect of nanomaterials on aquatic organisms in the natural water system which receive wastewater treated with nanotechnology. The investigation of adsorption affinity between nanomaterials and microorganisms in comparison to the elements and other materials that are necessary for cell growth might explain the toxicity mechanism of nanoparticles. It can be concluded from this study that the utilisation of nanotechnology in the treatment of wastewater might affect the biological diversity in the environment.

Based on the above literature review, it can be concluded that a novel alternative treatment system of greywater needs to be considered in relation to many aspects such as the nature of the materials used, their efficiency as well as the toxicity levels. This is because the main aim of treating greywater is to protect human health from the adverse effects of toxic compounds. To overcome the shortcomings of chemical and physical treatments, microbial processes might be an alternative for the degradation XOCs in greywater through enzymatic reactions. Furthermore, the combination of AOPs and biological methods has been paid more attention in recent years (He et al. 2014).

12.8 Conclusion

The Streeter–Phelps equation should be able to facilitate a proper assessment in the selection of equations for the coefficient method of each re-aeration (K_2), deoxygenation (k_d) and initial ultimate BOD (Lo) after streams or rivers and greywater have been mixed depending on the conditions of the location. Further studies are required for the estimation of greywater quantities, the evaluation of household practices and the variation of greywater pollutants. The application of the Streeter–Phelps model is expected to tabulate the greywater flow and to which extent it can contribute to river pollution. Besides, the model describes the decrease in dissolved oxygen (DO) levels in rivers or streams along a certain distance through the degradation of BOD.

Acknowledgements Special gratitude goes to the laboratory technicians at the Micropollutant Research Centre, Faculty of Civil and Environmental Engineering, Universiti Tun Hussein Onn Malaysia (UTHM) for providing the facilities for this research. The authors also wish to thank the Ministry of Higher Education (MOHE) for supporting this research under FRGS vot 1574 and also the Research Management Centre (RMC) UTHM for providing grant IGSP U682 for this research.

References

Ahmed M, Arora M (2012) Suitability of Grey Water Recycling as decentralized alternative water supply option for Integrated Urban Water Management. J Eng 2(9):31–35
Al-Gheethi AA, Ismail N, Efaq AN, Bala JD, Al-Amery RM (2015) Solar disinfection and lime stabilization processes for reduction of pathogenic bacteria in sewage effluents and biosolids for agricultural purposes in Yemen. J Water Reuse Desalin 5(3):419–429
Bansode S, Patil K (2016) Water balance assessment using Q-SWAT. Int J Eng Res 5(6):2319–6890

Baun A, Sørensen SN, Rasmussen RF, Hartmann NB, Koch CB (2008) Toxicity and bioaccumulation of xenobiotic organic compounds in the presence of aqueous suspensions of aggregates of nano-C_{60}. Aquat Toxicol 86(3):379–387

Beck MB (1982) Identifying models of environmental systems' behaviour. Math Model 3(5):467–480

Benson A, Zane M, Becker TE, Visser A, Uriostegui SH, DeRubeis E, Moran JE, Esser BK, Clark JF (2014) Quantifying reaeration rates in alpine streams using deliberate gas tracer experiments. Water (Switzerland) 6(4):1013–1027

Capodaglio AG (2017) Integrated, decentralized wastewater management for resource recovery in rural and peri-urban areas. Res 6(2):1–22

Cathey AM (2005) The calibration, validation, and sensitivity analysis of Do sag: an instream dissolved oxygen model. University of Georgia, Master of Science

Chirisa I, Bandauko E, Matamanda A, Mandisvika G (2016) Decentralized domestic wastewater systems in developing countries: the case study of Harare (Zimbabwe). Appl Water Sci 1–10

Chung S (1998) Modeling the fate and transport of agricultural pollutants and their environmental impact on surface and subsurface water quality. Ph.D. thesis, Iowa State University, Ames Iowa

Cowan DM, Kingsbury T, Perez AL, Woods TA, Kovochich M, Hill DS, Madl AK, Paustenbach DJ (2014) Evaluation of the California safer consumer products regulation and the impact on consumers and product manufacturers. Regul Toxicol Pharmacol 68:23–40

Davis ML, Cornwell DA (2013) Introduction of environment engineering, 5th edn. Open University Press, Mcgraw-Hill International

Dhinadhayalan M, Nema AK (2012) Decentralised wastewater management new concepts and innovative technological feasibility for developing countries. Sustain Environ Res 22(1):39–44

Feng C, Hu A, Chen S, Yu CP (2013) A decentralized wastewater treatment system using microbial fuel cell techniques and its response to a copper shock load. Biores Technol 143:76–82

Gabarrell X, Morales-Pinzón T, Rieradevall J, Rovira MR, Villalba G, Josa A, Martínez-Gasol C, Dias AC, Martínez-Aceves DX (2014) Plugrisost: a model for design, economic cost and environmental analysis of rainwater harvesting in urban systems. Water Practice Technol 9(2):243–255

Golmohammadi G, Rudra RP, Prasher SO, Madani A, Goel PK, Mohammadi K (2016) Modeling the impacts of tillage practices on water table depth, drain outflow and nitrogen losses using DRAINMOD. Comput Electron Agric 124:73–83

Gunady M, Shishkina N, Tan H, Rodriquez C (2015) A review of on-site wastewater treatment systems in Western Australia from 1997 to 2011. J Environ Public Health 2015:12

Guo X, Liu Z, Chen M, Liu J, Yang M (2014) Decentralized wastewater treatment technologies and management in Chinese villages. Frontiers Environ Sci Eng 8(6):929–936

Haider H, Al W, Haydar S (2013) A review of dissolved oxygen and biochemical oxygen demand models for large river. Pak J Engg Appl Sci 12:127–142

Haris H, Chow MF, Usman F, Sidek LM, Roseli ZA, Norlida MD (2016) Urban stormwater management model and tools for designing stormwater management of green infrastructure practices. In: IOP conference series: earth and environmental science, vol 32, pp 1–19

Hartmann NB, Baun A (2010) The nano cocktail: ecotoxicological effects of engineered nanoparticles in chemical mixtures. Integr Environ Assess Manage 6(2):311

He SJ, Wang JL, Ye LF, Zhang YX, Yu J (2014) Removal of diclofenac from surface water by electron beam irradiation combined with a biological aerated filter. Radiat Phys Chem 105:104–108

Hendrawan D, Widarnako S, Moersidik S, Triwenko WR (2013) Evaluation of centralized WWTP and the need of communal Wwtp in supporting community-based sanitation in Indonesia. Eur Sci J 9(17):1857–7881

Hendriarianti E, Karnaningroem N (2015) Rate of nitrification-denitrification Brantas River in the city of Malang. Appl Environ Biol Sci 5(12):978–979

Hjorth R, Hansen SF, Jacobs M, Tickner J, Ellenbecker M, Baun A (2017) The applicability of chemical alternatives assessment for engineered nanomaterials. Int Environ Assess Manage 13(1):177–187

Jacobs MM, Malloy TF, Tickner JA, Edwards S (2015) Alternatives assessment frameworks: research needs for the informed substitution of hazardous chemicals. Environ Health Persp 124:265–280

Jolankai G (2000). Description of the CAL programme on water quality modelling. Basic river and lake water quality models, 2nd edn. United Nations Educational Scientific and Cultural Organization

Kowalska B, Kowalski D, Widomski MK (2013) Modelling of hydraulics and pollutants transport in Sewer systems. Monografie—PolitechnikaLubelska. Lublin University of Technology

Kumarasamy M (2011) Simulation of spatial and temporal variations of dissolved oxygen of Bayne-spruit stream in South Africa. Water Res Manage 145:403–414

Lakshmi C, Ranjitha J, Vijayalakshmi S (2014) Waste water treatment using vermifilteration technique at institutional level. Int J Adv Sci Tech Res 1(4):581–590

Liang X, Dijk VMP (2012) Cost benefit analysis of centralized wastewater reuse systems. J Benefit-Cost Anal 3(3):1–30

Liguori B, Hansen SF, Baun A, Jensen KA (2016) Control banding tools for occupational exposure assessment of nanomaterials—Ready for use in a regulatory context? NanoImpact 2:1–17

Longe EO, Omole DO (2008) Analysis of pollution status of River Illo, Ota, Nigeria. Environmentalist 28:451–457

Ma XC, Xue X, Gonzalez-Mejia A, Garland J, Cashdollar J (2015) Sustainability (Switzerland) 7(9):12071–12105

Maroneze MM, Zepka LQ, Vieira JG, Queiroz MI, Eduardo JL (2014) De-oxygenation rate, re-aeration and potential for self-purification of a small tropical urban stream. An Inter J Appl Sci 9(3):445–458

Nas SS, Nas E (2009) Water quality modeling and dissolved oxygen balance in streams: a point source Streeter-Phelps application in the case of the Harsit stream. Clean—Soil, Air and Water 37(1):67–74

Nolte TM, Hartmann NB, Kleijn JM, Garnæs J, van de Meent D, Hendriks AJ, Baun A (2017) The toxicity of plastic nanoparticles to green algae as influenced by surface modification, medium hardness and cellular adsorption. Aquat Toxicol 183:11–20

Omole DO, Adewumi IK, Longe EO, Ogbiye AS (2012) Study of auto purification capacity of River Atuwara in Nigeria. Int J Eng Technol 2(2):229–235

Pak G, Lee J, Kim H, Yoo Yun Z, Choi S, Yoon J (2010) Applicability of Aquacycle model to urban water cycle analysis. Des Water Treat 19(1–3):80–85

Pantelakis D, Thomas Z, Baltas E (2013) Hydraulic models for the simulation of flow routing. Global NEST J 15(3):315–323

Pathirana A, Maheng MD, Brdjanovic D (2011) A two-dimensional pollutant transport model for sewer overflow impact simulation. In: International conference on urban drainage, pp 10–15

Peng H, Yao W, Huang P (2010) Application of modified Streeter-Phelps model and COD changing model to Xiangxi River in Three Gorges Reservoir Area. In: International conference on bioinformatics and biomedical engineering, ICBBE. University Yichang, China, pp 1–4

Qian SS (1999) ESR 202 Applied environmental studies: preparation for problem solving. Environmental Sciences and Resources Portland, pp 1–65

Raymond PA, Zappa CJ, Butman D, Bott TL, Potter J, Mulholland P, Laursen AE, McDowell WH, Newbold D (2012) Scaling the gas transfer velocity and hydraulic geometry in streams and small rivers. Limnol Oceanogr: Fluids & Environ 2:41–53

Simunek J, Jacques D, Langergraber G, Bradford SA, Sejna M, Genuchten VM (2013) Numerical modeling of contaminant transport using HYDRUS and its specialized modules. J Indian Inst Sci 93(2):265–284

Singh O, Sharma MK (2015) Measurement of dissolved oxygen and biochemical oxygen demand for The Hindon River, India. Indian Water Res Soc 35(1):42–50

Sinha D, Aggarwal S, Tyagi B (2014) Analysis of DO sag for multiple point sources. Math Theory Model 4(3):1–7

Skaggs RW, Youssef MA, Chescheir GM (2012) Drainmod: model use, calibration, and validation. Trans ASABE 551(4):1509–1522

Sudarsan JS, Roy RL, Baskar G, Deeptha VT, Nithiyanantham S (2015) Domestic wastewater treatment performance using constructed wetland. Sust Water Res Manage 1(2):89–96

Teh XY, Poh PE, Gouwanda D, Chong MN (2015) Decentralized light greywater treatment using aerobic digestion and hydrogen peroxide disinfection for non-potable reuse. J Cleaner Prod 99:305–311

Tsakiris G, Alexakis D (2012) Water quality models: an overview. Eur Water 37:33–46

Ugbebor JN, Agunwamba JC, Amah VE (2012) Determination of reaeration Coefficient K_2 for polluted stream as a function of depth, hydraulic radius, temperature and velocity. Niger J Technol 31(2):174–180

Uzoigwe LO, Maduakolam SC, Samuel C (2015) Development of oxygen sag curve: a case study of Otamiri River, Imo State. Sci Eng Appl Sci 12(2):133–150

Wang Q, Li S, Jia P, Qi C, Ding F (2013) A review of surface water quality models. Sci World J 1–7

Chapter 13
Removal of Nutrients from Meat Processing Wastewater Through the Phycoremediation Process

A. S. Vikneswara, Radin Maya Saphira Radin Mohamed, Adel Ali Saeed Al-Gheethi, Amir Hashim Mohd Kassim and Norzila Othman

Abstract Wastewater from slaughterhouses and meat processing factories are normally assessed in terms of mass parameters due to the specific amounts of wastewater and pollutant load parallel to the animals slaughtered or processed that differ from the meat processing industry. Normally, this type of wastewater contains significant amounts of total phosphorus (TP), total nitrogen (TN), total organic carbon (TOC), chemical oxygen demand (COD), total suspended solids (TSS) and biochemical oxygen demand (BOD). The present chapter aims to highlight the characteristics of meat processing wastewater and the effect of the direct disposal of these wastes on the environment. Moreover, the potential of the phycoremediation process using microalgae species to remove nutrients in terms of total nitrogen (TN) and total phosphorus (TP) from the meat processing wastewater was reviewed. The role of a biokinetic study in determining the specific removal rate of nutrient by microalgae was also discussed.

Keywords Biokinetic models · Characteristics · Removal · Treatment
Meat processing wastewater

13.1 Introduction

The commercial meat-based food industry generates huge quantities of wastewater. Discharge of untreated or partially treated wastewater into the environment may lead

A. S. Vikneswara · R. M. S. Radin Mohamed (✉) · A. A. S. Al-Gheethi (✉)
A. H. Mohd Kassim · N. Othman
Micro-Pollutant Research Centre (MPRC), Department of Water and Environmental Engineering, Faculty of Civil and Environmental Engineering, Universiti Tun Hussein Onn Malaysia (UTHM), 86400 Parit Raja, Batu Pahat, Johor, Malaysia
e-mail: maya@uthm.edu.my

A. A. S. Al-Gheethi
e-mail: adel@uthm.edu.my

© Springer International Publishing AG, part of Springer Nature 2019
R. M. S. Radin Mohamed et al. (eds.), *Management of Greywater in Developing Countries*,
Water Science and Technology Library 87,
https://doi.org/10.1007/978-3-319-90269-2_13

to the increase of nutrients and organic compounds in the water body which further causes eutrophication (Bustillo-Lecompte et al. 2015). This is due to the highest concentrations of nitrogen, 2743.6 mg/L and phosphate, 328.4 mg/L in meat processing wastewater compared to other types of wastewater resulting from food operations (Cristian 2010; Cai et al. 2013). The treatment of meat processing wastewater is either a chemical or physical process. The purpose of the treatment is to reduce suspended solids (SS), which can clog water channels as they settle as a response to gravity, biodegradable organic matter which reduces the content of dissolved oxygen (DO). The biodegradable organic compounds can serve as food for undesirable microorganisms which combine with oxygen to yield energy required for the growth. Hence, the other organisms in the water body will lack oxygen and nutrients. This will eventually lead to high concentrations of unwanted algae such as phytoplankton.

The common process of wastewater treatment needed in meat processing factories includes regulation, aeration and settling tanks. Normally, this mechanical aeration is very expensive (Crites et al. 2014). Therefore, the small meat food processing industries which basically have low production fail to treat wastewater appropriately subsequently causing river pollution. Several decomposing microorganisms have been applied in the wastewater treatment process. This includes the use of microalgae in the phycoremediation process. The application of microalgae for wastewater treatment is an innovative measure. The potential of microalgae to reduce the amount of nutrients and remove heavy metals from different types of wastewater such as acidic wastewater by *Botryococcus* sp., municipal wastewater by *Botryococcus* sp. and *S. obliquus*, agricultural wastewater by *Botryococcus* sp., industrial wastewater by *Botryococcus* sp. and *Chlorella saccharophila* and livestock wastewater by *Botryococcus* sp. and *Chlorella* sp. have been reported in the literature (An et al. 2003; Shen et al. 2008; Órpez et al. 2009; Chinnasamy et al. 2010; Areco et al. 2012). However, the potential and biokinetics of these microalgae in the phycoremediation of meat processing wastewater have not been investigated extensively. This gap offers researchers a great opportunity to explore the potential of microalgae and its removal mechanisms in the phycoremediation of wastewater.

The present chapter aims to highlight the characteristics of meat processing wastewater quality through BOD, COD, TSS, pH, TN, NH_4^+ and PO_4^{3-} values as well as to review the biokinetic models used for the investigation of coefficients, reaction rate constant (k), half-saturated constant (K_m) and yield coefficient (Y) of microalgae during the phycoremediation process of wastewater.

13.2   Food Processing Industry

The meat food processing industry in Malaysia mainly processes poultry meat (chicken), bovine meat (cattle), ovine meat (goat and sheep) and seafood (fish, prawn and crab) (Pazim et al. 2009). The meat food processing industry normally manufactures meat-based products such as sausages, meatballs, burgers, pickled meat, canned meat, dried, curried or spiced meat, meat floss, meatloaves and minced meat.

There are 32 main companies involved in food processing in Malaysia and more than 50 small-scale entrepreneurs are actively involved (Senik 1995).

The commercial meat-based food industry generates huge quantities of solid waste and wastewater. Solid waste is mainly composed of the whole animals which are rejected or unsatisfactory, carcasses, skeletons and animal waste, offal containing viscera and skin mostly produced from the evisceration and cleaning process of those animals. Wastewater mainly consists of the organic contaminants such as blood, animal faeces, fats, oils and grease, fine parts of eviscerated organs and flesh and other parts in the form of soluble, colloidal and particulates which are mostly produced from cleaning, washing and cutting of raw materials. This wastewater is supposed to be treated before it is being discharged into the drainage systems.

13.3 Environmental Effects Of Untreated Wastewater Disposal

Wastewater contain high organic and nutrient loads which include nitrogen, phosphorous and other elements (Fe, Mn, Cu, Zn, Pb, Cd and Hg), as well as oil and grease, salt and pathogens (*Salmonella* sp. and *Campylobacter* sp.) (Sunda and Susan 1998; Cai et al. 2013). Excessive nutrients and elements in the wastewater can cause eutrophication and disrupt the balance of the ecosystem. The discharge of untreated wastewater into drainage systems develops high nutrient and element loading into the aquatic environment which provides a favourable condition for the proliferation of undesirable phytoplankton (Cai et al. 2013).

Phytoplanktons are organisms which are competent in absorbing compulsory substances for growth and reproduction from surrounding waters (Kuroshi 2012; Sunda and Susan 1998). The two most vital nutrients needed for the growth of phytoplankton are nitrogen and phosphorous which are present in vast amounts of meat processing wastewater. The warm and humid weather in Malaysia also makes it conducive for phytoplankton blooms to take place (Cai et al. 2013; Nursuhayati et al. 2013). When the phytoplankton bloom takes place, there may be dangerous ramifications. It may cause hypoxia, also known as dissolved oxygen (DO) depletion, as the decay of a large number of dead phytoplankton will cause a large amount of DO to be consumed. This directly leads to 'fish kills' due to anoxia (Hallegraeff 2010). Depletion of DO occurs more at night as the phytoplankton use more oxygen than they give off during the day during photosynthesis. As phytoplankton bloom leads to 'fish kills', it may eventually disrupt the biodiversity of the ecosystem (De Silva 2012; Kangur et al. 2013).

Besides that, certain types of phytoplankton produce toxic chemicals when they die. These toxic chemicals may be hazardous to other aquatic organisms (Anderson et al. 2002). Toxicity can sometimes cause serious illnesses and even death to animals that consume the biotoxins present in water (Berdalet et al. 2016). When humans are subjected to this toxin, they might suffer severe symptoms such as memory loss, diarrhoea, gastroenteritis, lung irritations, paralysis and even death (Hallegraeff 1995). The toxins can alter zooplankton communities, decrease the growth of trouts and obstruct the growth of fish and amphibians. Therefore, it is obvious that the phytoplankton bloom caused by the meat processing industry can cause a loss of biodiversity (Cai et al. 2013; Berdalet et al. 2016).

Phytoplankton blooms may cause water bodies to have bad odour, colour and taste. In addition, phytoplankton blooms produce isoprene, the second most abundant biologically derived greenhouse gas that will react with the ozone layer, produce smog and emit methane during decay (Stephanie 2010). Smog particles may cause respiratory problems and cardiac diseases among human beings and damage the environment (Seaton et al. 2009). As a catalyst for phytoplankton blooms, untreated wastewater such as meat processing wastewater plays both direct and indirect roles in causing air pollution and global warming (Paerl and Paul 2012).

Other than the effect of the nutrients present in untreated wastewater from the meat processing industry, there are some other side effects of disposing untreated wastewater into natural water bodies. Meat processing wastewater may cause pollution due to dissolved pollutants such as blood which has a chemical oxygen demand (COD) of 375 000 mg/L (Tritt and Schuchardt 1992). Wastewater from the meat processing industry also contains excessive concentrations of suspended solids (SS), which include hair, feathers, grease, flesh, pieces of fat, grit, manure and undigested feed. About 50% of the pollution charge is represented by these insoluble and slowly biodegradable SS while another 25% originated from colloidal solids (Sayed et al. 1988). By producing a high amount of pollutants, meat processing wastewater precipitates water pollution and indirectly plays a role in environmental defects caused by water pollution (Bustillo-Lecompte et al. 2015).

13.4 Characteristics of Untreated Wastewater from Meat-Related Activities

The wastewater from slaughterhouses and meat processing factories are normally assessed in terms of mass parameters due to the specific amounts of wastewater and pollutant load parallel to the animals slaughtered or processed that differ from the meat processing industry. Normally, it contains significant amounts of total phosphorus (TP), total nitrogen (TN), total organic carbon (TOC), chemical oxygen demand (COD), total suspended solids (TSS), and biochemical oxygen demand (BOD) (Tritt and Schuchardt 1992; Johns 1995; Mittal 2006; Cao and Mehrvar 2011; Wu and Gauri 2011; Barrera et al. 2012; Bustillo-Lecompte et al. 2013, 2014, 2015).

The concentrations of nitrogen and phosphorous content in meat processing wastewater investigated in previous studies are shown in Table 13.1. The highest content of nitrogen and phosphorous recorded in meat processing wastewater was 2743.6 mg/L of nitrogen and 328.4 mg/L of phosphorus (Cristian 2010). This is followed by slaughterhouse wastewater with 1057 mg/L of TKN and 217 mg/L of

Table 13.1 Nitrogen and phosphorous content in wastewater from meat-related activity from previous studies

Source of wastewater	Nitrogen (mg/L)	Phosphorous (mg/L)	References
Meat processing	2743.6	328.4	Cristian (2010)
Meat processing	199.5	51	Bustillo-Lecompte et al. (2015)
Slaughterhouse	700 (NH_4^+)	200 (PO_4^{3-})	Ge et al. (2013)
Slaughterhouse	1057 (TKN)	217	Cassidy and Belia (2005)
Slaughterhouse	325	43	Jensen et al. (2014)
Slaughterhouse and meat packing	465 (NH_4^+)	176.5	Rajeshwari et al. (2000)
Slaughterhouse	51 ($N\text{-}NH^{-3}$)	17.5	Akan et al. (2010)
Meat processing	91	29	Lu et al. (2015)
Slaughterhouse	190	50	Keller et al. (1997)
Meat processing	145 (TKN)	34	Nagalingam (2002)
Poultry slaughterhouse	116 ($N\text{-}NH^{-3}$)	57	Zheng et al. (2013)

phosphorus (Cassidy and Belia 2005). The lowest amount of nitrogen and phosphorus was also found in slaughterhouse wastewater with 51 mg/L. of N-NH-3 and 17.5 mg/L of phosphorus in a study by Akan et al. (2010).

In one of the studies presented in Table 13.1 which was conducted in Romania by Cristian (2010), the results revealed that the wastewater produced by meat processing factories contained high amounts of COD (1683.6 mg/L), BOD (863.4 mg/L), TSS (640.2 mg/L) and chlorides (382.6 mg/L) which were comparatively lesser than the ones found in wastewater produced by factories manufacturing milk and dairy products. However, it was found that meat processing wastewater contained the highest concentrations of nutrients namely 2743.6 mg/L of nitrogen and 328.4 mg/L of phosphorus with an N:P ratio of 8:1 (Cristian 2010).

Bustillo-Lecompte et al. (2015) mentioned that 51% of the meat processing plants do not have onsite wastewater treatment. 17% use aerobic treatment, 32% utilise passive systems such as storage tanks to settle solids, and only 2% utilise grease traps for fat separation and blood collection. The content of nitrogen (60–339 mg/L) and phosphorus (25.7–75.9 mg/L) with N:P ratios of 2:1 to 4:1 was examined at selected provincially licensed plants from the survey.

Based on Table 13.1, another fact that can be proven is that the nitrogen and phosphorous content and the N:P ratio of meat processing wastewater varies due to the diverse kinds of meat used and various stages of meat processing. The nutrient content in meat processing wastewater from other countries is different from the nutrient content found in meat processing wastewater in Malaysia (Pazim et al. 2009).

13.5   Wastewater Treatment in the Meat Processing Industry

The presence of nutrients in the water body due to the discharge of untreated meat processing wastewater is a major concern as it might threaten the lives of humans and animals, as well as the biodiversity in aquatic ecosystems (Melnick 2005). Hence, in order to avoid these problems from occurring, it is a critical necessity to remove nutrients in wastewater systematically. This environmental problem has led to extensive studies on the development of effective alternative technologies to remove nutrients from wastewater. However, each alternative method has its own advantages and disadvantages.

In the conventional method used in Malaysia for the wastewater treatment process, there is a need for mechanical aeration to provide oxygen for aerobic bacteria so that it can consume organic compounds in wastewater. This mechanical aeration is a very expensive procedure (Crites et al. 2014).For an algae-based wastewater treatment, however, there is no need for mechanical aeration because the microalgae are able to provide oxygen for aerobic bacteria (Munoz and Guieysse 2006; Crites et al. 2014)

Microalgae offer a productive way to consume nutrients and heavy metals and at the same time provide required oxygen for aerobic bacteria through the photosynthesis process (Cai et al. 2013). Algae-based wastewater treatment is said to be a more cost-saving method compared to any other secondary wastewater treatment processes (Suad and Gu 2014).

Table 13.2 shows different alternative methods of meat processing wastewater treatment. These treatment methods mainly focus on the treatment of BOD, COD and TSS. Among these methods, the highest removal efficiency of BOD (97%), COD (96%) and TSS (95%)was achieved through the use of an anaerobic lagoon for slaughterhouse wastewater. The removal efficiency of BOD between 62.0 and 78.8% and the removal efficiency of COD between 74.6 and 79.5% were achieved through physicochemical treatment (Senna 2008). The lowest TSS removal efficiency (81–86%) was identified in a study by Caixeta and Magali (2002) using UASB treatment.

Senna (2008) also focused on the removal of oils and greases as well as TS which achieved up to 85% in terms of removal efficiency. Aguilar et al. (2005) claimed that the coagulation–flocculation using ferric sulphate in slaughterhouse wastewater achieved a removal efficiency of particles between 78 and 99%. Gong et al. (2008) indicated that a turbidity removal efficiency of 93.7% was achieved. Each alternative method has its own advantages and disadvantages in removing particular pollutants.

Table 13.2 Previous studies on different treatments for wastewater in the meat industry

Source of wastewater	Type of treatment	Treatment efficiency	References .
Slaughterhouse	UASB	BOD > 95% COD > 77–91% TSS > 81–86%	Caixeta and Magali (2002)
Meat packing	Microbial fuel cell (MFC)	More than 86% BOD and TOC removal	Heilmann and Logan (2006)
Meat processing industry	Physicochemical treatment	Oils and greases, and TS > 85%, BOD > 62.0–78.8% COD > 74.6–79.5%	Senna (2008)
Slaughterhouse and meat packing	Anaerobic fluidised bed reactor	COD > 94%	Rajeshwary et al. (2003)
Slaughterhouse	Coagulation–flocculation using ferric sulphate	Particles removal > 78–99%	Aguilar et al. (2005)
Meat processing	Bioflocculation by Serratiaficaria	COD > 76.3% Turbidity > 93.7%	Gong et al. (2008)

13.6 The Use of Microalgae for the Phycoremediation of Meat Processing Wastewater

Microalgae are unicellular organisms which exist individually, in chains or in the groups. The size of microalgae ranges from a few micrometres to hundreds of micrometres depending on the species. Even though microalgae do not have roots, stems and leaves like other plants, they are capable of performing photosynthesis (Rydin et al. 2013). Hence, microalgae are vital for life on earth as they generate about half of the atmospheric oxygen and also use carbon dioxide for their photoautotrophic growth (Rydin et al. 2013; Moreira et al. 2000). Microalgae represent immense biodiversity and are an almost unexploited resource (Pulz and Gross 2004).

There is a number of microalgae species generated in hatcheries for commercial use. Based on previous studies, it has been estimated that the dimension of the vessel or bioreactor where the microalgae are being cultured, subjection to light or irradiation and the concentration of cells within the reactor are the main factors that contribute to the success of a microalgae hatchery system. The open pond system is a method that has been used for microalgae cultivation since the 1950s (Cai et al. 2013). The advantage of this process is the simplicity in its construction and operation compared to the closed pond system. However, this method is not regularly used due to certain aspects such as evaporation, optimal growth temperature and preservation of the environment which are challenging to maintain as well as the low capacity of certain materialistically important strains such as *Arthrospira* sp. (Pandey et al. 2013).

A photobioreactor is a fully covered equipment which yields a controlled environment and enables greater efficiency in the production of algae and prevents contamination (Kunjapur and Eldridge 2010; Rawat et al. 2011). Closed pond systems cost more than open pond systems. An air-lift method is used during the outdoor cultivation and production of microalgae where air is transferred within a system for the circulation of water where microalgae are grown. Fermentation is carried out by bioreactors called fermenter-type reactors (FTR). FTRs have not advanced hugely in the cultivation of microalgae due to its low capability in utilising sunlight and a bad surface area to volume ratio (Pandey et al. 2013).

There are various commercial applications of microalgae in sectors such as aquaculture, biofuel production, cosmetic products, bio-fertilisers and so on (Priyadarshani and Biswajit 2012; Ravishankar et al. 2012). In the field of aquaculture, microalgae are a vital nutrition source. It is either a direct or an attached source of essential nutrition and is mostly used at farms breeding larvae of fish, echinoderms, molluscs and crustaceans. As microalgae like *Botryococcus braunii* are outstandingly abundant in oils up to 80% of the dry weight of biomass, it is suitable to be converted into fuel (Mofijur et al. 2012; Saharan et al. 2013). Moreover, microalgae have greater productivity than land-based agricultural crops and are thus more sustainable in the long run. Red microalgae such as *Laurencia pacifica* are categorised by pigments called phycobiliproteins that have innate colourants which are utilised in pharmaceuticals and cosmetic products (Wang et al. 2012). Forbio-fertilisers, blue-green algae such as *Cyanobacterium* sp. and *Microcystis aeruginosa* are used to fix nitrogen by allowing cyanobacteria to reproduce in the soil (Xu et al. 2017). Nitrogen fixation is a vital method which allows inorganic compounds like nitrogen to be converted into organic forms which can then be utilised by plants (Priyadarshani and Biswajit 2012).

The microalgae *Chlorella* sp. has been used for nutrient removal in meat processing wastewater, according to a study by Lu et al. (2015). It has achieved aremoval efficiency of 90.38 and 50.94% for NH_4^+ and TN respectively. In a study by Hernández et al. (2016), mixed algae which consisted of *Chlamydomonassub caudata*, *Anabaena* sp. and *Nitzschia* sp. were used to treat slaughterhouse wastewater in high rate algal ponds (HRAP) and achieved 92 and 91% in terms of total COD and soluble phosphorous removal. Selvarani et al. (2016) achieved 100% in terms of NH3-N removal and 81% in terms of PO4-P removal for seafood processing wastewater using mixed microalgae through phycoremediation.

There are many advantages of using microalgae for the phycoremediation of meat processing wastewater. It has been shown to be a cheaper method compared to the activated sludge process and other secondary treatment processes to remove biochemical oxygen demand, pathogens, phosphorus and nitrogen. Other than that, the traditional wastewater treatment processes need mechanical aeration to supply oxygen to aerobic bacteria to consume the organic compounds in wastewater which causes greater energy costs. Aeration is an energy exhaustive process which takes up about 45–75% of a wastewater treatment plant's entire energy costs (Crites et al. 2014). In an algae-based wastewater treatment, algae help to supply oxygen for aerobic bacteria (Alcantara and O'driscoll 2014).

Algae are able to consume nutrients and supply aerobic bacteria with oxygen through photosynthesis (Crites et al. 2014). In an activated sludge process, one kWh of electricity is required for aeration and generates one kg of fossil CO_2 from power generation to remove roughly one kg of BOD. In comparison, no energy is needed for photosynthetic oxygenation which removes one kg of BOD and produces enough algal biomass to produce methane that can generate one kWh of electric power.

Microalgae also have the good ability to reduce sludge formation during wastewater treatment. In conventional wastewater treatment systems, the primary aim is to reduce or remove sludge. A variety of dangerous chemicals such as chlorine are used for sludge removal, pH correction, colour removal and odour removal in the conventional treatment of industrial effluents (Kaushika et al. 2016). The large-scale usage of chemicals for effluent treatment produces large amounts of sludge which form the so-called dangerous solid waste produced by the industry. The waste is finally discarded or deposited in landfills (Furcas and Balletto 2014).In an algae-based wastewater treatment, sludge produced by algal biomass is energy-rich and can be further processed to make biofuel or other costly products such as fertilisers (Borowitzka 2013; Pandey et al. 2013). Algal technology reduces the need for chemicals and makes the whole procedure of effluent treatment easier. There is also a significant decrease in sludge formation (Nandeshwar and Satpute 2014; Mishra et al. 2014).

The emission of greenhouse gases caused by wastewater treatment can be reduced by using microalgae for wastewater phycoremediation. The US Environmental Protection Agency (EPA) has confirmed that conventional wastewater treatment plants are the primary contributors to greenhouse gases. CO_2 is released by algae-based wastewater treatment too but this does not pose a problem as the consumption of CO_2 by microalgae during growth is higher than the CO_2 it produces (Nandeshwar and Satpute 2014). Moreover, algae biomass is a source of beneficial products such as biodiesel. In the 1990s, the National Renewable Energy Laboratory (NREL) revealed that algae are able to produce up to 40 times the amount of oil for biodiesel per unit area of land under supervised conditions in contrast to terrestrial oilseed crops like soy and canola (Sheehan et al. 1998 and Nandeshwar and Satpute 2014). Nevertheless, their results concluded that large-scale algae cultivation for energy generation was overpriced at that time and suggested future research on waste-stream integration.

It is expected that the economic situation will eventually be enhanced by integrating biodiesel feedstock production with agricultural or municipal wastewater treatment and CO_2 fixation. Algae can be utilised to produce bioethanol and biobutanol. It may also have the potential to generate higher quantities of vegetable oil compared to terrestrial crops grown for the same reason (Raut et al. 2015). Algae can be grown to generate hydrogen and biomass which can be burnt to generate heat and electricity.

Table 13.3 Previous applications of *Botryococcus* sp. in the removal of nutrients from wastewater

Source of wastewater	Type of treatment	Microalgae sp.	Treatment efficiency	References
Domestic effluent before and after pretreatment	Removal of nutrients TN and TP as well as lipid productivity	*Botryococcus* sp.	TN (61–65%) TP (100%)	Rinna (2014)
Industrial	Nutrient and element (Fe, Mg, Ca, Mo, Mn and Zn) removal and lipid production	*Botryococcus* sp., *Chlorella saccharophila*, *Pleurochrysis-carterae*	TN (99.7–99.8%) TP (98.8–99.1%)	Chinnasamy et al. (2010)
Livestock	Removal of nutrients TN and TP and hydrocarbon productivity.	*Botryococcus* sp. and *Chlorella*	TN (88%) TP (98%)	Shen et al. (2008)
Piggery	Removal of nutrients TN and TP	*Botryococcus* sp.	TN (80%) TP (98%)	An et al. (2003) Korea
Wet market wastewater	Removal of nutrients TN and TP	*Botryococcus* sp.	>80%	Al-Gheethi et al. (2017)
Greywater	Removal of nutrients TN and TP	*Botryococcus* sp.	>90%	Mohamed et al. (2017)

13.6.1 Previous Applications of Microalgae in Phycoremediation

Water-intensive food industries are one of the major contributors of effluents and normally contain excessive concentrations of nitrogen and phosphorous as well as toxic metals, making their treatment more expensive (Gasperi et al. 2008). Under such conditions, microalgae can grow successfully by gathering nutrients and metals, making them sustainable and convenient for cheaper wastewater treatment (De-Bashan and Yoav 2010). Table 13.3 shows the previous applications of the microalgae *Botryococcus* sp. in the removal of nutrients from wastewater.

Findings from the previous studies revealed that the removal efficiency of *Botryococcus* sp. for nitrogen and phosphorous was the highest for industrial wastewater. Table 13.3 shows that the highest removal efficiency of nitrate is 99.7–99.8%, whereas the highest removal efficiency of phosphate is 98.8–99.1%, followed by the removal efficiency rate in livestock wastewater and then piggery wastewater. The highest removal efficiency rate of nitrogen and phosphate by *Botryococcus* sp. is achieved for industrial wastewater because it has the optimum amount of nutrients for the microalgae growth. There might be some other factors such as the environment and the climate which cause differences in temperature, salinity, light intensity, and cultivation time which play a major role in influencing algae growth (Ruangsomboon 2012).

In a study by Shen et al. (2008), growth media was prepared in eight 225 mL Corning cell culture flasks numbered A1–A8. One flask was used as the control sample with the inoculation medium while other flasks contained blends of autoclaved wastewater (120 °C for 20 min) and distilled water at autoclaved wastewater concentrations of 25, 50, 75 and 100%, respectively. The samples were used to study how algae grow in wastewater in the absence of bacteria or wild algae. 15 mL of algae inoculum with a concentration of about 1×10^6 cells mgL^{-1} was added to flasks A1–A7 containing 110 mL of growth medium each. Flask A8 contained 125 mL of raw wastewater and was used as an un-inoculated control sample. The cultivation was carried out batch-wise for 25 days until a stationary growth phase was reached. By day 14, 88% of TN and 98% of TN were removed. Therefore, this indicates that *Botryococcus* sp. is a type of microalgae that is useful for nutrient removal from livestock wastewater.

Órpez et al. (2009) carried out experiments in batch photobioreactors at a laboratory scale with pre-culture for 1 week. The pH of the culture medium was 8.0, whereas the temperature was maintained constantly at 25 °C. An air-supply volume between 0.5 and 1.5 v/v/min, 12 h light/12 h darkness cycle and mechanical stirring at 60 rpm were maintained only during the culture stage. The results showed that the use of secondarily treated sewage from domestic wastewater as a medium to grow *B.braunii* for the removal of nitrogen and phosphorus resulted in a rate of 11.9 mg/L/day and 11.5 mg/L/day, respectively by algal consumption.

According to Rinna (2014), the domestic effluents were collected after the pretreatment stage and at the discharge point from aeration tanks. Samples were collected from a wastewater treatment plant in the municipality of Salvador, Bahia, Brazil. The wastewater utilised was untreated and CHU 13 was used as a control medium. The experiments were carried out in triplicates using 1 L borosilicate Pyrex flasks. Two strains of *B. braunii* were used. The TN removal percentage for both strains was 62 and 65%, respectively whereas the TP removal percentage for both strains was 100%.

Based on Table 13.3, it was found that there is a huge variation in the types of wastewater, nutrients and microalgae that are capable of being used. There are differences in the efficiency of its usage under various environmental conditions. Therefore, there searchers involved in this study opine that the type of wastewater is an important factor that must be given priority in future research.

Based on an analysis of the existing literature, most microalgae applications have been done using industrial wastewater followed by municipal wastewater. Industrial wastewater itself can be divided into more groups and experiments should be done to more specific types of wastewater. The chemical and physical properties of the wastewater should also be taken into account. Furthermore, microalgae applications have not been done on meat processing wastewater. Therefore, it is important to do this research to find out the efficiency of microalgae in removing nutrient content from meat processing wastewater.

13.7 Biokinetics of Nutrient Removal from Wastewater

The main mechanisms in which the microalgae cells remove nutrients from wastewater include the uptake into cells through the assimilation process and ammonia stripping through elevated pH (Hoffmann 1998 and Bich et al. 1999). The removal of these nutrients mainly depends on their concentrations in the wastewater. Therefore, in order to test the efficiency of the microalgae species in removing nutrients from any wastewater sample, bioremediation kinetics should be studied. Conventionally, batch experiments are conducted to investigate the effect of the bioremediation performance of microalgae and to determine biokinetic coefficients such as k, reaction rate constant, Km, half saturation constant and Y, yield coefficient. Models which have commonly been used to study biosorption kinetics include Langmuir and Freundlich isotherm models (Mallikarjun et al. 2012). Table 13.4 shows a summary of previous studies on bioremediation kinetics with varying models, substrates, adsorbents and media.

According to previous studies, it was noted that wastewater microalgae bioremediation kinetics in the meat processing industry has not been investigated. The study which most related to this research is 'Organic Carbon and Nitrogen removal from slaughterhouse wastewater' done by Kundu et al. (2013) as it uses slaughterhouse wastewater which has almost similar properties and characteristics as meat processing wastewater. However, the adsorbents used, parameters studied and the type of wastewater involved in that particular study differ from the ones used in this study. Therefore, more research is needed to find out the bioremediation kinetics of microalgae in reducing nutrients in wastewater from the meat processing industry. Another research that is related to this study is 'Nitrogen and Phosphorous removal from synthetic wastewater' done by Aslan and Kapdan (2006) as the parameters studied for kinetics include ammonia and orthophosphate. The wastewater used was synthetic wastewater. As the parameters explored in both of these studies are the same, it is assumed that the Michaelis Menten model might be the best fit for the determination of the biokinetic coefficients. Michaelis–Menten kinetics is one of the most common mathematical models which is used for the study of enzyme kinetics. Table 13.5 shows a summary of the previous studies on bioremediation kinetics using the Michaelis–Menten model with varying substrates, adsorbents and media.

Corresponding to the previous applications of the Michaelis–Menten model in determining the biokinetics of the adsorbents as shown in Table 13.5, it can be concluded that there was a huge variation of adsorbents as many types of microorganisms were used, as well as a huge variation in the type of parameters studied. There were also differences in the efficiency of its usage under various factors that can be made constant such as environmental conditions which cover temperature, light intensity and pH value. According to the information found in previous studies, it is clear that the studies involving the Michaelis–Menten model were mostly applied in the removal of dye followed by the removal of heavy metals. This may be caused by the characteristics of wastewater with high dye content and thus, it is important to

Table 13.4 Previous studieson bioremediation kinetics

Media	Type of adsorbent	Parameters studied	Model used	Treatment efficiency	References
Organic carbon and nitrogen removal from slaughterhouse wastewater	Active microbial seeds	COD and TN	Lawrence and McCarty's modified Monod	COD > 96% TN > 90.12%	Kundu et al. (2013)
Phosphate adsorption characteristics on clay soil	Clay soil	Phosphorous	Langmuir and Freundlich	5 mg/L in first 60 min	Mallikarjun et al. (2012)
Nitrogen and phosphorous removal from synthetic wastewater	*Chlorella Vulgaris*	Ammonia and Orthophosphate	Michaelis–Menten	Ammonia > 100% Phosphorous > 78%	Aslan and Kapdan (2006)
Biodegradation kinetics of phenol and catechol	*Pseudomonas putida*	Phenol and Catechol	Linearized-Haldane's	Both approximately 100%	Kumar et al. (2005)
Kinetics of aerobic biodegradation of benzene and toluene in sandy aquifer material	Microbial	Benzene and toluene	Monod	Benzene > 80% Toulene > 80%	Alvarez et al. (1991)

A. S. Vikneswara et al.

Table 13.5 Previous studies on bioremediation kinetics using the Michaelis–Menten model

Title	Type of adsorbent	Model used	References
Removal of copper	*Scenedesmussubspicatus*	1.30×10^{-10} mol·[g dry wt algae]$^{-1}$·min^{-1}	Knauer et al. (1997)
Batch kinetics for the removal of TN and TP from synthetic wastewater	*Chlorella Vulgaris*	100% 78%	Aslan and Kapdan (2006)
Biodegradation of Malachite Green based on the reduction in COD	*Cosmarium*sp.	99 and 75%	Daneshvar et al. (2007)
Kinetic study for hydrolysis of wheat straw by cellulase	Cellulase enzyme solution (Novozymes A/S)	$V_{emax} = 0.55$ g/L min $K_e = 38.2$ g/L min	Carrillo et al. (2005)
Biodegradation of azo dyes by immobilised bacteria	*Pseudomonas luteola*	45%	Chang et al. (2001)
Removal of acid anthraquinone dye by bacteria	*Bacillus gordonae Bacillus benzeovoras Pseudomonas putida*	13, 18 and 19%	Walker and Weatherley (2000)
Kinetics study for the removal of Co, Mn and Zn by algae	*Chlorella salina*	19, 2, and 4%	Garnham et al. (1992)

identify the kinetics of the respective absorbents for dye removal. As information on the kinetics for nutrient removal is sorely lacking, more studies are needed in this field in order to better understand the efficiency of phycoremediation in the removal of nutrients from meat processing wastewater.

13.8 Conclusion

Based on the aforementioned studies, it can be concluded that the phycoremediation process for meat processing wastewater is applicable due to the high content of nutrient required for microalgae to survive and grow. The Michaelis–Menten model is one of the models which can be used to determine the growth rate of microalgae in response to the concentration of nutrients in wastewater as well as the removal rate of these pollutants. However, more studies are required to investigate the biokinetics of nutrient removal from meat processing wastewater using microalgae.

Acknowledgements The authors also wish to thank The Ministry of Science, Technology and Innovation (MOSTI) for supporting this research under E-Science Fund (02-01-13-SF0135) and also the Research Management Centre (RMC) UTHM under grant IGSP U682 for this research.

References

Aguilar MI, Saez J, Lloréns M, Soler A, Ortuno JF, Meseguer V, Fuentes A (2005) Improvement of coagulation–flocculation process using anionic polyacrylamide as coagulant aid. Chemosphere 58(1):47–56

Akan JC, Abdulrahman FI, Yusuf E (2010) Physical and chemical parameters in abattoir wastewater sample, Maiduguri Metropolis, Nigeria. Pac J Sci Technol 11(1):640–648

Alcantara D, O'driscoll M (2014) Congenital microcephaly. Am J Med Genet Part C: Semin Med Genet 166(2):124–139)

Al-Gheethi AA, Mohamed RM, Jais NM, Efaq AN, Halid AA, Wurochekke AA, Amir-Hashim MK (2017) Influence of pathogenic bacterial activity on growth of Scenedesmus sp. and removal of nutrients from public market wastewater. J Water Health 15(5):741–756

Alvarez PJ, Anid PJ, Vogel TM (1991) Kinetics of aerobic biodegradation of benzene and toluene in sandy aquifer material. Biodegradation 2(1):43–51

An JY, Sim SJ, Lee JS, Kim BW (2003) Hydrocarbon production from secondarily treated piggery wastewater by the green alga *Botryococcusbraunii*. J ApplPhycol 15(2):185–191

Anderson K, Blanchette CA, Broitman B, Cooper SD, Halpern BS (2002) A cross-ecosystem comparison of the strength of trophic cascades. Ecol Lett 5(6):785–791

Areco MM, Hanela S, Duran J, Afonso MS (2012) Biosorption of Cu(II), Zn(II), Cd(II) and Pb(II) by dead biomasses of green alga Ulvalactuca and the development of a sustainable matrix for adsorption implementation. J Hazard Mater 213–214:123–132 (Epub January 30, 2012)

Aslan S, Kapdan IK (2006) Batch kinetics of nitrogen and phosphorus removal from synthetic wastewater by algae. Ecol Eng 28(1):64–70

Barrera M, Mehrvar M, Gilbride KA, McCarthy LH, Laursen AE, Bostan V, Pushchak R (2012) Photolytic treatment of organic constituents and bacterial pathogens in secondary effluent of synthetic slaughterhouse wastewater. Chem Eng Res Des 90(9):1335–1350

Berdalet E, Fleming LE, Gowen R, Davidson K, Hess P, Backer LC, Moore SK, Hoagland P, Enevoldsen H (2016) Marine harmful algal blooms, human health and wellbeing: challenges and opportunities in the 21st century. J Marine Biol Ass U.K. 96(01):61–91

Bich NN, Yaziz MI, Bakti NAK (1999) Combination of *Chlorella vulgaris* and *Eichhornia crassipes* for wastewater nitrogen removal. Water Res 33(10):2357–2362

Borowitzka MA (2013) High-value products from microalgae—their development and commercialisation. J Appl Phycol 25(3):743–756

Bustillo-Lecompte CF, Mehrab M, Quiñones-Bolaños E (2013) Combined anaerobic-aerobic and UV/H_2O_2 processes for the treatment of synthetic slaughterhouse wastewater. J Environ Sci Health, Part A 48:1122–1135

Bustillo-Lecompte CF, Mehrab M, Quiñones-Bolaños E (2014) Cost-effectiveness analysis of TOC removal from slaughterhouse wastewater using combined anaerobic–aerobic and UV/H_2O_2 processes. J Environ Manage 134:145–152

Bustillo-Lecompte CF, Mark K, Mehrab M (2015) Assessing the performance of UV/H_2O_2 as a pretreatment process in TOC removal of an actual petroleum refinery wastewater and its inhibitory effects on activated sludge. Can J Chem Eng 93:798–807

Cai T, Park SY, Li Y (2013) Nutrient recovery from wastewater streams by microalgae: status and prospects. Rene Sust Energy Rev 19:360–369

Caixeta CET, Magali CC (2002) Alcina MF (2002) Slaughterhouse wastewater treatment: evaluation of a new three-phase separation system in a UASB reactor. Biores Technol 81(1):61–69

Cao W, Mehrvar M (2011) Slaughterhouse wastewater treatment by combined anaerobic baffled reactor and UV/H_2O_2 processes. Chem Eng Res Des 89:1136–1143

Carrillo F, Lis MJ, Colom X, López-Mesas M, Valldeperas J (2005) Effect of alkali pretreatment on cellulase hydrolysis of wheat straw: kinetic study. Process Biochem 40(10):3360–3364

Cassidy DP, Belia E (2005) Nitrogen and phosphorus removal from an abattoir wastewater in a SBR with aerobic granular sludge. Water Res 39(19):4817–4823

Chang JS, Chou C, Chen SY (2001) Decolorization of azo dyes with immobilized *Pseudomonas luteola*. Process Biochem 36(8):757–763

Chinnasamy S, Bhatnagar A, Claxton R, Das KC (2010) Biomass and bioenergy production potential of microalgae consortium in open and closed bioreactors using untreated carpet industry effluent as growth medium. Biores Tech 101(17):6751–6760

Cristian O (2010) Characteristics of the untreated wastewater produced by food industry. Analele Universității din Oradea, Fascicula: Protecția Mediului, vol 15

Crites RW, Middlebrooks J, Robert KB (2014) Natural wastewater treatment systems. CRC Press, Boca Raton

Daneshvar N, Ayazloo M, Khataee AR, Pourhassan M (2007) Biological decolorization of dye solution containing Malachite green by microalgae *Cosmarium* sp. Biores Technol 98(6):1176–1182

De Silva SS (2012) Aquaculture: a newly emergent food production sector and perspectives of its impacts on biodiversity and conservation. Biodivers Conserv 21(12):3187–3220

De-Bashan Luz E, Yoav B (2010) Immobilized microalgae for removing pollutants: review of practical aspects. Biores Technol 101(6):1611–1627

Furcas C, Balletto G (2014) Increasing the value of dimension stone waste for a more achievable sustainability in the management of non-renewable resources. J Solid Waste Technol Manage 40:185–196

Garnham GW, Codd GA, Gadd GM (1992) Kinetics of uptake and intracellular location of cobalt, manganese and zinc in the estuarine green alga *Chlorella salina*. Appl Microbiol Biotechnol 37(2):270–276

Gasperi J, Garnaud S, Rocher V, Moilleron R (2008) Priority pollutants in wastewater and combined sewer overflow. Sci Total Environ 407(1):263–272

Ge H, Batstone DJ, Keller J (2013) Operating aerobic wastewater treatment at very short sludge ages enables treatment and energy recovery through anaerobic sludge digestion. Water Res 47(17):6546–6557

Gong WX, Wang SG, Sun XF, Liu XW, Yue QY, Gao BY (2008) Bioflocculant production by culture of *Serratia ficaria* and its application in wastewater treatment. Biores Technol 99(11):4668–4674

Hallegraeff GM (1995) Harmful algal blooms: a global overview. Manual on harmful marine microalgae, vol 33, pp 1–22

Hallegraeff GM (2010) Ocean climate change, phytoplankton community responses, and harmful algal blooms: a formidable predictive challenge. J Phycol 46(2):220–235

Heilmann J, Logan BE (2006) Production of electricity from proteins using a microbial fuel cell. Water Environ Res 78(5):531–537

Hernández D, Riaño B, Coca M, Solana M, Bertucco A, Garcia-Gonzalez MC (2016) Microalgae cultivation in high rate algal ponds using slaughterhouse wastewater for biofuel applications. Chem Eng J 285:449–458

Hoffmann JP (1998) Wastewater treatment with suspended and nonsuspended algae. J Phycol 34:757–763

Jensen PD, Sullivan T, Carney C, Batstone DJ (2014) Analysis of the potential to recover energy and nutrient resources from cattle slaughterhouses in Australia by employing anaerobic digestion. Appl Energy 136:23–31

Johns MR (1995) Developments in wastewater treatment in the meat processing industry: a review. Biores Technol 54(3):203–216

Kangur K, Kangur P, Ginter K, Orru K, Haldna M, Möls T, Kangur A (2013) Long-term effects of extreme weather events and eutrophication on the fish community of shallow Lake Peipsi (Estonia/Russia). J Limnol 72(2):30

Kaushika ND, Reddy KS, Kaushik K (2016) Sustainable energy and the environment: a clean technology approach. Springer

Keller J, Subramaniam K, Gösswein J, Greenfield PF (1997) Nutrient removal from industrial wastewater using single tank sequencing batch reactors. Water Sci Technol 35(6):137–144

Knauer K, Behra R, Sigg L (1997) Adsorption and uptake of copper by the green alga *Scenedesmus subspicatus* (Chlorophyta). J Phycol 33(4):596–601

Kumar A, Kumar S, Kumar S (2005) Biodegradation kinetics of phenol and catechol using *Pseudomonas putida* MTCC 1194. Biochem Eng J 22(2):151–159

Kundu P, Debsarkar A, Mukherjee S (2013) Treatment of slaughter house wastewater in a sequencing batch reactor: performance evaluation and biodegradation kinetics. BioMed Res Int 2013

Kunjapur AM, Eldridge RB (2010) Photobioreactor design for commercial biofuel production from microalgae. Ind Eng Chem Res 49(8):3516–3526

Kuroshi LA (2012) Onshore ballast water treatment stations: a harbour specific vector management proposition/by Lawrence A. Kuroshi

Lu Q, Zhou W, Min M, Ma X, Chandra C, Doan Y, Ma Y, Zheng H, Cheng S, Griffith R, Chen P (2015) Growing *Chlorella* sp. on meat processing wastewater for nutrient removal and biomass production. Biores Technol 198:189–197

Mallikarjun BODA, Ignor MM, ProciwPJ, Sutarwala TSH (2012) US Patent No. D669,427. US Patent and Trademark Office, Washington, DC

Melnick D (2005) Environment and human well-being: a practical strategy. Earthscan

Mishra SK, Suh WI, Farooq W, Moon M, Shrivastav A, Park MS, Yang JW (2014) Rapid quantification of microalgal lipids in aqueous medium by a simple colorimetric method. Biores Technol 155:330–333

Mittal GS (2006) Treatment of wastewater from abattoirs before land application—a review. Biores Technol 97(9):1119–1135

Mofijur M, Masjuki HH, Kalam MA, Hazrat MA, Liaquat AM, Shahabuddin M, Varman M (2012) Prospects of biodiesel from Jatropha in Malaysia. Renew Sustin Energy Rev 16(7):5007–5020

Mohamed RM, Al-Gheethi AA, Aznin SS, Hasila AH, Wurochekke AA, Kassim AH (2017) Removal of nutrients and organic pollutants from household greywater by phycoremediation for safe disposal. Int J Energy Environ Eng 8(3):259–272

Moreira D, Le Hervé G, Hervé P (2000) The origin of red algae and the evolution of chloroplasts. Nature 405(6782):69–72

Munoz R, Guieysse B (2006) Algal–bacterial processes for the treatment of hazardous contaminants: a review. Water Res 40(15):2799–2815

Nandeshwar SN, Satpute GD (2014) Green technical methods for treatment of waste water using microalgae and its application in the management of natural water resources—a Review. Curr World Environ 9(3):837

Nursuhayati AS, Yusoff FM, Shariff M (2013) Spatial and temporal distribution of phytoplankton in perak estuary, Malaysia, during monsoon season. J Fish Aquatic Sci 8(4):480

Órpez R, Martínez ME, Hodaifa G, El Yousfi F, Jbari N, Sánchez S (2009) Growth of the microalga *Botryococcus braunii* in secondarily treated sewage. Desalination 246(1–3):625–630

Paerl HW, Paul VJ (2012) Climate change: links to global expansion of harmful cyanobacteria. Water Res 46(5):1349–1363

Pandey VD, Pandey A, Sharma V (2013) Biotechnological applications of cyanobacterial phyco-biliproteins. Int J Curr Biol Appl Sci 2:89–97

Pazim O, Irfan S, Wan S (2009) Malaysia as an international halal food hub: competitiveness and potential of meat-based industries. ASEAN Econ Bull 26(3):306–320

Priyadarshani I, Biswajit R (2012) Commercial and industrial applications of micro algae. J Algal Biomass Utln 3(4):89–100

Pulz O, Gross W (2004) Valuable products from biotechnology of microalgae. Appl Microbiol Biotechnol 65(6):635

Rajeshwari KV, Balakrishnan M, Kansal A, Lata K, Kishore VV (2000) State-of-the-art of anaerobic digestion technology for industrial wastewater treatment. Renew Sustain Energy Rev 4(2):135–156

Raut N, Al-Balushi T, Panwar S, Vaidya RS, Shinde GB (2015) Microalgal biofuel. In: Biofuels-status and perspective (InTech)

Ravishankar GA, Sarada R, Vidyashankar S, VenuGopal KS, Kumudha A (2012) Cultivation of micro-algae for lipids and hydrocarbons, and utilization of spent biomass for livestock feed and for bio-active constituents. Biofuel Co-Prod As Livestock Feed, 423

Rawat I, Kumar RR, Mutanda T, Bux F (2011) Dual role of microalgae: Phycoremediation of domestic wastewater and biomass production for sustainable biofuels production. Appl Energy 88:3411–3424

Rinna F (2014) Microalgae biomass production at different growth conditions assessing the lipid content and fatty acid profile for feed, food and energy applications. Agraria, Scienzeagrarie e agro-alimentari

Ruangsomboon S (2012) Effect of light, nutrient, cultivation time and salinity on lipid production of newly isolated strain of the green microalga, *Botryococcus braunii*. Biores Technol 109:261–265

Rydin H, Jeglum JK, Jeglum JK (2013) The biology of peatlands, 2nd edn. Oxford university press, Oxford

Saharan BS, Sharma D, Sahu R, Sahin O, Warren A (2013) Towards algal biofuel production: a concept of green bio energy development. Innovative Rom Food Biotechnol 12:1

Sayed S, Zanden JVD, René W, Gatze L (1988) Anaerobic degradation of the various fractions of slaughterhouse wastewater. Biol Wastes 23(2):117–142

Seaton A, Tran L, Aitken R, Donaldson K (2009) Nanoparticles, human health hazard and regulation. J Royal Soc Interface (rsif.20090252)

Selvarani AJ, Padmavathy P, Srinivasan A, Jawahar P (2016) Bioremediation of municipal wastewater, sewage water and seafood processing plant wastewater using mixed micro algae. Environ Ecol 34(4B):2134–2138

Senik G (1995) Small-scale food processing enterprises in Malaysia. ASPAC Food & Fertilizer Technology Center

Senna JPM (2008) Biotechnology and biosecurity in the production of vaccines and diagnostic kits: biosecurity in the development of vaccines and diagnostic kits. Ciência Veterinária nos Trópicos 11(suplemento 1):119–122

Sheehan J, Dunahay T, Benemann J, Roessler P (1998) A look back at the US Department of Energy's aquatic species program: biodiesel from algae. Natl Renew Energ Lab 328

Shen Y, Yuan W, Pei Z, Mao E (2008) Culture of microalga botryococcus in livestock wastewater. Am Soc Agric Biol Eng 1395–1400

Stephanie M (2010) Treatment of sewage water to reduce pollutant load. Am J Sci Res 2(4):33–39

Suad J, Gu S (2014) Commercialization potential of microalgae for biofuels production. Renew Sustain Energy Rev 14:2596–610

Sunda WG, Susan AH (1998) Processes regulating cellular metal accumulation and physiological effects: phytoplankton as model systems. Sci Total Environ 219:165–181

Tritt WP, Schuchardt F (1992) Materials flow and possibilities of treating liquid and solid wastes from slaughterhouses in Germany. Biores Technol 41:235–245

Walker GM, Weatherley LR (2000) Biodegradation and biosorption of acid anthraquinone dye. Environ Poll 108(2):219–223

Wang H, Xiong H, Hui Z, Zeng X (2012) Mixotrophic cultivation of *Chlorella pyrenoidosa* with diluted primary piggery wastewater to produce lipids. Biores Technol 104:215–220

Wu PF, Gauri SM (2011) Characterization of provincially inspected slaughterhouse wastewater in Ontario. Can Biosyst Eng 53:6

Xu R, Wang L, Zi J, Hilt S, Hou X, Chang X (2017) Allelopathic effects of *Microcystis aeruginosa* on green algae and a diatom: evidence from exudates addition and co-culturing. Harmful Algae 61:56–62

Zheng W, Li X, Kelly WR (2013) Occurrence and removal of pharmaceutical and hormone contaminants in rural wastewater treatment lagoons. Sci Total Environ 445:22–28

Printed in the United States
By Bookmasters